AF870187

CERTIFICATION OF AUTHENTICATION

YOU ARE THE OWNER OF LIMITED EDITION VOLUME NUMBER 114
OF THE HISTORICAL SERIES OF "DOWN THE OLD SPANISH TRAIL:
THE HISTORIES OF THE PLANTATIONS FROM MINTMERE TO GRAVIER"

THIS CERTIFIES THAT LIIMITED EDITION VOLUME NUMBER 114
IS A TRUE AND AUTHENTIC COPY OF THE ORIGINAL FIRST
VOLUME OF THE HISTORICAL SERIES OF "DOWN THE OLD
SPANISH TRAIL."

Kitty Counts

Mr. Bill Isher,
Good Luck
+
Happy Reading,
Kitty Counts

Courts, Kitty 1952 -

The Histories of the Old Sugar Plantations Along the Old Spanish Trail -------- Non Fiction, Historical Fact

ISBN 09658756-0-1

Originally published by Research Services

More Copies of this book can be
ordered from: Kitty Courts,
Post Office Box 11411,
New Iberia, Louisiana 70562-1411

The research material used in this books was compiled from a very careful research and study of the respective records, such as conveyance records, mortgage records, succession records, civil lawsuits, probate civil suits, book of sheriff sales, and the miscellaneous records of these Offices of the Clerk of Court in St. Mary Parish, St. Martin Parish, Iberia Parish, and St. Landry Parish.

To My Readers,

I hope you will find this book both a learning experience and a reading experience. I have enjoyed researching and writing this book, and my aim is to share this information with as many people as possible through readership.

The history of the Spanish Trail is very interesting as well as colorful. In researching this books, I became aware of so many landmarks that were really right out my back door that I never knew about. It is fascinating to learn that this area was not forgotten during the other times of history that are made so well known to us. My aim is to enrich all of our lives by making us aware of the historical times and landmarks that are so to speak right out of our back door and down the Old Spanish Trail.

Enjoy your reading and I look forward to being able to share and teach you all this new knowledge.

Sincerely,

Kitty Courts

Kitty Courts

DOWN THE OLD SPANISH TRAIL

VOLUME I

The Histories of the Plantations From Mintmere Plantation to the Gravier Plantation

INTRODUCTION

The Old Spanish Trail was the pathway used by the early settlers and Spanish explorers to push westward across the Southwestern part of the northern New World. Today, this section of the Southwestern part of the norther New World is known as the States of Florida, Alabama, Mississippi, Louisiana, Texas, New Mexico, Arizona, and California. The Old Spanish Trail begins in St. Augustine, Florida and ends in San Diego, California. Also, the Old Spanish Trail was one of the earliest public roads in this country.

Down the Old Spanish Trail came many early settlers from Spain, France, Canada, England, and the northern United States. These early settlers came to find new places to make their homes and farms, and they settled mainly upon the west side of Bayou Teche. Many of the early settler's land faced Bayou Teche, which was used a waterway, and their backdoors were on the public road, known as the Old Spanish Trail. These early settlers began to farm corn, cotton, indigo, and sugarcane. The crops of sugarcane proved to be very profitable. After gaining wealth, these early settlers gave up their original cabins for more splendid homes of the antebellum period.

The early settlers that came down the Old Spanish Trail were Louis le Pelletier de la Houssaye, and his brother, Alexander, le Pelletier de la Houssaye, Francoise Cezar Boutte', pere, Charles Napoleon Olivier Devezin, Jean Baptiste Despanets DeBlanc, and Louis Charles DeBlanc, Jacques Gravier, and Nicolas Hebert. Louis and Alexander le Pelletier de la Houssaye claimed land in the area of present day Wade Gajan farm, Kyle Acres and as far down the Old Spanish Trail as the property of Martin Cassia. Charles Napoleon Olivier Devezin claimed land in the present day Orange Grove area back, and Francois Cezar Boutte' claimed land from the present day Mintmere plantation to the land of Iberia General Medical Center. The land of the present day Iberia General Medical Center and the western part of the Wade Gajan farm was claimed by Jean Baptiste Despanets DeBlanc. This land was later sold to Stephen Owens Nelson of the City of New Orleans, Louisiana. Jacques Gravier claimed the land where the present day Cloverdale, Deepwood, and Ravenswood plantations are located.

Also, down the Spanish Trail came peddlers and traders. These early peddlers and traders had stores in the City of New Orleans, but they came to sell their wares to the wealthy plantation owners. At this time, along the Old Spanish Trail, there were no stores or markets. Many of these early peddlers were of German, and Italian ancestry, and many decided to settle in the neighborhood of the Old Spanish Trail.

Now the Spanish Trail, is known as the Old Spanish Trail, and it comprises East and West Main Street in the town of New Iberia, Louisiana. Also, the Old Spanish Trail leads westward from the town of New Iberia to the villages of Cade and Broussard and to the City of Lafayette, Louisiana.

The Old Spanish Trail as it appears today.

MINTMERE

On the date of July 26, 1913, Richard S. McMahon purchased the property situated in the 1300 block of East Main Street, in the town of New Iberia, Louisiana, on which the antebellum home, known as Mintmere is located from John R Taylor at a sale and mortgage.

In the year of 1913, the property was described as a certain lot of ground, which was unimproved, and it was situated in the lower limits of the City of New Iberia, Louisiana. The property boundary began at a point of intersection of the Bayard property on Main Street in the City of New Iberia, or the public road, which was known as the Old Spanish Trail. The public road or the Old Spanish Trail leads from the town of Jeanerette to the town of New Iberia. The lot where Mintmere is located runs in a northerly direction along the Main Street of the town of New Iberia to a distance of two hundred and fifty feet, and then resuming east a distance of six feet, then resuming east again to Bayou Teche, and then running east along the Bayou Teche a distance of one hundred and seventy two feet, and then running in a westerly direction to the Main Street of the town of New Iberia to the point of beginning. Also, the Mintmere property was bounded on the north side by land of John R. Taylor, on the east side by Bayou Teche, on the south side by land belonging to Bayard, and on the west side by Main Street of the town of New Iberia, Louisiana.

On the date of January 2, 1902, William Edward Satterfield conveyed to Edmund C. DeGeneris a tract of land, known as Bellehaven Place, which is situated in the Parish of Iberia. Before the formation of Iberia Parish in the month of October of the year of 1869, Bellehaven Place was situated partly in St. Martin Parish and partly in St. Mary Parish. In the year of 1902, Bellehaven Place was situated below the corporate limits of the town of New Iberia, Louisiana. Bellehaven Place consisted of two tracts of land, one tract of land was situated on the west side of Bayou Teche, and the second tract of land was situated on the east side of Bayou Teche. The tract on the west side of Bayou Teche had a frontage of three arpents along Bayou Teche, with a depth of forty arpents, or one hundred and twenty arpents in total area. The tract of land on the east side of Bayou Teche had a frontage along Bayou Teche of three arpents, with a depth of four arpents. Both tracts of land were bounded above by lands belonging to the estate of Cordelia D. Lewis, and below by land belonging to Baron Benoit Bayard.

Also, it was agreed and stipulated that the purchaser, Edmund C. DeGeneris, is to keep insured from the risk of fire, the residence home, known as Mintmere, and the furniture, which is contained in the residence home. The fire insurance policy is to be of an amount of at least $3,000.00 to cover any loss from fire. This fire insurance policy is payable to the holder or the holders of the mortgage notes. In the event of loss by fire to the residence home, Mintmere, and the furniture in the home, the money from the fire insurance policy is to be used toward the rebuilding and refurnishing of the new residence home on the property.

The purchase price of the property, known as Bellehaven Place, and the home, Mintmere, was the amount of $20,000.00. Edmund C. DeGeneris, the purchaser, has furnished twenty promissory notes. These twenty promissory notes are made to his own order and endorsed by him in the blank. Also, the twenty promissory notes are dated January 2, 1902. The first promissory note is for the amount of $800.00, and it falls due on December 1, 1902, the second promissory note is form the amount of $800.00, and the second promissory note falls due on December 1, 1903. The third promissory note through the nineteenth promissory note are for the amount of $1,000.00, and these promissory notes become due on December 1 of each year. The twentieth promissory note is for the amount of $1,400.00, and this final promissory note is due on the date of December 1, 1921. All twenty promissory notes are conditioned to bear interest at the rate of 8% until maturity.

The privilege is granted to the purchaser, E. C. DeGeneris, that he can sell the property on which the home, Mintmere, is situated with the approval of the seller, William Satterfield. The property is warranted to a clear title. The proceeds of the selling price shall be applied to the liquidation of the mortgage notes. This act of conveyance was executed with the presiding notary public, Fred W. Bauman, serving in his capacity, and the two witnesses to the act of conveyance were Robert Perry, and Edward Delhome.

On the date of November 16, 1902, Edmund C. DeGeneris conveyed at a sale and mortgage a tract of land, which was situated on both the east and the west sides of Bayou Teche, to Richard S. McMahon, of the Parish of Iberia. This tract of land, which was situated on the west side of Bayou Teche, was the location of the plantation home, Mintmere. The execution of this act of conveyance took place at the office of William G. Weeks, who served in his capacity as notary public for the execution of the act of conveyance. Also, the execution of the act of conveyance was passed in the presence of William Bauman, and Norwood Satterfield, who served as witnesses to the execution of the act of conveyance.

The tract of land that was conveyed was situated below the corporate limits of the town of New Iberia, Louisiana, and was known as Bellehaven Place. The tract of land was situated on both the east and the west sides of Bayou Teche. The tract on the west side of Bayou Teche had a depth of forty arpents, whereas, the tract of land on the east side of Bayou Teche had a depth of four arpents. This tract of land was bounded above by land belonging to the estate of the late Cordelia Lewis, and below by land belonging to Baron Benoit Bayard.

At the execution of the act of conveyance, it was agreed that Richard S. McMahon, the purchaser, was to insure and keep insured against the risk of fire until the full repayment of the mortgage notes the buildings and the improvements, which were erected on the property. Richard S. McMahon was to buy a fire insurance policy with a good and solvent insurance company, and he was to insure the buildings and the improvements, which were situated on the property, for the sum of at least $3,000.00. In the event of default of the mortgage notes, Richard S. McMahon was to transfer to the holder or the holders of the mortgage notes, the fire insurance policy or policies. Any holder or holders of the mortgage notes were authorized and empowered to cause a fire insurance policy to be taken out at the cost, charge, and expense of any purchaser, but this is not to be construed as obligatory upon the holder or the holders of the mortgage notes to do so. Also, the holder or the holders of the mortgage notes shall not be able to hold any purchasers responsible for any loss or damage by fire which may result from the non-insurance of the buildings and improvements, which are situated on the property.

Also, it was understood between the parties, Richard S. McMahon, the purchaser, and Edmund C. DeGeneris, the seller, that Richard S. McMahon has the right to sell any portion or portions of the property of the tract of land. Upon the payment to the holder or the holders of the mortgage note of the full amount of the purchase price in settlement of the promissory note that the holder or holders of the promissory notes will release the property sold from the mortgage and the vendor's lien. The purchase price of the tract of land, which was known as Bellehaven Place, was the amount of $18,000.00. Richard S. McMahon furnished seventeen promissory notes, which notes were to his own order, signed by him, and endorsed in the blank. All seventeen promissory notes were dated the date of November 17, 1902. The first mortgage note was due in sixty days from the sale, and it was for the amount of $2,000.00. Then, each of the remaining sixteen promissory notes were for the amount of $1,000.00, and these promissory notes were falling due in one, two, three, four, five, six, seven, eight, nine, ten, eleven, twelve, thirteen, fourteen, fifteen, and sixteen years after the date of November 19, 1902, which was the date of the conveyance. All seventeen promissory notes were conditioned to bear interest at the rate of 8% per annum from maturity until the final payment. Also, the seventeen promissory notes were paraphed, "Ne Varietur," by William G. Weeks, who served as notary public. The reason that these promissory notes were paraphed, "Ne Varietur," was to identify these seventeen promissory notes with the particular act of conveyance. These seventeen promissory notes were delivered to the seller, Edmund C. DeGeneris, who acknowledged and granted acquittal.

Also, it was expressly understood and agreed that Richard S. McMahon was to have the right and privilege at any time as he (Richard S. McMahon) may see fit of discounting or paying any one or more or all of the promissory notes. Also, Richard S. McMahon may select the promissory notes to be paid by paying to the holder or the holders of the notes such amount or amounts as would if placed at 8% interest per year. The interest bearing on the promissory notes is compounded annually for the remaining unexpired term of the notes, which Richard S. McMahon desires to take up and pay the amount to the face value of the promissory notes taken up and paid by Richard S. McMahon.

Now, coming and intervening was William Satterfield, a former owner of the place, known as Bellehaven Place, and he was the seller of the property to Edmund C. DeGeneris. Also, William Satterfield was a resident of the town of New Iberia, Louisiana. The reason that William Satterfield intervened in this act of conveyance was to take cognizance of the foregoing act of conveyance between the parties Richard S. McMahon, the purchaser, and Edmund C. DeGeneris, the seller. William Satterfield acknowledged that he ratified the same in each and every particular, and that he recognizes it as binding on him in all of its stipulations as fully and to all interests and purposes as effectively as though William Edward Satterfield was the seller of the property.

On the date of May 13, 1881, Sarah Cade Smeedes, who was the widow of Evert Bunker Smeedes, sold the property on which the antebellum home, known as Mintmere, was situated to William Edward Satterfield, of the town of New Iberia, Louisiana.

In the year of 1881, the property was described as a certain tract of land, which was situated in the Parish of Iberia. Also, the tract of land had a frontage of three arpents on both the east and the west sides of Bayou Teche. The property which was situated on the west side of Bayou Teche had a depth of forty arpents, and the property, which was situated on the east side of Bayou Teche, had a depth of four arpents. This tract of land was bounded on both the east and the west sides of Bayou Teche on the upperside by lands belonging to Cordelia D. Lewis, and on the lowerside by land belonging to Baron Benoit Bayard. Also, sold with the property, were all the buildings and improvements situated on the property.

At this act of conveyance, Sarah Smeedes also sold one horse carriage and harness, and a herd of five cattle. Besides conveying these articles, Sarah Cade Smeedes also conveyed the agricultural contracts, which she may have made with persons to lease and cultivate the land of her plantation.

This act of conveyance was passed and accepted for the purchase price of $7,500.00. William Edward Satterfield paid the purchase price of $7,500.00 in cash money at the execution of the act of conveyance. Also, Sarah Cade Smeedes acknowledged the receipt of the payment and granted acquittal to William Edward Satterfield for the payment of the purchase price.

The antebellum home, known as Mintmere, was purchased by Evert Bunker Smeedes on the date of April 10, 1860, from the sellers Millington M. Hartman and his brother, Stanley C. Hartman at a sale and mortgage. Also, William Robertson, who was a notary public for the Parish of St. Martin, presided at the execution of this act of conveyance.

At the time, when Evert Bunker Smeedes purchased the antebellum home, Mintmere, he (E. B. Smeedes) and his family were residents of the City of New Orleans. After the purchase of this property, the home Mintmere, became the residence house of Evert Banker Smeedes and his wife, Sarah Cade Smeedes, and their children, Henry D., Robert Cade, Charles Edwin, and Mary Cade Smeedes.

In the year of 1860, the property was described as a tract of land, which was situated partly in the Parish of St. Martin and partly in the Parish of St. Mary. This tract of land had a front of three arpents on the west side of Bayou Teche with a depth of forty arpents. The boundaries of this tract of land were on the upperside by the lands belonging to Cordelia D. Lewis, and on the lowerside by land belonging to Baron Benoit Bayard. Also, sold to Evert Bunker Smeedes were all the buildings and improvements, which were situated on the land. Also, there was another tract of land, which was situated on the east side of Bayou Teche that was contained in this act of conveyance. The tract of land that was situated on the east side of Bayou Teche was situated opposite the tract of land that was situated on the west side of Bayou Teche. The tract measured three arpents frontage along Bayou Teche with a depth of forty arpents. Also, the tract, which was situated on the east side of Bayou Teche was bounded above by lands belonging to Baron Benoit Bayard, and below by land belonging to Baron Benoit Bayard. The tract of land that was situated on the west side of Bayou Teche was subject to a reservation of forty feet in width on the upper side from Bayou Teche backward to the depth of the land.

Stanley C. Hartman, of the Parish of St. Mary, declared that there are no mortgages bearing or existing on the property sold, except an original mortgage in favor of Francois Optat Darby. Also, the debt for which the amount of the mortgage was given has been paid. Both Stanley and Millington Hartman stated to William Robertson, who served as notary public for this act of conveyance, that they promise and obligate themselves to raise all mortgage payments now existing on the property, which is being sold before the maturity of the first note and before demanding payment of the note. Also, Stanley and Millington Hartman promise to give to Evert Bunker Smeedes, the purchaser, a certificate from the Recorder of St. Mary Parish stating that Stanley and Millington Hartman promise and obligate themselves to pay all mortgage payments on existing mortgages.

The purchase price of this act of conveyance was made and accepted for the purchase price of $6,000.00. One third of the purchase price or $2,000.00 was paid in cash by Evert Banker Smeedes at the execution of this act of conveyance. The balance of the amount of the purchase price or $4,000.00 is to be paid in two equal annual installments. These installments are to be paid in one year or on the date of April 10, 1861, and in two years or on the date of April 10, 1862.

On the date of April 29, 1861. James G. Parkerson, who was a notary public in the Parish of St. Mary, was authorized and empowered to take an inventory of all the immovable and movable property belonging to the community of acquets and gains, which existed between the late Evert Bunker Smeedes and his surviving widow, Sarah Cade Smeedes. Also, James G. Parkerson was authorized to make a return to the probate court of the Third Judicial District Court, which was holding court in the Parish of St. Mary, of the property consisting of immovable and movable property that belonged to the community of acquets and gains.

Prior to this date of April 29, 1861, Evert B. Smeedes had passed away at his residence home, Mintmere, which was situated in upper St. Mary Parish, Louisiana.

Joshua Baker and Wilson McKerall were solemnly sworn that they will and faithfully appraise the property and effects, rights, and credits shown and designated as belonging to the estate of the late Evert B. Smeedes. Also, this property and effects belonged to the community of acquest and gains, which existed during the marriage of Evert Bunker Smeedes and his surviving widow, Sarah Cade Smeedes. In the State of Louisiana, the community of acquets and gains exists between a husband and a wife during their marriage, but if one passes away or a divorce takes effect, then the community of acquets and gains is dissolved. When one partner dies, the community of acquets and gains becomes a part of the estate of the deceased partner. According to the laws of the State of Louisiana, the community of acquets and gains consists of all property that has been acquired during the years of their marriage. Also, in the State of Louisiana, when a married couple purchases a home together, that home will be considered as belonging to the community of acquets and gains. Unless, one married partner files a petition with the clerk of court stating that a certain tract of land or a home is being purchased as their separate paraphernal property, all property purchased during marriage can be considered as belonging to the community of acquets and gains.

On the date of April 29, 1861, James G. Parkerson made an inventory of the immovable and movable property that belonged to the estate of the late Evert B. Smeedes and to the community of acquest and gains, which existed during the marriage of the late Evert B. Smeedes and his surviving widow, Sarah Cade Smeedes. The immovable property that belonged to the estate of the late E. B. Smeedes was a certain tract of land, which was situated in upper St. Mary Parish, Louisiana, on the west side of Bayou Teche. This tract of land, which was located on the west side of Bayou Teche had a frontage along Bayou Teche of three arpents, with a depth of forty arpents. Also, there was another tract of land, which was situated on the east side of Bayou Teche, directly opposite the tract, which was situated on the west side of Bayou Teche. This tract of land that was located on the east side of Bayou Teche also had a frontage of three arpents along Bayou Teche, with a depth of forty arpents. The two tracts of land were bounded on the upperside by land belonging to John F. Miller, and on the lowerside by land belonging to Baron Benoit Bayard. Also, included in this inventory was the dwelling house, known as Mintmere, the buildings that were situated on the land, and all the improvements existing on the land. The total valuation of this property was $15,000.00.

These slaves were considered movable property and these slaves belonged to the estate of the late Evert Bunker Smeedes. The slaves that were living on the Smeedes Plantation or at the place known as Mintmere were Anne, a colored slave woman, who was age 45 years old, and her six children. The six children of the slave woman, Anne, were Nathan, a slave man, and he was 20 years of age, Ephriam, a teenage boy slave, and he was 16 years of age, Cripy, a slave adolescent child, and Cripy was 12 years of age, Maria, a girl slave child, and she was 10 years of age, Ned, a slave child of 5 years of age, and Nat, a two years old toddler slave child. The slave woman, Anne, and her six children, Nathan, Ephriam, Cripy, Maria, Ned, and Nat were valued at the amount of $4,200.00.

Sam, a slave man, who was about 23 years of age, and his wife, Easther, who was also 23 years of age. The slave man, Sam, and his wife, Easther, were the parents of two children, Aaron, a slave boy, who was 3 years old, and Daniel, a slave toddler, who was one years old. The slave man, Sam, and his wife, Easther, and their two children, Aaron, and Daniel, were valued at the price of $3,000.00.

Sandy, a slave man, and his wife, Nelly, a slave woman, were also residing on the Smeedes Plantation. The slave man, Sandy, was about 65 years of age, and Nelly, his wife, was about 45 years of age. Sandy and Nelly had four children, and their children were Lorenzo, who was a teenager age 15 years of age, Sandy, another teenage slave child, who was age 13 years of age, Randall, a slave boy, who was age 12 years of age, and Fray, who was a slave child, age 10 years old. Sandy and his wife, Nelly, and their four children, Lorenzo, Sandy, Randall, and Fray were all valued at the price of $4,000.00.

Juluis, a slave man, age 30 years of age, and he was valued at $400.00.

Peter, a slave man, and he was age 65 years old. The slave man, Peter, was valued at the price of $200.00.

Susan, a slave woman, and her child, Jim, who was age 2 years of age, were also residing on the Smeedes Plantation. The slave woman, Susan, and her child, Jim, were valued at the price of $1,200.00.

The livestock that were on the Smeedes Plantation at the time of the inventory was taken,in the month of April of the year of 1861, consisted of three match American horses, which were valued at $200.00 each horse, and the total valuation was $600.00. One American horse, which was valued at the price of $100.00. One pair of mules, which were valued at $200.00. One milk cow and calves, which were valued at $60.00. This was all the livestock that was owned by Evert Bunker Smeedes on his farm, and this livestock was considered movable property. Also, this livestock was considered as belonging to the community of acquets and gains, which existed during the marriage of Evert Bunker Smeedes and his wife, Sarah Cade Smeedes. After the death of Evert Bunker Smeedes, this livestock was considered as belonging to both the estate of Evert Bunker Smeedes and the community of acquets and gains of the marriage of E. B. Smeedes and his wife, Sarah Cade Smeedes.

This movable property also belonged to the estate of the late E. B. Smeedes and to the community of acquets and gains, which existed during the marriage between the late E. B. Smeedes and his surviving widow, Sarah Cade Smeedes. The movable property consisted of one buggy and harness, which was valued at the price of $150.00, one Jersey wagon with a double harness, which was valued at $100.00, one horse cart, which was valued at the price of $60.00, two plows and gearing and harrows, which was valued at the price of $30.00, household furniture, which was contained in the plantation home, Mintmere, and was valued at the price of $1,000.00., one library, which was valued at $350.00, one lot of wines and liquors, which was valued at $250.00, and one double trunk, which was valued at the price of $25.00. The total valuation of the immovable and movable property that belonged to the estate of the late E. B. Smeedes, and to the community of acquets and gains, which existed between the late E. B. Smeedes and his surviving widow, Sarah Cade Smeedes, amounted to the total amount of $30,925.00.

On the date of May 7, 1861, Sarah Cade Smeedes, the widow of the late E. B. Smeedes, solemnly swore that she would truly perform the duties incumbent upon her as natural tutrix of her four minor children, Mary Cade, Robert Cade, Charles Edward Cade, and

Henry D. Cade.

On the date of April 21, 1860, Francois Optat Darby granted a release of mortgage to Leonie Hartman, and this mortgage resulted from the sale of the property on which the residence home, known as Mintmere, was situated. Francois Optat Darby was the seller,of the two tracts of land to Leonie Hartman, and the act of conveyance that existed between these two parties was a sale and mortgage. This mortgage that was being released by this act was the original mortgage that existed from the act of conveyance.

The mortgage was dated February 13, 1857, and this mortgage was drawn by Leonie Hartman, of St. Mary Parish, in favor of Francois Optat Darby for the sum of $3,000.00. Also, this act of mortgage was paraphed to the act of conveyance, which act of conveyance bore the date of February 13, 1857. This mortgage was secured by an act importing a mortgage on the date of February 13, 1857, and this mortgage was bearing on the property, which consisted of two tracts of land, which were situated on both the east and the west sides of Bayou Teche. The property was sold by Francois Optat Darby to Leonie Hartman on the date of February 13, 1857. Then, Leonie Hartman sold these two tracts of land to her two sons, Stanley C. Hartman, and Millington Michael Hartman on the date of September 9, 1859. Also, this act that imported the mortgage, resulted in Francois Optat Darby retaining an special mortgage and vendor's privilege on the two tracts of land, which were situated on both the east and the west sides of Bayou Teche. The reason that Francois Optat Darby was able to obtain a special mortgage and a vendor's privilege was to secure the payment of the promissory note. Also, William Robertson, who was a notary public in the Parish of St. Martin, certified that the signature on the reverse side of the promissory note are genuine and that the record of the mortgage, which is recorded in the office of the clerk of court in the courthouse of the Parish of St. Mary may be cancelled in full. This cancellation of this mortgage was given at the office of William Robertson, which office was situated in the Parish of St. Martin and the town of New Iberia.

On the date of September 9, 1859, Leonie Hartman, who was the widow of the late Michael Hartman, conveyed to her son, Stanley C. Hartman, of the Parish of St. Mary, three slaves. James G. Parkerson, a notary public, presided at the execution of the conveyance of these three slaves. This conveyance was a sale and mortgage.

The three slaves that were conveyed were a certain colored slave boy, named Sylvania, who was a teenage slave, only age 14 years old. Sylvania, the teenage slave, was sold for the purchase price of $1,000.00. Anne, a colored woman slave, and she was 30 years of age. Also, Anne, the slave woman, was a sickly person. Anne, the slave woman, was sold for the purchase price of $500.00. Frederick, who was an orphan and a slave child of the tender age of four years old was also sold. Frederick, the slave child, was sold for the purchase price of $500.00. All three slaves, Sylvania, Anne, and Frederick, were all sold for the total purchase price of $2,000.00. Also, all three slaves, Sylvania, Anne, and Frederick, were all slaves for life unto their new owner, Stanley C. Hartman. Stanley C. Hartman knew all three slaves very well, as they were the slaves, who were working for his mother, Leonie Theriot Hartman, at the plantation, now known as Mintmere. At this time, Stanley C. Hartman took all three slaves Anne, Sylvania, and Frederick, at his (Stanley C. Hartman's) own peril and risk, as regards to vices, defects, and redhibitory prescribed by the laws of the State of Louisiana.

The purchaser, Stanley C. Hartman, obligates himself to pay over to his sister, Rosalie Silman Hartman, her share of the purchase price of the three slaves. At the time of this sale of the three slaves, Rosalie Silman Hartman was a minor under the age of majority and under notarial tutorship of her mother, Leonie Theriot Hartman, who was the widow of Michael Hartman, of St. Mary Parish. Also, Leonie Theriot Hartman swore that her minor daughter, Rosalie S. Hartman, will attain her majority age or her otherwise being emancipated before the promissory note is due. The promissory note is made for the principal of $2,000.00, and the promissory note is conditioned to bear interest at the rate of 5% per annum.

Also, the three slaves, Sylvania, Anne, and Frederick, will remain specially mortgaged, affected, and hypothecated in favor of Leonie Theriot Hartman, the seller of the slaves, until the full and entire payment of all the principal and interest. The full payment of the promissory note is $2,100.00.

On the date of August 4, 1859, Madame Leonie Hartman sold the property to her two sons, Stanley C. Hartman and Millington Michael Hartman. Madame Leonie Theriot Hartman was the widow of the late Dwight Michael Hartman, of the Parish of St. Mary.

At this time, the property consisted of two tracts of land. The first tract of land was situated on the west side of Bayou Teche, and this tract had a frontage of three arpents along Bayou Teche with a depth of forty arpents. Also, the tract situated on the east side of Bayou Teche had a frontage of three arpents along Bayou Teche with a depth of forty arpents. The property was bounded on the upperside by the lands of Cordelia D. Lewis, and on the lowerside by the lands of Baron Benoit Bayard. Also, this property was partly situated in the Parish of St. Mary and the Parish of St. Martin.

The purchase price of these two tracts of land was $2,150.00. Also, there was a mortgage bearing on the property. Originally, this mortgage was in favor of Francois Optat Darby. Also, the amount of the mortgage was $1,900.00. At the time of the conveyance of the property, in the year of 1859, Henry Knight was the owner or the holder of the mortgage.

Also, conveyed with the two tracts of land was these slaves. A colored slave woman, named Charlotte, who was age 21 years of age, and her child, Maria, who was two years of age. Both the slave woman, Charlotte, and her child, Maria, were sold for the purchase price of $1,250.00. Also, a colored slave man, named Moses, was sold. The colored man slave, Moses, was sold for the purchase price of $1,200.00. Another colored man slave, Nat, was also sold. Nat, the colored slave man, was sold for the purchase price of $1,600.00.

Also, these slaves were to remain mortgage by the original mortgage, which was now in favor of Henry Knight. The slaves that were affected by the mortgage were a slave man, Jim, age 21 years old, and a slave woman, who was about 35 years of age, Calvin, a slave child, who was age 12 years old, Hamitt, a slave child, who was 11 years of age, and another slave child, Frederick, who was age 4 years old. All these slaves were to remain mortgaged to Henry Knight until the payment and satisfaction of the mortgage by Madame Leonie Theriot Hartman, of the Parish of St. Mary.

On the date of February 18, 1859, James Parkerson, a notary public for the Parish of St. Mary, certified that a certain promissory note, which was signed by Francois Optat Darby, and made payable to Hortense Berard Duperier, who was the widow of Frederick Duperier, has been presented to me (James G. Parkerson) by the drawer, Francois Optat Darby, and the promissory note has been fully satisfied. This promissory note originated from the sale of the property by Hortense Berard Duperier to Francois Optat Darby, which act of conveyance was held on the date of July 28, 1853, and this act of conveyance was executed by William Robertson. Also, this promissory note was paraphed, "Ne Varietur," by William Robertson, the presiding notary public, to identify this promissory note with this act of conveyance. The promissory note was made for the sum of $3,000.00, and it was conditioned to bear interest at the rate of 8% per annum from the date until paid in full. This mortgage was secured by a vendor's privilege and a vendor's lien which was bearing on the property. Since the mortgage has been fully satisfied, now the promissory notes may be cancelled, annulled, and erased on the books of record, which are recorded in the office of the clerk of court in the Parish of St. Mary.

On the date of August 10, 1857, Leonie Theriot Hartman, who was the widow of the late Dwight Michael Hartman, sold to her sons, Stanley C. and Millington Michael Hartman, a slave. The slave that was sold was named, Nat, who was age 26 years of age. The slave man, Nat, was sold for the purchase price of $1,600.00.

Also, on this same date, Leonie Theriot Hartman declared that whereas the two tracts of land, which was sold by her to her two sons, Stanley C. and Millington Michael Hartman, is affected by a special mortgage. This special mortgage originally was in favor of Francois Optat Darby. Now, on the date of August 10, 1857, William Henry Knight is the present holder of the mortgage, and the mortgage represents the two tracts of land, which are situated on both the east and the west sides of Bayou Teche. Also, with the interest agreed, this promissory note amounts to the total of $1,900.00. In order for Leonie Theriot Hartman to keep her two sons, Stanley C. Hartman and Millington Michael Hartman, harmless from the liability of the mortgage, and from having to pay the full amount or any amount of the promissory note, Leonie Hartman specially mortgages in favor of her two sons, Stanley C. and Millington M. Hartman, a certain slave man, named Jim, who is age 21 years old, and a colored slave woman, named, Anne, who is age 35 years of age, another slave named, Calvin, who is a child of 12 years of age, Hewitt, a slave child, who is age 11 years of age, and Frederick, a slave child, who is four years old. These five slaves are to remain mortgaged until the final payment and the satisfaction by Leonie Theriot Hartman of the mortgage.

On the date of February 13, 1857, Francois Optat Darby sold to Leonie Theriot Hartman, who was the widow of Dwight Michael Hartman, of St. Mary Parish, a tract of land, which tract of land was situated in upper St. Mary Parish. Actually, this tract of land was situated partly in St. Mary Parish and partly in St. Martin Parish. Also, this tract of land had a frontage of three arpents on the right bank of Bayou Teche by a depth of forty arpents. This tract of land was bounded above by land belonging to John F. Miller, and below by land belonging to Baron Benoit Bayard. Also, this tract of land was situated on the west bank of Bayou Teche. During the spring of 1857, construction was started on the plantation home, known as Mintmere.

Francois Optat Darby reserves a strip forty feet wide on the upper side of the land for the purpose of a public road, which road will lead from Bayou Teche to the prairie along the whole depth of the forty arpents. Also, Francois Optat Darby sells a certain other tract of land, which is situated on the east side of Bayou Teche. This tract of land had a frontage of three arpents along Bayou Teche with a depth of forty arpents. Also, the boundary of this tract of land, which was situated on the east bank of Bayou Teche were on the lowerside by land belonging to Baron Benoit Bayard.

The purchase price of these two tracts of land was the sum of $3,500.00, which was payable as a sale and mortgage. The sum of $500.00 was paid in cash at the execution of this act of conveyance, and the balance of $2,900.00 was to be paid in installments of one or two years at the option of the purchaser, Leonie Theriot Hartman. Also, this mortgage was to bear interest at the rate of 8% per annum from the date until paid. Francois Optat Darby acknowledges the receipt of the sum of the downpayment of $500.00 in cash money and the promissory note. The promissory note was a drawn by Leonie Theriot Hartman and signed by her, and made to the order of Francois Optat Darby. Also, the promissory note was paraphed, "Ne Varietur," to identify the promissory note with this act of conveyance. Francois Optat Darby retains a mortgage on the land conveyed to Leonie Theriot Hartman, and to secure the payment of the promissory note there exists a vendor's lien and a vendor's privilege.

The parties, Francois Optat Darby and Leonie Theriot Hartman, dispense with the certificate required by Article 3328 of the Civil Code of the State of Louisiana. Article 3328 of the Civil Code of the State of Louisiana exonerates the notary public or lawyer from all liability in the premises of the land being conveyed. Therefore, the parties to an act of conveyance are able to dispense with Article 3328, which releases the notary public or the lawyer from all liability.

This act of conveyance was executed in the town of New Iberia, on the St. Martin Parish side, on the date of February 13, 1857. Also, the two witnesses to the execution of the act of conveyance were Dorisons Deblanc and Jean B. Mistrot.

After the signing of the act of conveyance, it was discovered that an error had been committed in not making the interest that may accrue on the promissory note. The interest was payable annually, and it was expressly understood that as a part of the conditions of the sale that the interest shall be paid annually. The payments of the promissory note shall extinguish the principal between one and two years.

After purchasing the two tracts of land, in the spring of the year of 1857, Leonie Theriot Hartman started constructing the plantation home, which is known as Mintmere. The plantation home, Mintmere, was constructed with slave labor. After the plantation home, Mintmere, was completed, it was occupied with the family of either Stanley C. or Millington Michael Hartman.

Francois Optat Darby purchased at a sale and mortgage two tracts of land, which were situated partly in St. Martin Parish and St. Mary Parish from Hortense Berard Duperier, who was the widow of Frederick Duperier. Hortense Berard was the daughter of Marie Hortense Boutte, by her first marriage to Achille Berard. Marie Hortense Boutte Berard Bayard originally owned these two tracts of land, and gave these two tracts of land to her daughter, Hortense Duperier. Also, this tract of land, which the present day plantation home, Mintmere, is situated on, was once a part of a plantation, which was owned by Francois Cesear Boutte. Francois Cesear Boutte was a very wealthy planter, and an extensive landowner, and was also the father of Marie Hortense Boutte, the mother of Hortense Berard Duperier. Marie Hortense Boutte was married first to Achille Berard and secondly to Baron Benoit

Bayard. Francois Cesear Boutte owned this tract of land, and he gave it to his daughter, Marie Hortense Boutte prior or just after her marriage to her first husband, Achille Berard.

The tract of land that Francois Optat Darby purchased consisted of two tracts of land, which were directly opposite each other on both the east and the west sides of Bayou Teche. The tract of land, which was situated on the west side or the right bank of Bayou Teche had a frontage of three arpents along Bayou Teche, with a depth of forty arpents. This tract of land was bounded on the upperside by land belonging to John F. Miller, and on the lowerside by land belonging to Baron Benoit Bayard. Also, the tract of land, which was situated on the east side or the left bank of Bayou Teche had a frontage of three arpents with a depth of forty arpents. This tract of land, which was situated on the east side of Bayou Teche was bounded on the upperside by land belonging to John F. Miller, and on the lowerside by land belonging to Baron Benoit Bayard.

This sale was made for the purchase price of $3,000.00, which was payable in twelve months or on the date of July 29, 1854. Also, this mortgage was conditioned to bear interest at the rate of 8% per annum from the date of July 29, 1853 or until paid in full. At the execution of the act of conveyance, Hortense Berard Duperier acknowledged the receipt of a promissory note of even date, which promissory note was drawn by Francois Optat Darby. The land is to remain specially mortgaged to Hortense Berard Duperier, the seller, until the full and final payment of the promissory note. Hortense Berard Duperier had a vendor's lien and a vendor's privilege existing on the land until the full payment of the promissory note.

Both, Francois Optat Darby and Hortense Berard Duperier, dispense with Article 3328 of the Louisiana Civil Code of the State of Louisiana, which article exonerates the notary public from all liability in the premises of the land conveyed. In this case, William Robertson, who served as notary public, was exonerated by both Francois Optat Darby and Hortense Berard Duperier, who was the widow of Frederick Duperier, from all liability in the premises of the land being conveyed.

A view of East Main Street in the town of New Iberia, Louisiana, which once formed a part of the old Bayard plantation

THE BAYARD PLANTATION

The Bayard Plantation was once located on what is now presently known as East Main Street in the town of New Iberia, Louisiana. The property line on the west side of Bayou Teche of the Bayard Plantation was situated on the property, now known as the Gonsoulin properties, and owned by the Gonsoulin Brothers. This tract of land includes the former family home of the Ulysse Gonsoulin, Senior family, the site of the old Harry Broussard Grocery Store, which is situated at 2210 East Main Street, the site of Teche Business Supplies, which is situated at 1985 East Main Street, the Davis Alignment Center, which is situated at 1612 East Main Street, the Teche Motel, which is situated at 1830 East Main Street, the site of the office of the architectural firm of Beyt, Rish, Robbins, Inc., Architects and Planners, which is situated at 1805 East Main Street, the former site of Clay's Boat's and Motor's, and formerly the site of Main Motors, which was a former Dodge and Chrysler automobile dealership, which was once owned by Hawly Gary, which is situated at 1925 East Main Street, the site of Foster Auto Repair, which is situated at 1801 East Main Street, the site of the old Caulking grocery store, the site of Iberia Professional Center, which is situated at 1612 East Main Street, the site of the former Kroger Store and the former Canal Villere Store, now known as Simoneaud's East and the Iberia Bowl, which is situated at 1501 East Main Street.

On the east side of Bayou Teche, the Bayard Plantation was situated and ran along what is now known as the Old Jeanerette Road. On the east side of Bayou Teche, the Bayard Plantation was located on the property, which is now presently known as Palmland Drive, Stockstill Street, Belle Haven Drive, and Grand Prix Avenue.

On the date of January 20, 1817, Francois Cesear Boutte' sold to his daughter, Marie Hortense Boutte', who was the widow of Achille Berard, two slaves. These two slaves resided on the former Boutte' plantation, but now the plantation was owned by Marie Hortense Boutte', who was the widow of the late Achille Berard. Judge John Wilkerson, who was serving as judge in the Parish of St. Mary, presided at the execution of this act of conveyance.

The two slaves that were sold was a colored man slave, named, Paul who was about 20 years of age, and a colored woman slave, named Euphrosine. The slave man, Paul, was an African slave by birth of the Congo nation, meaning that the slave, Paul, was a pure African slave. The slave, Paul, was not considered a mulatto, a quadroon, or a gringo, which means that the slave's bloodlines were not pure African, but the blood lines had been mixed with another blood line of another race. Also, the slave woman, Euphrosine, was about 20 years of age. The slave woman, Euphrosine, was considered a creole, meaning that her bloodline had probably been mixed with Spanish blood.

On November 17, 1821, Marie Hortense Boutte', who was the widow of the late Achille Berard, was confirmed by the Judge of the Probate Court, in St. Mary Parish, Louisiana, as natural tutrix for her two minor children, Achille Berard and Achille Camille Berard. A family meeting was held previous to the date of November 17, 1821, and the reason for this family meeting was the consequence of the marriage of Marie Hortense Boutte, the widow of Achille Berard, to Baron Benoit Bayard, a native of France.

Also, on the date of November 17, 1821, Marie Hortense Boutte', co-jointly with her husband, Baron Benoit Bayard as principal, and Colonel Charles Napoleon Olivier as surety, appealed to the Parish Judge of St. Mary Parish for the sum of $10,250.00. This sum of $10,250.00 was owed to Marie Hortense Boutte', the widow of Achille Berard, and now the wife of Baron Benoit Bayard, for her faithful administration as tutrix of the estate of the late Achille Berard for her two minor children, and her children of the age of majority.

Marie Hortense Boutte', who was now married to Baron Benoit Bayard, gave Colonel Charles Napoleon Olivier a mortgage on a certain tract of land as security. Also, Hortense Boutte' Bayard was wishing to release her property from the decree and tacit mortgage, which resulted from her appointment by the Judge of the Probate Court in St. Mary Parish as natural tutrix of her two minor sons. Hortense Bayard prayed that the judge would cause a family meeting to be commenced at an early date as possible in order to deliberate on the property. Hortense Bayard wished to offer a certain tract of land to be specially mortgaged to her two minor sons, Achille Berard and Achille Camille Berard. The reason that this property had to be mortgaged was to secure that the person appointed natural tutrix would handle the administration of an estate in an honest and proper way for minor children or orphans. Therefore, the natural tutor or tutrix had to mortgage their plantations, property, or other property of value to insure against any mishandlement. Also, the legislature of the State of Louisiana had passed an act that persons appointed natural tutor or tutrix had to mortgage property to insure against the mishandlement of estates involving minors of which that person had been appointed natural tutor or tutrix.

On the date of November 14, 1822, Baron Benoit Bayard, who was the second husband of Marie Hortense Boutte', purchased a slave woman and her two children from Madeline Borel. The slave woman that was purchased was named Celeste, and she was 24 years of age, and her two children were Cola, a slave boy of six years of age, and Jenny, a slave girl of two years of age. The total purchase price paid by Baron Benoit Bayard for the slave woman, Celeste, and her two children, Jenny and Cola, was the sum of $900.00. Also, the purchase price of $900.00 was paid in cash money at the execution of this act of conveyance of the slaves. Now, Celeste, the slave woman, and her two children, Jenny and Cola, were slaves for life to Baron Benoit Bayard.

Also, on the date of December 20, 1825, Baron Benoit Bayard purchased five slaves from J. Frederick. The five slaves were Charles, a colored man slave, Suzanne, a colored woman slave, Jess, a mulatto slave man, Issac, a colored man slave, and Pour, a slave. All these five slaves together with the slave woman, Celeste and her two children, Jenny and Cola, resided in the slave quarters of the Bayard Plantation.

On the date of October 19, 1827 the last testament and will of Francois Cesear Boutte' was read and probated at the plantation home of his son-in-law, Baron Benoit Bayard I, and this plantation home was located on the Bayard Plantation. The Bayard Plantation was situated on both the east and the west sides of Bayou Teche, on land that presently comprises East Main Street in the town of New Iberia, Louisiana. The Bayard Plantation home was located on that part of the plantation, which presently comprises East Main Street, and this plantation home was situated directly across East Main street from Simoneaud's Grocery Store and the Iberia Bowl.

Charles Napoleon Olivier and Frederick Pellerin went to the plantation home of Baron Benoit Bayard I for the purpose of making an inventory of the property belonging to the estate of the late Francois Cesear Boutte'. Also, Charles Napoleon Olivier and Frederick Pellerin were making an appraisal of the property, by estimating the property belonging to the estate. Charles Napoleon Olivier and Frederick Pellerin estimated and appraised the property belonging to the estate of the late Francois Cesear Boutte' in the presence of Claude Frilot and Louis DeBlanc.

The property belonging to the estate or the succession of the late Francois Cesear Boutte' consisted of two pairs of oxen, at the price of $500.00, three pairs of work oxen at the price of $90.00, one steer, which was three years old, at the price of $9.00, three bulls, which are three years old, at the price of $18.00, six bulls, which are one year old, at the price of $24.00, two hefters, which are one years old, at the price of $12.00, one hefter, which is two years old, at the price of $7.00, two cows and calves at the price of $12.00 each or the total price of $24.00, a herd of twenty nine head of sheep, the sheep were appraised at $1.00 each, and the total price of the herd of sheep was $29.00, ten cows and calves at the price of $100.00, six cows at the price of $48.00, four 1 year old hefter at the price of $16.00, one hefter, which is two years old, at the price of $5.00, five mares and colts at the price of $75.00, six mares at the price of $60.00, two colts, which are one years old, at the price of $10.00, one stud, which is four years old, at the price of $25.00, one old mule at the price of $6.00, one mule at the price of $25.00, and three horses at the price of $60.00. This was all of the livestock that belonged to the estate of the late Francois Cesear Boutte', and this livestock was situated on the Bayard Plantation.

The equipment that belonged to the estate of the late Francois Cesear Boutte' consisted of one cotton pump, which pump is old and worn out, and the cotton pump is worth nothing, one old horse collar, for Jim, is worth $70.00, one magazine is worth $30.00, one negro cabin is worth $15.00, one old house is worth $12.00, and one small magazine is worth $10.00. The total amount of the equipment, which is situated on the Bayard Plantation, and belongs to the estate of the late Francois Cesear Boutte' is worth the sum of $99.30. Francois Cesear Boutte' was the father of the wife of Baron Benoit Bayard, I, who was Marie Hortense Boutte'.

Also, Francois Cesear Boutte' owned property in the Parish of Iberville, which is situated in the State of Louisiana. On the date of November 9, 1827, Judge John Dutton of the Parish of Iberville, authorized the probate courts in the Parishes of St. Martin and St. Mary to proceed with the sale of the property belonging to the succession of the late Francois Cesear Boutte'.

At the succession sale of the property belonging to his late father-in-law's estate, Baron Benoit Bayard I purchased the oxen for the price of $36.00, two young steers for the price of $23.00, twenty four head of sheep for the price of $24.00, one stud horse for the price of $30.50, one house for the cotton gin at the price of $70.00, one magazine at the price of $30.00, one old house at the price of $12.00, and one small magazine for the price of $10.00. A magazine is a place where the gunpowder was stored on the plantation.

A family meeting to discuss the interests and rights of the minor children of the late Achille Berard and his surviving widow, Marie Hortense Boutte' Berard, now the wife of Baron Benoit Bayard, was held on the date of September 15, 1832. This family meeting was held at the plantation home of Baron Benoit Bayard, the step father of the minor children. On Friday, September 15, 1832, at 5:00 P.M. in the afternoon, a family meeting was held. The family meeting was composed of Francois Cesear Boutte, pere, the Grandfather of the minor children, Francois Cesear Boutte', fils, a maternal uncle of the minor children, Jean B. Berard, an older brother of the minor children, and he was of the age of majority, and Joseph V. Boutte', a cousin of the minor children. All the persons attending the family meeting were the nearest relatives, who were residing in St. Mary Parish of the Berard minor children. Francois Cesear Boutte', who was the maternal Grandfather and also the father of Marie Hortense Boutte', who was the widow of Achille Berard, and now the wife of Baron B. Bayard, was appointed under tutor of the Berard minor children. At this family meeting, Judge Donelson Caffery, of the Parish of St. Mary, presided.

On the date of March 11, 1830, the legislature of the State of Louisiana approved an act, which act was entitled, "An Act in Addition to the Laws now in force to Tutors and Curators of Minors." At the family meeting, which family meeting was held to discuss the rights and interests of the two minor children of the late Achille Berard, was read to the tutors. Also, this act of the legislature of the State of Louisiana was read to the under tutors of the minor children, and this was done in compliance with the provisions of the seventh section of the act.

Also, on the fifteenth day of the month of September of the year of 1832, Marie Hortense Boutte', the widow of Achille Berard, and now the wife of Baron B. Bayard, and Baron B. Bayard appeared before Judge Donelson Caffery of the Parish of St. Mary, in regards to the tutorship of the two minor Berard children. Hortense Boutte' Bayard was the mother and the natural tutrix of the two minor Berard children, Achille and Camille Berard. In the presence of the judge, Baron B. Bayard acknowledged that to the best of his knowledge and belief that the amount due to Hortense Boutte' Berard Bayard and himself (Baron B. Bayard) as tutors of the two minor Berard children, as well as for the inherited amount due to the minor Berard children from the succession of their late paternal Grandfather, Jean Baptiste Beard, is the sum of $8,700.00 in capital and interest. The two minor Berard children, Achille and Camille, inherited from the estate of their late Grandfather, Jean B. Berard, in representation of their deceased father, Achille Berard. Also, Hortense Boutte' Bayard inherited from her late father-in-law's estate in representation of her late husband's, Achille Berard, as her share of the community of acquets and gains which existed during her marriage to the late Achille Berard.

Hortense Bayard and her husband, Baron B. Bayard, had presented the certificate of the recorder of mortgages for the Parish of St. Mary to Judge Donelson Caffery. There was only one mortgage, which was bearing on the Bayard plantation. This was a legal and tacit mortgage that was in the favor of the two minor Berard children, Achille and Camille.

The family meeting that was held at the plantation home of Mr. and Mrs. Baron B. Bayard duly and masterfully deliberated that the property of the two minor Berard children that was offered for sale was sufficient to secure the two Berard minor children in their rights, interest, and to faithfully discharge Marie Hortense Boutte' Berard Bayard, and Baron B. Bayard, who was the step father of the two minor children, in their function as a tutor and tutrix of the two minor children.

Also, on the date of September 15, 1832, the under tutor of the two minor Berard children, Louis DeBlanc, fils, (junior) having approved himself satisfied with the deliberations of the family meeting unanimously agrees that the aforesaid immovable property be received in a special mortgage. The balance of the property, which belongs to Marie Hortense Boutte' and Baron B. Bayard be released from the legal and tacit mortgage. This legal and tacit mortgage that was bearing on the Bayard property in favor of the two minor Berard children was also released.

Then on the date of September 15, 1832, Colonel Charles Napoleon Olivier, of the Parish of St. Mary, granted a release of a mortgage to Marie Hortense Boutte, who was the widow of the late Achille Berard and now the wife of Baron Benoit Bayard. The releasing of Hortense Bayard from the mortgage was done in the presence of Judge Donelson Caffery of the Parish of St. Mary.

At this time, Colonel Charles Napoleon Olivier came and declared in front of Judge Donelson Caffery, that on the seventeenth day of the month of November of the year of 1821, by act of mortgage, which was passed before former Judge John Wilkerson, that he (Colonel Charles Napoleon Olivier) signed as security in a bond given by Marie Hortense Boutte' Beard Bayard. This bond was given to secure that Marie Hortense Boutte, who was the widow of the late Achille Berard, and authorize administration in the functions of Marie Hortense Boutte' Berard as natural tutrix of her two minor children, Achille and Camille Berard. Achille and Camille Berard were the legal heirs and representatives of their late father, Achille Berard, for the sum of $10,250.00. This amount of $10,250.00 was payable to the Parish Judge of the Parish of St. Mary. Also, this bond, in the amount of $10,250.00, was to indemnify and secure Colonel Charles Napoleon Olivier, Marie Hortense Boutte' Berard, who is now the legal wife of Baron B. Bayard, and mortgage unto Colonel Charles Napoleon Olivier, a certain tract of land, which was situated in the Parish of St. Mary.

The tract of land, which was mortgaged in favor of Colonel Charles Napoleon Olivier was situated on both the west and the east sides of Bayou Teche, and this tract of land was bounded on both the west and the east sides of Bayou Teche on the upperside by land belonging to Robert Kyle McCarty, and on the lowerside by land belonging to Francois Cesear Boutte', pere. The french term, pere, means Senior in English. The tract of land, which was situated on the west side of Bayou Teche had a frontage along Bayou Teche of sixteen arpents with a depth of forty arpents. Also, seven slaves were included in this mortgage in favor of Colonel Charles Napoleon Olivier.

Colonel Charles Napoleon Olivier declared the release of the legal and tacit mortgage, and discharge of the property and the seven slaves from his favor. Now, the property and the seven slaves were free from all mortgages, privileges, and liens.

On the fourteenth day of the month of September of the year of 1832, Judge Donelson Caffery, of the Parish of St. Mary, appointed Jacques Fontenette and Jean B. Perett as experts to appraise the lands that belonged to the estate of the late Achille Berard, who was the late husband of Marie Hortense Boutte', now the wife of Baron Benoit Bayard. These lands may be offered by the Hortense Boutte' Berard Bayard and Baron Benoit Bayard, who is the stepfather of the minor children, for sale. Also, Louis DeBlanc, fils, was named and appointed under tutor to the two Berard minor children, Achille Berard and Achille Camille Berard.

Hortense Boutte' Bayard and Baron B. Bayard offered the immovable property that belonged to the estate of the late Achille Berard to be specially mortgaged in order to secure the two minor children of the late Achille Berard, in their rights both in capital and interest. The land that belonged to the estate was appraised at the sum of $7,100.00.

The immovable property that belonged to the estate of the late Achille Berard was a tract of land, which was situated in the Parish of St. Mary. This tract of land contained twenty one arpents of land fronting Bayou Teche with a depth of forty arpents. Also, this tract of land was situated on the west side of Bayou Teche. Another tract of land, which was situated on the east side of Bayou Teche, and this tract of land measured four arpents in depth with a twenty one arpent frontage along Bayou Teche.

In the year of 1832, the plantation of Francois Cesear Boutte' was located in upper St. Mary Parish on the east side of Bayou Teche. This plantation had a frontage along the east side of Bayou Teche of fifteen arpents of land with the depth of thirty six arpents of land. The total land area of this Boutte' plantation was five hundred and forty arpents of land or four hundred and fifty nine American acres of land. The boundaries of the plantation of Francois Cesear Boutte' was on the upperside by lands belonging to Francois Duplechin, and below by land belonging to Robert Kyle McCarty.

The slaves on the plantation of Francois Cesear Boutte' was a slave man, named Linon, who was 40 years of age, another slave man, Jo, who was 25 years of age, a slave man, Ned, who was 22 years of age, a slave man, T-bright, who was age 26 years old, a slave woman, Pierre, age 23 years of age, a slave man, Fontuen, who was 32 years of age, a slave woman, Josephine, who was 28 years of age, and a slave woman, Elina, who was 22 years of age. All of these slaves were slaves for life, and the property of Francois Cesear Boutte'.

On the date of May 19, 1832, Francois Cesear Boutte' sold to his son, Francois Cesear Boutte', fils, (which means junior) his plantation and slaves. Francois Cesear Boutte, fils, was a resident of the Parish of St. Martin, and he was the husband of the former, Marie Celeste Gonsoulin. At this act of conveyance, the two witnesses were Baron Benoit Bayard, and his wife, Marie Hortense Bayard. Marie Hortense Boutte' Bayard and Francois Cesear Boutte'', fils, were brother and sister. At this time, Francois Cesear Boutte'', pere, was appointing his son, Francois Cesear Boutte', fils, as his attorney in fact with full power to substitute any other person in his place and stead, to borrow in the City of New Orleans, Louisiana from any bank, or individual the amount of $3,500.00. This amount of $3,500.00 was to secure the debt with Francois Cesear Boutte', pere, and to authorize Francois Cesear Boutte', fils, his attorney in fact to appoint a person to mortgage the plantation.

On the date of June 11, 1832, Baron Benoit Bayard, I purchased a tract of land, which was seized in regards to the civil lawsuit of Jean Baptiste Delepere versus Francois Cesear Boutte'. Francois Cesear Boutte' was the father-in-law of Baron Benoit Bayard, I.

Benjamin Hudson, who was serving as Sheriff of the Parish of St. Mary, by virtue of a writ of seizure and sale, did seize an eight arpent tract of land. This writ of seizure and sale was issued out of the 5th District Court, which was holding court in the Parish of St. Mary. This writ of seizure and sale was issued against the goods, chattels, and lands of Francois Cesear Boutte'.

The tract of land had a frontage of two arpents of land on the west side of Bayou Teche with a depth of forty arpents. The boundaries of this tract of land were on the upperside by the lands of Francois Cesear Boutte' and below by the lands of Mager Duplechin.

The public sale was held on the date of May 25, 1832. At the sheriff sale, Sheriff Benjamin Hudson spoke in a loud and audible voice. First, he stated that there existed a certificate of mortgage on this tract of land, which is the property of Francois Cesear Boutte', and this certificate of mortgage is dated the date of May 24, 1832. Baron Benoit Bayard became the purchaser of this eighty arpent tract of land for the purchase price of $166.66. At the sheriff sale, Baron B. Bayard had to pay the amount of $166.66 in cash money to Sheriff Benjamin Hudson of the Parish of St. Mary.

Baron Benoit Bayard, the native of France, and the legal husband of Marie Hortense Boutte', died at the Bayard plantation in the month of October of the year of 1852. Marie Hortense Bayard was now a widow of her second marriage, and she was left with three sons, Baron Benoit Bayard, II, Hypolite Bayard, and Alfred Bayard.

The estate left by the late Baron Benoit Bayard I was opened and probated in the probate court of the Parish of St. Mary with Judge Cornelius Voorhies presiding. Also, the inventory of the estate of the late Baron Benoit Bayard, I, was taken by John Dumartrait, who was assisted by Eugene B. Olivier and Alexander Hebert.

The real estate property of the estate of the late Baron Benoit Bayard consisted of a tract of land being the second or double concellion of the land belonging in proper to Marie Hortense Boutte' Berard Bayard. This tract of land had a twenty one arpentage frontage along the backline of the Bayard plantation with a depth of forty arpents. The Bayard plantation contained a total of eight hundred and forty arpents of land. An arpent of land of the Bayard plantation was appraised at the sum of $4.00 an arpent. The total value of the Bayard plantation was the sum of $3,360.00.

Also, situated on the Bayard plantation were one dwelling house, which was the home of the family of the late Baron Benoit Bayard, and considered the Bayard plantation home. The total value of the Bayard dwelling house was the value of $500.00. Situated on the Berard plantation were twenty one cabins with outhouses, and the cabins were valued at $12.00 a cabin. The total valuation of the twenty one cabins and outhouses was the total sum of $252.00. The slave labor resided on the Bayard plantation in these cabins. Also, on the Bayard plantation was one small corn crib, and this corn crib was valued at the sum of $20.00.

There was another building on the Bayard plantation, which was the large cotton gin building, and the cotton gin building was valued at the price of $900.00. Also, one thousand and two hundred (1,200) fence pieux were valued at four cents a pieux. The total valuation of these fence pieuxs was $4.80. There were forty one sugar kettles on the Bayard plantation, and these sugar kettles were used to process the cane juice into raw, brown sugar on the Bayard plantation. The forty one sugar kettles were valued at the price of $27.00. Also, there were four broken sugar kettles, and these broken sugar kettles were valued at the sum of $5.00. On the Bayard plantation were fifty bales of cotton, which was the present crop of cotton that was raised and made on the Bayard plantation. The fifty bales of cotton were valued at $1,708.50. The livestock that was being raised on the Bayard plantation consisted of a herd of three hundred horned cattle. Each of the horned cattle were valued at the price of $7.00 each, and the total valuation of the herd of three hundred horned cattle was the price of $2,100.00.

The slaves that were residing on the Bayard plantation at the time of the death of Baron Benoit Bayard I, were Jeffe, a colored slave man, who as age 25 years old, and he was valued at the price of $1,000.00, John, a colored man slave, and he was 40 years old, and John was valued at the price of $1,000.00, Davis, a colored slave man, age 41 years old, and Davis was valued at the price of $800.00, Thomas, a colored man slave, age 45 years of age, and Thomas was valued at the price of $700.00, William, a colored man slave, age 19 years of age, and William was valued at the price of $900.00, Anthony, a slave man, age 40 years of age, and Anthony was valued at the price of $500.00, also, Anthony was considered a ruptured slave meaning that he was crippled, Touissant, a colored man slave, and he was 40 years of age, Touissant was valued at the price of $500.00, Etienne, a teenage boy slave, who was 16 years of age, and Etienne was valued at the price of $800.00, Johnny, a slave man, who was 20 years of age, and Johnny was valued at the price of $1,000.00, Lewis, a slave boy of 12 years of age, and Lewis was valued at the price of $700.00, Tom, a slave man, who was

50 years of age, and Tom was valued at the price of $500.00, Jean, a slave man, who was age 25 years of age, and Jean was valued at the price of $1,100.00, Charles, a slave man, who was 30 years of age, and Charles was valued at the price of $1,000.00, Colas, a slave man, who was age 30 years old, and Colas was valued at the price of $1,000.00, Christine, a colored woman slave, and she was a teenager of the age of 16 years old, and Christine was valued at the price of $750.00, Frances, a colored woman slave, who as 20 years old, and Frances was valued at the price of $700.00, Heneritte, a colored teenage slave girl, who was 14 years of age, and Heneritte was valued at the price of $600.00, Celestine, a colored teenage slave girl, and Celestine was valued at the price of $700.00, Delphine, a colored slave woman, who was 27 years of age, and Delphine was valued at the price of $700.00, Louise, a colored woman slave, who was 55 years of age, and Louise was valued at the price of $500.00, Charity a colored woman slave, who was 50 years of age, and Charity was valued at the price of $500.00, Suzanne was an old woman slave, who was 70 years of age, and Suzanne was valued at the price of $500.00, Suzanne, a slave woman, who was 20 years of age, and her infant were valued at the price of $700.00, Eugenee, a colored woman slave, who was 28 years of age, and her three children, Joe, a boy slave child, age 7 years of age, Mary, a slave girl child, who was age 4 years old, and Charles, a toddler slave child, who was age 16 months old, were valued at the price of $1,300.00, Panella, a colored slave woman, who was age 25 years old, and her three slave children, Nellie, a slave child of 6 years of age, Michael, a slave boy child, age 3 years of age, and an infant slave, and Panella and her three children were valued at the price of $1,300.00, Louise, a slave woman, age 30 years old, and Louise was valued at the price of $700.00, Felice, a colored slave child of 5 years of age, and Felice was valued at the price of $300.00, Cyrus, a slave boy child, who was 3 years of age, and Cyrus was the son of the slave woman, Louise, also Cyrus was valued at the price of $200.00, Ulzeide, a slave child, who was 1 years old, and a daughter of the slave woman, Louise, and Ulzeide was valued at the price of $100.00, Marie, a colored slave woman, who was 30 years of age, and her five children, Poupome, a slave child of 11 years of age, Annette, a girl slave child of 9 years of age, Marie, a girl slave child of 7 years of age, Jacques, a slave boy child who was 3 years of age, and Rose, a girl slave child, who was 2 years of age. Marie, the slave woman, and her five children were valued at the price of $2,000.00. Eliza, a slave woman and her slave child, Nicholas, a boy slave, were valued at the price of $1,200.00, Celeste, a colored slave woman, who was age 55 years old, and her child, Rose, a girls slave child of 10 years of age, were valued at the price of $600.00. These were the slaves that were residing on the Bayard plantation in the year of 1852. All these slaves were slaves for life.

The livestock that was situated on the Bayard plantation consisted of a herd of fifty one creole horses, five creole stallions, sheep, cattle, and oxen. The herd of fifty one creole horses were valued at the total valuation of $1,020.00, and each creole horse was valued at the price of $20.00. Each creole stallion was valued at the price of $20.00. There was a large herd of creole mares and colts, and this herd of mares and colts were valued at the price of $276.00. Each creole mare and colt were valued at the price of $4.00. Also, there was one creole saddle horse, which was valued at the price of $5.00. There was a herd of one hundred and forty sheep, and each sheep was valued at the price of $1.50. The total valuation of the one hundred and forty herd of sheep was the price of $210.00. The five oxen were valued at the price of $12.50 each oxen. Oxen were used on the plantations for plowing and pulling wagons and carts, and pulling barges of loads of sugar cane down Bayou Teche. Also, there was a herd of one hundred and sixty two cattle, and each cow, calf, or bull was worth the price of $10.00. The total valuation of the herd of one hundred and sixty two cattle was the price of $1,620.00.

There were two ox carts that were valued at the price of $20.00, and the total valuation of the two ox carts was $40.00. Also, there was one gig, which was valued at the price of $50.00.

There was a judgment against Joseph D. Babineaux for the amount of which it was not known at this time. This is a doubtful claim, and Joseph D. Babineaux has surrendered his property to his creditors. The amount was later discovered to be the sum of $840.00.

The land and the slaves that belonged to the estate of the late Baron Benoit Bayard was sold at a public auction. The terms of the sale were one, two, and three years after the date of the sale. The movable property for all sums under $100.00 was sold for cash, and for all sums over $100.00 was sold on a credit basis. All purchasers, who purchased on credit, had to give good security in soldio. Also, the promissory notes resulting from the sale on credit of the Bayard immovable and movable property were to be made to the order of Alfred Duperier, who was serving in his capacity as administrator of the estate of the late Baron Benoit Bayard. All property sold on credit was susceptible of mortgage, and the property was to remain specially mortgaged until the full and final payment.

On the date of February 16, 1869, the heirs and legal representatives of the late Marie Hortense Boutte', who was the widow of her first marriage to Achille Berard, and the widow of her second marriage to Baron Benoit Bayard, filed and probated the estate of their late mother. The estate of the late Mrs. Bayard was probated in the probate court of the newly created Parish of Iberia. The heirs and legal representatives of the late Marie Hortense Boutte' Berard Bayard were the children of her first marriage, Camille Berard, Hortense Berard Duperier, who was the widow of the late Frederick Duperier, Achille Berard, Ernest Berard, and Mathilda Berard Duperier, who was the wife of Doctor

Frederick Duperier, and the children of her second marriage were Baron Benoit Bayard, II, Hypolite Bayard, and Alfred Bayard. At this time the legal representatives wanted the property left by their late mother to be divided among themselves.

The tract of land that was to be divided among the heirs of the late Marie Hortense Boutte Berard Bayard, consisted of a tract of land, which was situated about three miles below the town of New Iberia, Louisiana. This tract of land had a frontage along Bayou Teche of twenty one arpents, and a depth of forty arpents. Also, this tract of land was bounded on the east by the lands of Camille Jean Baptiste Berard, Hortense Berard, and Hypolite and Alfred Bayard, on the north by land belonging to the estate of Henry Peebles, on the south by lands belonging to Romero, and on the west by lands formerly belonging to Henry Peebles, and by land belonging to the heirs of Bernard Migues. This property was appraised at the price of $10.00 per arpent, which gave a total valuation of $400.00.

Lot one was a seventh part of the tract of land, which was inherited by the heirs of the late Hortense Boutte' Berard Bayard, and this tract of land was allotted to Hortense Berard Duperier, who was the widow of Frederick Duperier. This was the portion that Hortense Duperier received from the estate of her late mother, Hortense Bayard. This tract of land had a frontage of three arpents along Bayou Teche, with a depth of forty arpents. Also, this tract of land was bounded on the east side by land belonging to Hortense Berard, and it was the same tract of land that was received from her of her late mother, Hortense Bayard, by legal notarial act before John A. Dumartrait, on the date of February 16, 1839. This tract of land was bounded on the north side by lands belonging to the estate of Henry Peebles, and by land allotted this day to Baron Benoit Bayard, Junior, and in the rear by lands belonging to the heirs of Bernard Migues. The appraisal value of lot one was $1,200.00.

Lot two was another seventh portion of the tract of land, that belonged to the estate of the late Marie Hortense Boutte' Berard Bayard. Lot two was allotted to Baron Benoit Bayard, Junior, as his share of the estate of his late mother, Hortense Bayard. Lot two had a frontage along Bayou Teche of three arpents, with a depth of forty arpents. Also, lot two was bounded on the east side by land belonging to Baron Benoit Bayard, which was received by him from his mother, Hortense Bayard, in agreement and transaction between his mother, Hortense Bayard and her children. Also, lot two was bounded on the north side by land belonging to Hortense Berard, who was the widow of the late Frederick Duperier, on the south side by land belonging to the heirs of Jean Baptiste Berard, and on the west side by land belonging to the estate of Henry Peebles, and by land belonging to the heirs of Bernard Migues. The appraisal value of lot two was the sum of $1,200.00.

Lot three was composed of another seventh part of the tract of land that belonged to the estate of the late Hortense Bayard. Also, lot three was allotted to Alfred Bayard, who was a son of the late Hortense Boutte' and her second husband, Baron Benoit Bayard. This tract of land had a frontage of three arpents along Bayou Teche with a depth of forty arpents. Also, this tract of land was bounded on the east side by land belonging to Alfred Bayard, and received from his mother, Hortense Bayard, in agreement and transaction, on the north side by lands belonging to the heirs and legal representatives of Jean Baptiste Berard, and this day allotted to them, on the south side by land belonging to Hypolite Bayard, a brother of Alfred Bayard, and another son of the late Hortense Boutte' Bayard and her second husband, Baron Benoit Bayard, and on the west by land belonging to Bernard Migues. The appraisal value of this tract of land was the amount of $1,200.00.

Another part of lot three was allotted to Hypolite Bayard, who was another son of the late Hortense Boutte' and her second husband, Baron Benoit Bayard. This lot had a frontage along Bayou Teche of three arpents with a depth of forty arpents. Also, this lot was bounded on the east side by land belonging to Hypolite Bayard, and this land was received of him from his mother, Hortense Bayard in agreement and transaction, on the north side by land belonging to Alfred Bayard, and this day allotted to him from the estate of the late Hortense Boutte' Bayard, on the south side by land belonging to Achille Berard, and this day allotted to Achille Berard. The appraisal value of this lot was the amount of $1,200.00.

Lot four was another seventh portion of the tract of land that belonged to the estate of the late Hortense Boutte', who was the widow of her first marriage of Achille Berard, and the widow of her second marriage of Baron Benoit Bayard. Lot four was allotted to Alfred Bayard, a son of the late Hortense Boutte' and her second husband, Baron Benoit Bayard. Also, lot four had a frontage along Bayou Teche of three arpents with a depth of forty arpents. Lot four was bounded on the east side by land belonging to Alfred Bayard, and this land was received by Alfred Bayard from his mother, Hortense Boutte' Bayard by agreement and transaction, on the north side by land belonging to the heirs of Jean Baptiste Berard, and this day allotted to them, on the south side by land belonging to Hypolite Bayard, and this day allotted to him. The appraisal value of lot four was the sum of $1,200.00.

Lot five was another seventh of the portion of the tract of land that belonged to the estate of the late Hortense Boutte' Bayard. Also, lot five was allotted to Hypolite Bayard, who was a son of the late Hortense Boutte' Bayard, and her second husband, Baron Benoit Bayard. Also, Hypolite Bayard and Alfred Bayard and Baron Benoit Bayard were full brothers. Lot five had a frontage along Bayou Teche of three arpents with a depth of forty arpent. Also, lot five was bounded on the east side by land belonging to

Hypolite Bayard, and received of him from his mother, Hortense Boutte' Bayard, by agreement and transaction, on the north side by land belonging to Alfred Bayard, and this day allotted to him, on the south side by land belonging to Achille Berard, and this day allotted to him, and on the west side by land belonging to Bernard Migues. Also, the appraisal value of this lot was the sum of $1,200.00.

Lot six was another seventh portion of the tract of land that belonged to the estate of the late Hortense Boutte' Bayard. Lot six was allotted to Achille Berard, who was a son of the late Hortense Boutte' by her first marriage of Achille Berard. Also, lot six had a frontage of three arpents along Bayou Teche with a depth of forty arpents. Lot six was bounded on the east side by land belonging to Camille Berard, on the north side by land belonging to Hypolite Bayard, on the west side by land belonging to Bernard Migues. Also, the appraisal value of lot six was the sum of $1,200.00.

On the date of September 1, 1869, Achille Berard sold a tract of land to Frederick Duperier and Ernest Berard. This deed of conveyance was executed before Patrick Burke, who was a notary public in the town of New Iberia, Louisiana.

Achille Berard was the half brother of Baron Benoit Bayard II, Alfred "Alphe" Bayard, and Hypolite B. Bayard. Achille Berard was the son of Marie Hortense Boutte' by her first marriage to Achille Berard. After the death of Achille Berard, Marie Hortense Boutte' Berard was left a young widow. Also, Frederick was the brother-in-law of Achille Berard.

This tract of land was lying on the west side of Bayou Teche. Also, this tract of land measured three arpents fronting the west side of Bayou Teche with a depth of forty arpents. The total land area of this tract of land was one hundred and twenty arpents of land. The boundaries of this tract of land were on the east side by land belonging to Camille Berard, on the north side by land belonging to Hypolite Berard, on the south side by land belonging to Camille Berard, and on the west side by land belonging to Henry W. Peebles, and land belonging to the heirs of Bernard Migues. This tract of land was the same tract of land that was allotted to Achille Berard in the partition made between the heirs of the late Mistress Marie Hortense Boutte', who was the widow of Achille Berard, and the widow of Baron Benoit Bayard, I. This act of partition was made on the date of February 16, 1869, by an act of partition, which was passed by Joseph Breaux, a notary public in the town of New Iberia, Louisiana.

This sale was made for the purchase price of the sum of $800.00. At the execution of this act of sale, Frederick Duperier and Ernest Berard paid Achille Berard the sum of $300.00 in cash money. The sum of 250.00 was paid on the date of March 1, 1870, and the sum of $250.00 was paid on the date of March 1, 1871. The tract of land will remain specially mortgaged unto Achille Berard until the final payment of the sum of $250.00 on the date of March 1, 1871. Also, to secure the repayment of the loan, a vendor's privilege also exists on the tract of land. A vendor's privilege is really a mortgage that the vendor has on a piece of immovable or movable property that is sold at a sale and mortgage, and if the purchaser cannot pay in part or whole the sum of the purchase price, then the seller can take back the property sold, and resale the property.

On the date of December 24, 1869, Baron Benoit Bayard was serving as tutor to Orite Theriot. Orite Theriot was a minor child of the late Charles Theriot and his wife, Marie Teryile Babineaux Theriot. Orite Theriot was the ward of Baron Benoit Bayard II, and the taking care of the money that was inherited by Orite Theriot was the responsibility of Baron Benoit Bayard. Ever so often, the tutor was required to give an account or a statement of monies received by them to the expenses and investment of their ward.

Marie Teryile Babineaux was the sister of the wife of Baron Benoit Bayard, II, who was Josephine Rosalie Babineaux. Josephine Roseline and Marie Teryile Babineaux were the daughters of Joseph Babineaux and his wife, Rosalie Segura.

On the date of April 1, 1861, Baron Benoit Bayard II, received $600.00 from the estate of Charles Theriot for administrative duties of the estate.

Also, on the date of April 1, 1861, Baron Benoit Bayard II, received from the estate of Charles Theriot, the sum of $600.00 as a fee for his handling of the administrative duties of the estate.

Also, on the date of December 13, 1860, Baron Benoit Bayard II, received the sum of $1,830.05, on the date of July 19, 1859, Baron Bayard II received the sum of $300.00, on the date of May 25, 1860, Baron Bayard received the sum of $1,100.00, on the date of May 18, 1860, Baron Bayard received the sum of $700.00, and on the date of May 1, 1860 Baron Bayard received a promissory note for the amount of $120.00. On the date of July 1, 1861, Baron Bayard received the sum of $815.00, on the date of June 19, 1861, Baron Bayard also received the sum of $125.00. Also, on the date of July 19, 1859, Baron Bayard received the sum of $410.00, and on the date of April 27, 1861, Baron Bayard received the sum of $300.00. The total amount of money coming into the hands of the tutor, Baron Benoit Bayard II, was the sum of $830.00. To this amount must be added the initial sums and the interest from the date at which this money came into the hands of Baron B. Bayard II.

The interest on the sum of $600.00 from the date of April 2, 1861, which was a period of five years and four months, was the sum of $160.00. Also, the interest on the sum of $2,000.00, from the date of April 20, 1860, which was a period of six years and three months, was the sum of $625.00. The interest on the sum of

$1,100.00, from the date of May 25, 1860, which is a period of six years and two months, was the sum of $337.00. The interest on the sum of $1,300.00, from the date of May 18, 1859, which was a period of seven years, was the sum of $105.00. The interest on the sum of $700.00 from the date of May 18, 1859, which was a period of seven years and two and one half months, was the sum of $250.00. The interest on the sum of $1,120.00 from the dates of May 1, 1860 to May 1, 1861 was the sum of $37.50. Also, the interest on the sum of $815.80 from the date of July 1, 1860 was the sum of $248.12. The interest on the sum of $125.00 from the date of June 19, 1861, which was for a period of five years and one month, was the sum of $31.77. The interest on the sum of $410.00 from the date of July 19, 1859, which was a period of seven years, was the sum of $143.57. Also, the interest on the sum of $300.00 from the date of April 27, 1861, which was a period of five years and two months, was the sum of $78.50.

Then the deduction of the commission of the tutor, Baron Benoit Bayard II, at the rate of 10% on the enumerated amount of those monies incurred, and the expenses of the minor child during her minority. The first commission was the sum of $160.00, and the second commission was the sum of $419.36. The third commission was the sum of $625.00, and the fourth commission was the sum of $337.50. Also, the fifth commission was the sum of $105.00, and the sixth commission was the sum of $250.00. The seventh commission was the sum of $37.50, and the eighth commission was the sum of $248.12. The ninth commission was the sum of $31.77, and the tenth commission was the sum of $143.57. Also, the eleventh commission was the sum of $78.50, and the twelfth commission was the sum of $124.20.

These were the expenses of the minor child, Orite Theriot, during the years of her minority, when she was the ward of Baron B. Bayard, II. Also, Orite Theriot lived in the Bayard plantation home with the family of her maternal Aunt Josephine and Uncle Baron B. Bayard, II. Baron B. Bayard sent his ward, Orite Theriot, to Bucky's School, and the schooling account was the sum of $187.10. The account of F. C. Breaux was the sum of $75.00. Also, Baron B. Bayard, II sent his ward, Orite Theriot, to Miss Camille Fenny's School, and the cost of this schooling was the sum of $100.00 per year. Also, the board and lodging of the minor child, Orite Theriot, was calculated at the rate of $175.00 per year during the space of six years, and the total sum for board and lodging in the Bayard home was the sum of $1,050.00.

On the date of December 31, 1869, Josephine Rosalie Babineaux, who was the legal and lawful wife of Baron Benoit Bayard, II, stated in the presence of Patrick Burke, who was a notary public, in the town of New Iberia, Louisiana, that her husband was legally indebted to her. This was done in the presence of two witnesses, who were Alfred Veazey and Viel E. Darby. This act of Dation en paienment was executed for the purpose of

Iberia Parish Created Oct. 29, 1869
Combining Portion of St. Mary w/ Portion St Martin.

Aug. 16, 1853 – St Marys Plantation Home and Acreage located By Wade Bayou ~~Home and Acreage~~ sold by Louis Toussaint De Blanc to Stephen Owens Nelson and partners of New Orleans La. for $140,000.00. 2209 arpents

Bounded North by Baron Benoit Bayard and South by Col. Chas. Napoleon Olivier (Orange Grove)

June 21 1854 – Stephen Owens Nelson ~~H~~ Husband of Martha Ann Turner Nelson Purchased the interests of all his partners in the Ownership of St. Marys Plantation becoming the sole Owner of the ~~Property~~ Plantation and of all the slaves horses mules livestock and other movable property ~~attached thereto~~. located thereon.

Tract #1 – 38 arpents Bayou X 40 arpents 1,520
Tract #2 – 11 arpents ✓ X 4 arpents – 44 arpents
Tract #3 – 3 arpents Bayou X 15 arpents – 45 arpents
Tract #4 – 5 arpents Bayou X 40 arpents – 200 arpents
5 arpents Bayou X 80 arpents – 400 arpents
2,209

preserving the mortgage or the privilege accorded to Josephine Roseline Babineaux Bayard, who was the wife of Baron B. Bayard, II, by the laws of the State of Louisiana. This law was in favor of married women in conformity with the Act # 95 of the legislature of the State of Louisiana, and Act #95 was approved by the legislature on the date of March 8, 1869. Act # 95 was entitled, " A previous act to carry into effect Article 123 of the Constitution of the State of Louisiana, and to provide for the reasoning of all mortgages and privileges."

Josephine Rosalie Babineaux and Baron Benoit Bayard, II were married on the date of October 17, 1842. She was the daughter of Joseph Babineaux and his wife, Rosalie Segura. Baron Benoit Bayard, II was the oldest son of Baron Benoit Bayard, I and his wife, Marie Hortense Boutte'. Also, Baron B. Bayard, II was the grandson of Francoise Cesear Boutte' and his wife, Marie Therese' DeGruy, of St. Martin Parish. The father of Baron Benoit Bayard II, who was Baron Benoit Bayard, I, was an immigrant of the Kingdom of France, and he was from the area of Greenoble, France. Shortly, after arriving on the shores of the New World, in the French province of Louisiana, Baron B. Bayard, I, married Marie Hortense Boutte'. Marie Hortense Boutte' was a young widow of Achille Berard, and the owner of a very large sugar plantation, fell in love with the young Baron B. Bayard, I. Baron B. Bayard, I helped his wife, Hortense Boutte', run her large plantation, which was situated in upper St. Mary Parish.

Josephine Rosalie Babineaux brought into her marriage with Baron Benoit Bayard, II, on the date of October 17, 1842, the sum of $2,000.00, which was received by her in the year of 1843, form her mother, Rosalie Segura. This amount of the sum of $2,000.00 was received as a donation from Rosalie Segura Babineaux, the mother, to her daughter, Josephine Rosalie Babineaux, who was the wife of Baron B. Bayard, II. This sum of $2,000.00 was remitted to Baron B. Bayard, II, who was the husband of Josephine R. Babineaux.

Also, in the year of 1846, Josephine Rosalie Babineaux, who was the wife of Baron B. Bayard, II, received from her father, Joseph O. Babineaux and her mother, Rosalie Segura Babineaux, a large vachery or herd of cattle. The value of this vachery or large herd of cattle was the sum of $3,000.00. Baron B. Bayard, II, the husband of Josephine R. Babineaux, sold this large vachery of cattle, and he received the proceeds of the sale of the vachery of cattle. This incurred an indebtness on the part of her husband, Baron B. Bayard, II, to his wife, Josephine Rosalie Babineaux Bayard, in the sum of $5,000.00. This whole amount of $5,000.00, Baron B. Bayard, II received in cash money and he appropriated this money for his own use and benefit. For the faithful reimbursement of which his wife, Josephine Bayard is accorded by law, it is now considered that Josephine Babineaux Bayard, who is the wife of Baron B. Bayard, II, hold a legal mortgage on all the property owned by her husband, Baron Benoit Bayard, II.

Also, conveyed to Josephine Rosalie Babineaux Bayard, through this act of Dation en paienment was current money and other articles. A total sum of $2,600.00 in current money and other provisions, such as corn, rice, sugar, molasses, and other goods were given to Josephine Babineaux Bayard. Also, a herd of thirty hogs, horses, mules, mares, and colts, which were valued at the sum of $900.00. A herd of fifteen sheep together with four carts, plows, harnesses, and hoes, which were valued at the sum of $50.00. The household furniture, including kitchen utensils, was valued at the sum of $30.00. The Bayard family carriage and harnesses was valued at the price of $20.00. The total amount of goods, livestock, and provisions that was conveyed to Josephine Rosalie Babineaux Bayard through the Dation en paienment from her legal husband, Baron Benoit Bayard, II, was the sum of $395.00.

Baron B. Bayard, II gave through this Dation en paienment to his wife, Josephine Bayard, all the real estate, which comprised the Bayard plantation, and the improvements, which were situated on the Bayard plantation, the machinery in the sugarhouse, and the farm machinery, and all the movable property with full warranty of title and subrogation to all the rights and actions in warranty which Baron B. Bayard, II, had against all present owners. Josephine Bayard accepts this sale through the Dation en paienment of property and movables at the value set forth of the sum of $3,950.00. Also, Josephine Bayard is reimbursed by her legal husband, Baron B. Bayard, II, to her (Josephine Bayard's) account of the amount of money due by Baron B. Bayard, II to his legal wife, Josephine Bayard, and for which the sum of $3,950.00 is owed. Also, Josephine Bayard grants unto Baron Benoit Bayard, II acquittance, but Josephine Bayard still holds Baron Benoit Bayard, II liable for the balance due to her interests and costs.

On the date of February 16, 1872, Baron Benoit Bayard, II sold a tract of land, some livestock, and a brand to his son, Baron Benoit Bayard, III. Baron Benoit Bayard III was the grandson of Baron Benoit Bayard, I, and his wife, Marie Therese' Boutte', who were the original owners of the Bayard Plantation. The execution of this deed of conveyance took place in the presence of Patrick E. Burke, who was a notary public in the town of New Iberia, Louisiana.

The tract of land sold to Baron B. Bayard III was situated on the east side of Bayou Teche. This tract of land was lying a little below the city limits of the town of New Iberia. This tract of land measured six arpents fronting the east side of Bayou Teche with a depth of forty arpents. The total land area of this tract of land was two hundred and forty arpents of land. The boundaries of this tract of land were on the north side by lands belonging to the widow of E. B. Smeedes, or Mintmere Plantation, on the east side by lands belonging to the estate of Jean Baptiste

Berard, and on the west side by Bayou Teche, and below by the lands of Alfred "Alphie" Bayard.

The livestock that was sold consisted of eighty two head of horned cattle. This herd of eighty two heads of horned cattle consisted of bulls, cows, and calves. Also, the brand of the cattle was sold. The brand is the mark that identifies a cow, calf, or bull as belonging to that particular herd or owner. Along the Bayou Teche, the owners of livestock branded their cattle in order to tell what herd that particular cow, calf, or bull belonged to, and as to who was the lawful and legal owner. It was understood between Baron B. Bayard II, and his son, Baron B. Bayard, III, that only the horned cattle bearing this brand were sold. No mule, horse, or mare that is branded with this brand or the Bayard brand is sold.

The purchase price of the tract of land, the herd of eighty two horned cattle, and the Bayard brand was the sum of $3,500.00. Baron B. Bayard III paid to his father, Baron B. Bayard, II, the sum of this purchase price at the execution of this act of conveyance.

On the date of February 13, 1872, Baron B. Bayard, II conveyed by act of Dation en paienment to his wife, Josephine Babineaux Bayard, land, money, animals, and other provisions. Baron Bayard II acknowledged that he was truly and justly indebted to his wife, Josephine Babineaux, in the full sum of $5,000.00. This amount of $5,000.00 was the amount that was received by Baron Bayard II for his wife's account prior of January 1, 1870. Baron Bayard II is desirous of paying up his debt to his wife, Josephine, so for as it is in the favor of Baron Bayard II and that availing himself of Article 2441 of the Civil Code of the State of Louisiana.

The tract of land that was conveyed to Josephine Babineaux Bayard by her husband, Baron Benoit Bayard II, was a tract of land together with the sugarhouse, and all the other buildings and improvements. Also, this tract of land was situated on the west side of Bayou Teche in the newly created Parish of Iberia, and it is situated a little below the town of New Iberia, Louisiana. This tract of land had a frontage along Bayou Teche of six arpents with a depth of forty arpents. Also, this tract of land was bounded on the north side by the property of Mistress Evert Bunker Smeedes, the former Sarah Cade, on the east side by Bayou Teche, on the south side by the property of Alfred "Alphie" Bayard, and on the west side by the pubic road, which road is known as the Old Spanish Trail. The Old Spanish Trail is the road that leads from the town of New Iberia to the town of Jeanerette, Louisiana. This tract of land forms a part of the sugar plantation, which is cultivated in sugarcane and owned by Baron B. Bayard II. Also, the present act of Dation en paienment includes all the sugarcane on the plantation, whether the sugarcane has been planted or not planted.

The tract of land with the sugarhouse and all the other buildings is valued at the sum of $2,600.00.

On the date of March 26, 1874, Judge Train rendered a judgement against Baron Benoit Bayard II in the civil lawsuit of Mrs. B. Martin, Administrator versus Baron Benoit Bayard, II, and bearing docket number 289. Mrs. B. Martin was serving as administrator of the succession of Ballhazard Martin. This judgment was for the sum of $4,176.40 and conditioned to bear 8% interest from the date of March 1, 1862.

Hypolite Bayard was the brother of Baron Benoit Bayard II, and Alfred "Alphie" Bayard. They were the children of their mother, Marie Hortense Boutte's second marriage to Baron Benoit Bayard I of Greenoble, France. The Bayard plantation was originally the lands belonging to Francoise Cesear Boutte', pere, who was the father of Marie Hortense Boutte'. Also, Francoise Cesear Boutte', pere, gave this tract of land that was to become the Bayard plantation to his daughter, Marie Hortense Boutte', before her first marriage to Achille Berard. Francoise Cesear Boutte', pere, was a very wealthy planter, who owned a plantation in St. Martin Parish.

Hypolite Bayard was raised at the Bayard plantation, but after his marriage he moved to a plantation near the Isle de Piqueant, which was the present day Patoutville, Louisiana. Hypolite Bayard married Irma Azelie Patout, who was the daughter of Madame Appoline F. Patout and Simeon Patout, who were the owners of Enterprise Plantation. The plantation that Hypolite Bayard owned and lived on in the Isle de Piqueant area was situated on the east side of Enterprise plantation. The lands of Hypolite Bayard were bounded on the east side by Enterprise plantation, which was the home of Madame Appoline Patout, on the south side by the swamp land, which was owned by Madame Patout, on the west side by the plantations of Junca and Gautreaux, and on the north side by the plantation of Leufroy Bonvillian.

On the date of May 27, 1872, Hypolite Bayard borrowed money from Pierre Pecot, of the Parish of St. Mary, for the sum of $940.00. For the reimbursement of this promissory note, Hypolite Bayard gave his promissory note, which was payable to the order of Pierre Pecot. This promissory note was due in one year from the date, or on the date of May 27, 1873 at the domicile of Pierre Pecot, which was situated in the Parish of St. Mary. Also, this promissory note was conditioned to bear interest at the rate of 8% per annum from the date of maturity. The execution of this promissory note took place in the town of New Iberia, Louisiana, and the two subscribing witnesses were Narcise Landry and Jerome Napoleon Lyles.

In the month of September of the year of 1876, Alfred B. Bayard declared that he was legally indebted to his wife, Felicina

Broussard, to whom he was married to on the date of July 27, 1857. Alfred Bayard was indebted to his wife, Feliciana, for the sum of $525.00. This sum of $525.00 was the value of thirty head of gentle cattle. At the time of the marriage, Feliciana Broussard brought into the marriage with Alfred Bayard the herd of thirty gentle cattle. Also, at this time, fifteen of the cows had calves. Alfred Bayard has disposed of the cattle and he kept the purchase money of the cattle. Alfred Bayard spent this purchase money at diverse times between the date of his marriage on July 27, 1857, and the breaking out of the Civil War in the year of 1861. Felicina Broussard Bayard held a mortgage, which is granted to her by the laws of the State of Louisiana, to secure the repayment of this mortgage. Alfred Bayard was called by his family, friends, and peers by the name, "Alphie."

On the date of March 30, 1877, Alfred Bayard executed a promissory note in the presence of notary public, Leonce de la Houssaye, and this promissory note was payable to Jane B. Hopkins. Jane B. Hopkins, who was the widow of the late Harvey Hopkins, loaned Alfred Bayard the sum of $1,000.00. Alfred Bayard had subscribed and made this promissory note, which was due in twelve months, or on the date of March 30, 1878. Also, this promissory note was conditioned to bear interest at the rate of 8% from the date of March 30, 1877 or until paid in full. Jane H. Hopkins advanced money in the sum of $1,000.00 to Alfred Bayard. After this promissory note had been paraphed, "Ne Varietur," by Leonce de la Housssaye, the notary public, Alfred Bayard handed this promissory note to William Robertson, who was representing Jane H. Hopkins. At the execution of this act of mortgage, William Robertson was present and acting in the behalf of Jane H. Hopkins, who was not present. Also, the wife of Alfred Bayard, Feliciana Broussard Bayard, acknowledged that she was aware of the foregoing act of mortgage. Feliciana Broussard Bayard, who was the wife of Alfred Bayard, made this renunciation to protect her matrimonial, dotal, and paraphernal rights, and also under the laws of the State of Louisiana she had a legal mortgage on the property of her husband, Alfred Bayard.

The reason that this promissory note was paraphed, "Ne Varietur," by the notary public was to identify this particular promissory note with this particular act of mortgage.

To secure the payment of the promissory note, and the interest due at maturity, Alfred Bayard specially mortgaged two tracts of land in favor of Jane B. Hopkins, who was the widow of the late Harvey Hopkins. This tract of land had a front of three arpents on the west side of Bayou Teche with an eighteen arpent depth. The total area of this tract of land was fifty four arpents or forty five acres and nine hundredths of an acre of land. Also, this tract of land was bounded above by the lands of Baron Benoit Bayard II, and below by the lands of Hypolite Bayard. Alfred Bayard also specially mortgaged another tract of land to Jane H.

Hopkins. This second tract of land was lying opposite the above tract of land, but this second tract of land was lying on the east side of Bayou Teche. Also, this second tract of land had a frontage along the east side of Bayou Teche of three arpents with a depth of forty arpents. The total land area of the second tract of land was one hundred and twenty arpents of land.

On the date of May 4, 1877, Robert S. Perry and the firm of Stott and Lutzenberger, which was represented by Francis Lutzenberger declared that for the price and sum of $2,072.00, they have sold Baron B. Bayard II, a steam engine. This steam engine was manufactured at the foundry of Stott and Lutzenberger, which was situated on the west side of the City of New Iberia. This steam engine had a twelve inch barrel and a four foot stoke. Also, sold was a sugar mill, which was four and one half feet long. All this equipment was lying in the foundry of Stott and Lutzenberger in the City of New Iberia, Louisiana.

The foundry firm of Stott and Lutzenberger binds and obligates themselves to have the sugar mill and engine repaired, and put in complete running order. The equipment must be removed from the foundry by Baron Benoit Bayard II before a certain day in the year of 1877. The boilers and connections which included chains and fixtures for one hundred feet of carrier were to remain at the foundry.

Before the removal of the equipment, Baron Benoit Bayard II is to name an engineer to inspect the machinery so that everything may be satisfactory before the removal of the machinery. It being understood that after such an inspection and removal of the machinery, the seller, Stott and Lutzenberger, are to be released from any further responsibility in connection with this equipment.

The purchase price of this equipment was the sum of $2,072.00. On the date of May 4, 1877, Baron Benoit Bayard II has drawn and delivered to Stott and Lutzenberger, the sellers, four promissory notes. These four promissory notes are payable to the order of Baron Benoit Bayard II, and are endorsed in the blank by Baron Benoit Bayard, II. The first promissory note is for the amount of $200.00, and it is payable thirty days after the date of sale, or on the date of June 4, 1877. The second promissory note is for the sum of $100.00, and it is due and payable in sixty days, or on the date of July 4, 1877. The third promissory note is for the sum of $100.00, and it is due and payable on the date of January 1, 1878. The fourth and last promissory note is for the sum of $600.00, and it is due and payable on the date of January 1, 1878. All four promissory notes are conditioned to bear interest at the rate of 8% from the date of maturity.

There are other promissory notes, one note is for the sum of $472.00, and it is payable on the date of January 1,1878. Also, there is a sum of $400.00, which is represented by two promissory notes, and these two promissory notes are payable on the date of January 1, 1879. These two promissory notes are conditioned to bear interest at the rate of 8% from the date of January 1, 1878. To secure the payment of these notes, Stott and Lutzenberger, the seller of the machinery, reserves the privilege and vendor's lien on the machinery and the sugar mill, which the laws of the State of Louisiana accord to vendors selling on a credit basis.

It is the intention of Baron Benoit Bayard II to put up the machinery, and the sugar mill on the land belonging to Clemarie Olivier, which is situated in the Grand Marais area. To further secure the payment of these promissory notes, Baron B. Bayard II pledges the foundry firm of Stott and Lutzenberger, the whole of the sugar mill establishment, including the mill house, the sheds, the purveying, the boiling room, the machinery, the furnaces, and all apparatus of the sugar establishment. Also, as a further securement of this loan, Baron Benoit Bayard II pledges the foundry firm of Stott and Lutzenberger twenty hogsheds of sugar, which is to be grinded at the sugar establishment of Baron Benoit Bayard II next fall or the fall of the year of 1877. Baron B. Bayard II will deliver these twenty hogsheds of sugar to the firm of Stott and Lutzenberger by shipping these hogsheds of sugar on or before the date of the 15th day of December of the year of 1877. The sugar is to be shipped to the merchants of Baron Benoit Bayard II in the City of New Orleans, and the seller, Stott and Lutzenberger, reserves the privilege guaranteed by law, such as it may bear upon the machinery and the sugar house.

On the date of August 6, 1877, Baron B. Bayard II of the Parish of Iberia, declared before Leonce de la Houssaye, a notary public, that he was legally indebted to Hypolite Solomn of the City of New Orleans, Louisiana for the sum of $300.00. This was cash money that Hypolite Solomn had advanced to Baron Benoit Bayard II. Also, during the year of 1877, Hypolite Solomn made advances of money to Baron Benoit Bayard II for the purpose of harvesting his crop of sugarcane. This crop of sugarcane had been raised at a place called, Grand Marais, by various planters. Also, this crop of sugarcane was to be processed into sugar products at a sugarhouse, which Baron Benoit Bayard II is buildings on the lands of Clemarie Olivier. The share of Baron Benoit Bayard II is two fifths of all the sugar and molasses manufactured at this sugarhouse in the Grand Marais area.

To secure all the money advanced, such as any liens and privilege, Baron Benoit Bayard II is obligated to ship to Hypolite Solomn all of his (Baron B. Bayard II) two fifths share of the manufactured sugar and molasses.

On the date of May 14, 1878, Alfred B. Bayard sold a tract of land to Nicholas Muller, of the Parish of Iberia, at a cash sale. The purchase price of this tract of land was the sum of

$1,194.94, which Nic Muller paid at the execution of this act of cash sale.

This tract of land was lying on both the east and the west sides of Bayou Teche. The upper portion of this tract of land was situated on the west bank of Bayou Teche, and it measured three acres fronting Bayou Teche. The lower portion of this tract of land was located on the east side of Bayou Teche. The boundaries of this tract of land were on the north side by the lands of Baron Benoit Bayard II, on the east side by land belonging to the heirs of Achille Berard, on the south side by the lands of Baron Benoit Bayard II, and on the west side by Bayou Teche.

On the date of November 7, 1878, Baron Benoit Bayard II entered into an act of mortgage with Dr. Bernard Suberbille. This was to borrow money to put the steam engine sugar mill in the Grand Marais area. This act of mortgage was executed by Leonce de la Houssaye, who was a notary public, in the town of New Iberia, Louisiana. The total amount of money that was loaned was the sum of $350.00.

Baron Benoit Bayard II had to pay the amount of the fire insurance premium, which was insuring his sugar mill against loss by fire. Baron Bayard II had to pay the sum of $21,25 for fire insurance policy number 1729.

Dr. Bernard Suberbille advanced $68.00 in cash money to Baron B. Bayard II. Then Baron B. Bayard received a draft, which was in favor of Gust Gonsoulin for the sum of $42.00. Also, Baron Bayard II paid Severin Romero his wages in the sum of $28.00, Joe Boutte'' his wages in the sum of $5.00, Alexis Derouen his wages in the sum of $24.00, Simon, who was the engineer of the sugarhouse, his wages in the sum of $42.00, and Baron Bayard II paid this commission to himself in the sum of $245.00. The balance of the money left from the $350.00 loan was the sum of $131.25.

Dr. Bernard Suberbille declared that the balance in the amount of $131.25 is due for work done on a certain sugar mill engine, and work done on other machinery built in the sugar house of Baron Bayard II. This sugar house was situated on the lands belonging to the succession of Clemarie P. Olivier, which was at a place called, Grand Marais. Also, the premium of the fire insurance policy is also due.

On the date of May 28, 1879, the succession of Irma Patout, who was the wife of Hypolite B. Bayard was probated in the courtroom of the Iberia Parish Courthouse. The inventory of the property belonging to the succession of Irma Patout Bayard was made on the date of May 1, 1877. All the land belonging to this succession was located in Patoutville, Louisiana. The property belonging to the community of acquets and gains existing of the marriage of Hypolite B. Bayard and Irma Patout Bayard was estimated at the sum of $8,320.20.

Hypolite B. Bayard was a son of the late Marie Hortense Boutte' Bayard and the late Baron Benoit Bayard I. He was raised at the Bayard Plantation, and after his marriage to Irma Patout, he moved to a plantation in the Patoutville area.

The children of Hypolite B. Bayard and Irma Patout Bayard were Irma, Ernest, Blanche Marie, Zubine, Alfred, and Clarie Pauline Bayard. These children had a tacit mortgage bearing on the property belonging to the succession of Irma Patout Bayard. This tacit mortgage is granted by the laws of the State of Louisiana upon all of the property belonging to their father, Hypolite B. Bayard.

On the date of February 28, 1879, a civil lawsuit of E. Frere versus Baron Benoit Bayard II was tried in the 3rd District court, which was holding court in the Parish of Iberia. In the case and the law, the evidence being in favor of E. Frere, and for the written reasons assigned by his honor, the Judge presiding on the bench for the trial of this case. It was ordered that a judgement be obtained in the case of Leonard J. Smith versus Baron Benoit Bayard II, which bears the docket number 6307. This judgement was signed by Judge Fred Gates, who was serving as judge of the 3rd District Court. The presiding judge of the court was Judge Theo Fontelieu.

This judgment was for the sum of $4,000.00 with interest at 8% per annum from the date of the first day of March of the year of 1864. The costs are subject to a credit. Also, to secure the payment on a certain tract of land, which is situated in the Parish of Iberia, upon which Baron Benoit Bayard II resides. This tract of land had a frontage along Bayou Teche of six arpents with a depth of forty arpents. The total land area is two hundred and forty arpents of land. This tract of land ran the depth back to forty arpents. Also, this tract of land was bounded above by the lands belonging to Leonie Theriot Hartman, which is known as Mintmere Plantation, and below by the lands of Alfred Bayard. The property of Baron Benoit Bayard II will be seized and sold to satisfy the debt of the judgment. This tract of land was located on the tracts of land that presently front East Main Street.

On the date of May 21, 1881, at a sheriff sale was sold to Emile Guidry and Company the sugar mill of Baron Benoit Bayard II, which was situated at a place called, Grand Marais. This sheriff sale resulted from a writ of seizure and sale, which was issued in the civil lawsuit of A. J. Decuir versus Baron Bayard II. The terms of this sale were cash.

Deputy Sheriff S. Daquesseau of the Parish of Iberia, did seize and take into possession the property of Baron Bayard II, which was a sugar mill and a steam engine. The Sheriff of the

Parish of Iberia had been issued a citation by the Honorable Court directing the Sheriff to take this property into possession.

The property seized consisted of a steam engine and a sugar mill with all the attachments, except the boiler. This sugar mill was situated on the land belonging to the succession of C. P. Olivier, which was around the Grand Marais area. This equipment was sold to Baron Bayard II by the Lutzenberger Foundry on the date of May 4, 1877.

After complying with all the legal formalities for the public auction or the sheriff sale, an advertisement was placed in the newspaper called, "the Louisiana Sugar Bowl." This newspaper, the Louisiana Sugar Bowl, was published in Iberia Parish. Also, this advertisement appeared on the date of July 7, 1881, and this advertisement stated that the sale of the sugar mill, steam engine, and all the apparatus of the sugar mill would take place on the date of July 9, 1881 at the doorsteps of the Iberia Parish Courthouse during the legal sale hours. Before the sheriff sale, the Sheriff of the Parish of Iberia, had to have this property appraised.

The Sheriff of the Parish of Iberia did offer this sugar mill, steam engine, and apparatus belonging to the sugar house for sale at a public auction. The sheriff sale was held at the doors of the Iberia Parish Courthouse on Saturday, the 9th day of the month of July of the year of 1881 during the legal sale hours in the morning. The sheriff first read the advertisement of the sheriff sale, then secondly read the certificate of mortgage, which was bearing on the sugar mill, and thirdly the writ of seizure and sale. The last and the highest bidder was the representatives of the Emile Guidry and Company.

The Emile Guidry Company purchased the sugar mill, the steam engine, and the apparatus for the purchase price of $1,500.00. The full amount of the purchase price was paid in cash. Also, the purchaser was charged with these other charges, a charge of $4.55 for the fees of the office of the clerk of court, a charge of $25.88 fee for the Iberia Parish Sheriff Department, and a charge of $6.11 for the fee of advertisement,w which was due to the Louisiana Sugar Bowl newspaper. Also, a $35.00 fee was added for the judgment obtained in the civil case of A. J. Decuir versus Baron Bayard II. The balance of the purchase price was the sum of $1,094.79, and this amount was retained by the purchaser to be applied in payment of the mortgages or privileges bearing on the property.

LEITMEYER

On June 7, 1880, John Ermant Leitmeyer purchased at a cash sale four hundred and nineteen arpents or three hundred and fifty six American acres and fifteen hundredths of an acre of land from the Citizens Bank of Louisiana. These four hundred and nineteen arpents of land were once a part of tract one or the plantation proper, which was the main tract of St. Mary's Plantation, which was owned by Stephen Owens Nelson of New Orleans, Louisiana. Today, at the present time on this tract, now stands the Iberia Parish General Medical Center, the medical offices of Drs. Haile, Ellias, Accardo, Ritter, Falterman, Bernard, and Hebert, the Consolata Nursing Home, the Diamond Shamrock Service Station, and the Iberia Broadcasting Center, which are all located on the western side of Bayou Teche.

On July 18, 1868, the estate comprising the lands owned by Stephen Owens Nelson in the Parishes of St. Mary, St. Martin, St. Landry, and Iberia were sold to the Citizens Bank of New Orleans, and the Crescent City Bank of New Orleans, Louisiana.

On September 25, 1866, the Citizens Bank of New Orleans and Crescent City Bank of New Orleans filed a petition against Stephen Owens Nelson. This civil suit was brought by the President and the Board of Directors of both banks, because Stephen Owens Nelson was legally indebted to both banks for the sum of $40,000.00. The two banks, the Citizens Bank and the Crescent City Bank were the holders or owners of four promissory notes, that were drawn, signed, and endorsed by Stephen Owens Nelson on February 1, 1855. These notes were in the amount of $4,000.00 each, and these notes were conditioned to bear interest at eight per cent per annum after maturity until paid. These four notes were payable to the order of Stephen Owens Nelson, or to his own order, at the counting house of S. O. Nelson & Company, which was situated in the City of New Orleans, Louisiana.

On July 1, 1862, the interest on two of the four promissory notes had been paid by Stephen Owens Nelson, and the banks extended the credit note. Also, on February 25, 1862, the two other notes were paid, and the credit was extended.

The two banks secured the payment of this mortgage, which was evidenced in these promissory notes, with principal and interest, by a special mortgage on the property comprising St. Mary's Plantation, and other lands owned by Stephen Owens Nelson in the Parishes of St. Martin, and St. Landry, and also the slaves were mortgaged.

The reason the banks were able to institute a civil suit

against Stephen Owens Nelson, and hold a mortgage on the real property and slaves of S. O. Nelson was because the banks were the owners of the promissory notes. The banks, the Citizens Bank and the Crescent City Bank, were both corporations, which were created by law.

As a result of the civil lawsuit issued out of the District Court of St. Mary Parish, Stephen Owens Nelson was cited to appear in court to answer the petition of the civil law suit. The petition was filed at the courthouse in Franklin, Louisiana, or St. Mary Parish, in the recorder's office or the office of the clerk of court, on September 25, 1866.

As a result of these two civil lawsuits, the Citizens Bank and the Crescent City Bank obtained a judgement in the favor of the banks, and against Stephen Owens Nelson. So the lands of Stephen Owens Nelson, which were situated in upper St. Mary Parish, now Iberia Parish, St. Martin Parish, and St. Landry Parish were seized by the two banks. Not being able to pay the $40,000.00 that he owed to these two banks, Stephen Owens Nelson lost all ownership, rights, privileges, and titles to all of his land. Therefore, the ownership of St. Mary's Plantation and the lands of S. O. Nelson was passed to the Crescent City Bank and the Citizens Bank of Louisiana, and both banks were located in the City of New Orleans, Louisiana.

Today, in Iberia Parish, the Nelson Canal and the Nelson Canal Road are the only reminders of the interest and ownership of the commercial firm of S. O. Nelson & Company, and Stephen Owens Nelson in this Southwest sugar cane region of Louisiana. Actually, the Nelson Canal was originally a drainage ditch, which was located on the property of St. Mary's Plantation or the property of Stephen Owens Nelson.

John Ermant Leitmeyer purchased a part of the property of which the Citizens Bank of Louisiana had acquired a two thirds undivided share from Emory E. Norton, who served as the assignee of bankruptcy in the estate of Stephen Owens Nelson and Thomas A. Nelson, bankrupt as individuals, and as members of the firm of S. O. Nelson & Company, by an act passed on the date of July 18, 1868. Also, the Crescent City Bank of New Orleans had acquired a one third share in the property formerly owned by Stephen Owens Nelson, individually, and as a member of the commercial firm of S. O. Nelson & Company. On January 4, 1876, Louis E. Dupuy, acting for the Crescent City Bank, conveyed the bank's one third interest and share in the title, rights, and privileges of the property formerly belonging to Stephen Owens Nelson, both individually and as a member of the firm of S. O. Nelson & Company to John Gaines, who was acting in his capacity for the Citizens Bank of Louisiana, and purchasing and accepting this one third share. Also, conveyed by the Crescent City Bank to the Citizens Bank of Louisiana was a one third undivided share of all the buildings, such as the sugar house, dwelling house, out houses, barns, cabins, fences, sugar works and sugar kettles used in the processing of cane juice into sugar. Louis E. Dupuy, was a single man, and a resident of the City of New Orleans, Louisiana. Also, Louis E. Dupuy purchased this one third share from the Crescent City Bank, on the date of February 28, 1872, by an act of conveyance before John Bederia, a notary public. This sale of the former property of Stephen Owens Nelson from the Crescent City Bank to the Citizens Bank of Louisiana was made and accepted by the Crescent City Bank for the amount of $11,000.00 This amount of eleven thousand dollars was represented by three promissory notes that were drawn by Louis E. Dupuy, and these three promissory notes were drawn to the order of Louis E. Dupuy, and endorsed by him (Louis E. Dupuy). The first promissory note was due in twelve notes from the date, the second promissory note was due in eighteen months from the date, and the third promissory note was due in twenty four months or two years from this date.

The property that was purchased by John Ermant Leitmeyer was leased for a term expiring on December 31, 1888. It was well understood that the property is to be delivered and possession given to John Ermant Leitmeyer on January 1, 1882. It was agreed that John Ermant Leitmeyer may obtain permission from Mr. M. Bonin, the present tenant of the conveyed property, for the immediate possession of the property. John Leitmeyer is able to use the property for the purpose of building structures on the property for his own use, and preparing the property for his own use. Also, it was well understood that included in the rights and privileges sold to John Leitmeyer, is the right of drainage into the Nelson land, now the Nelson Canal, and also the right to deepen any part of the land, so drainage can be improved.

On June 7, 1881, at their annual meeting the Board of Directors and the President of the Citizens Bank of Louisiana adopted a resolution to sell the part of the property, which was situated on tract one of the land formerly belonging to Stephen Owens Nelson, both individually and as a member of the commercial firm of S. O. Nelson & Company. Leon Carriere, was President of the Citizens Bank of Louisiana, and would represent the Citizens Bank of Louisiana in the powers vested in him as President of the Citizens Bank of Louisiana of New Orleans, Louisiana, to make a conveyance of a part of tract one of the land formerly comprising St. Mary's Plantation. At this meeting, which was held on June 7, 1881, the Board of Directors of the Citizens Bank of Louisiana agreed to sell this property with all legal warranties and with subrogation to all the rights and actions of warranty to which the Citizens Bank of Louisiana was entitled to against all the former owners, such as Stephen Owens Nelson, both individually and as a member of the commercial firm of S. O. Nelson & Company, and Messrs Louis Despanet DeBlanc and Charles DeBlanc.

John Ermant Leitmeyer purchased a tract of land, situated on the right bank of Bayou Teche, in the Parish of Iberia, and the State of Louisiana, and this tract of land was composed of the southwest lots of the Nelson Place, which was another name for the land, formerly belonging to S. O. Nelson, and other names were St. Mary's Plantation, DeBlanc Planation, and the Nelson Tract. These lots were the original lots numbers nineteen through twenty five, and this tract of land was divided into lots and given lot numbers, after a land survey made on the date of April 26, 1876, by a land surveyor of New Iberia, Louisiana. This tract of land measured eleven arpents or nine acres and thirty five hundredths of an acre fronting Bayou Teche by a depth of forty arpents or thirty four American acres, which embraces four hundred and nineteen arpents of land. This land was bounded above or on the upperside by the lands of B. Berard, and below or on the lowerside by the balance of the lands comprising the Nelson tract, or situated on the same side of Bayou Teche, and in the rear of lands belonging now or late to Terrence Boutte.'

The sale of this tract of land, comprising four hundred and nineteen arpents of land, which was sold to John Ermant Leitmeyer was made and accepted by the Citizens Bank of Louisiana for the consideration of the price and sum of $6,285.00. John Ermant Leitmeyer paid the sum of $1,000.00 in cash lawful American money at the execution of the act of conveyance, on the date of June 7, 1881, and the receipt was acknowledged by the President of the Citizens Bank of Louisiana, Leon Carriere. And for the balance, the purchaser has produced and furnished a certain promissory note. This promissory note bears the date of June 7, 1881, which is the date of the execution of the act of conveyance, and has been made to the order of John Ermant Leitmeyer, and endorsed by John Ermant Leitmeyer. This certain promissory note was due at the Citizens Bank of Louisiana, in the City of New Orleans, Louisiana, on the date of January 1, 1882. Also, the promissory note was conditioned to bear interest at 8% per annum. The eight per cent interest charged on the promissory note was charged from maturity, if the promissory note was not punctually paid, until the final payment of the note.

On August 11, 1894, the Jury of Freeholders of Iberia Parish appointed the then Iberia Parish Police Jury to trace and lay out a public road. This new public road would lead from Jeanerette to New Iberia, Louisiana, through the lands of John Ermant Leitmeyer and Alonzo E. Hall to the forty arpents road, now Admiral Doyle Drive. This road now bears the name of the Nelson Canal Road. The police jury, which consisted of Messrs. Ovide Landry, Robert A. Olivier, Adolphe B. Romero. Alfred Romero, Frank Gauthaux, and Thomas Romero were sworn that they would lay out a road to the greater ease and advantage of the inhabitants, and with as little prejudice to enclosures as maybe without favor of affection, malice, or hatred.

John Ermant Leitmeyer and Alonzo E. Hall were given notice of the meeting, three days previous to the meeting, on the date of August 8, 1894, and they were told the time and the place of the meeting. The meeting was to be held at the brick bridge across the Nelson Canal, which was the place designated in the notice. On August 10, 1894, John Ermant Leitmeyer and Alonzo E. Hall met with the Iberia Parish Police Jury for the purpose of discussing the tracing and laying down of the future Nelson Canal Road. It was determined at this meeting that the future Nelson Canal Road would begin at a point just above the Nelson Canal at the brick bridge, about ten or fifteen feet from the brick bridge, and running thorough the lands of John Ermant Leitmeyer and Alonzo E. Hall. There would be a distance of eight and one half arpents or seven acres and two hundred and twenty five hundredths of an acre taken out of the lands of John Ermant Leitmeyer. A triangle of land, which would contain ninety hundredths of an arpent or seventy seven hundredths of an acre of land would also be taken out of the lands of John Ermant Leitmeyer. Then the road would continue through the lands of Alonzo E. Hall for a distance of thirty arpents or twenty five acres and five hundredths of an acre taking sixteen arpents of the land belonging to Alonzo E. Hall. The road would consume a strip forty five feet wide along the Nelson Canal, as the least injurious place to lay out a road to the proprietors, Alonzo E. Hall and John Ermant Leitmeyer. The boundary was mutually agreed upon and showed to the Iberia Parish Police Jury by both John Ermant Leitmeyer and Alonzo E. Hall. Also, the boundary was designated to the Iberia Parish Police Jury by placing stakes in the ground and plowing furrows, so as to make the boundary easily recognizable to the termination of the road. The Nelson Canal Road was to be forty feet wide throughout its entire length, and was so traced and staked out as to be plainly visible throughout its entire course. A plat was made of the proposed road, showing the location and course of the Nelson Canal Road. This plat also showed the location of the lands of both Alonzo E. Hall and John Ermant Leitmeyer, whose properties the future Nelson Canal Road would run through, and the distance and quantity of the land expropriated from both John Ermant Leitmeyer and Alonzo E. Hall for the future Nelson Canal Road.

The Jury of Freeholders of Iberia Parish did on oaths taken, assess the following damages to both Alonzo E. Hall and John Ermant Leitmeyer in compensation for their land so taken and expropriated. John Ermant Leitmeyer was paid $25.00 per arpent or $22.50 for the land expropriated from him. Alonzo E. Hall was paid a fixed price of $22.50 for land expropriated from him. the reason that Alonzo E. Hall was paid a fixed price was due to the fact that Alonzo E. Hall had already worked the road and drained and hedged the road in the field.

John Ermant Leitmeyer died at his residence on the Leitmeyer Plantation, which was situated in Iberia Parish, on September 9, 1899, leaving his widow, Esther Heelum Leitmeyer, four minor children, and a stepson, William Newman. Dr. George Sabatier was the physician who attended John Leitmeyer in his last days. The residence home of the Leitmeyer family was situated on the tract of land, that presently the Consolata Nursing Home is situated on, and the front of the house was in the backyard of the Consolata Nursing Home. The home was a two story, white painted wooden home, and it was a very stately looking home. The house had a widow walk on the rooftop. Actually, the home was built closer to Bayou Teche than the Consolata Nursing Home. John Ermant Leitmeyer left a will and last testament, which was probated in Iberia Parish.

On October 12, 1899, by a virtue issued by the Honorable Court of Iberia Parish, which gave Mrs. Esther Leitmeyer, the widow of John Ermant Leitmeyer, the natural tutorship of her four minor children, Iraneus Joseph, Frank T., Frederick M., and Apolonia.

On October 21, 1899, a family meeting of the Leitmeyer family was held, with Frederick Bauman, a notary public, presiding and the purpose of this meeting was to discuss what would be done with the immovable and movable property left by the deceased, John E. Leitmeyer, and how the four minor children's interest and share in their late father's estate would be handled. Also, on October 26, 1899, an appraisement of the movable and immovable property left by John Ermant Leitmeyer was made, and the movable property was valued at $1.619.00. The two tracts of land that comprised the immovable property was valued as Lot 1 was valued at $10,475.00, and Lot 3 was valued at $1,320.00, which gave a total valuation of immovable property at $11,795.00. The movable property was sold to Mrs. Esther Heelum Leitmeyer, the widow of John Ermant Leitmeyer.

The movable property consisted of twenty two arpents of old fencing, two planks and three wires, which were appraised at $33.00, seven sets of gears, which was appraised at $17.00, three cane carts, which were appraised at $25.00 for each cane cart, or $75.00 for the three cane carts, inner and outer fencing around the main dwelling house, being the Leitmeyer home, with wiring pieuxs and pickets, which was appraised at $25.00, one old hen house, which was appraised at $5.00, one old store room and shed, which was appraised at $15.00, one lot of old tools, which was appraised at $5.00, two single plows, which was appraised at $5.00, one buggy, which was appraised at $10.00, one 7 foot by 9 foot cistern, which was appraised at $18.00, one old empty scale house, which was appraised at $2.50, one mowing machine, which was appraised at $30.00, one horse raker, which was appraised at $2.50, two old dung scrapers, which was appraised at $2.00, one wheel cultivator, which was appraised at $2.50, and one stubble cane digger, which was appraised at $2.50. A stubble cane digger was an agricultural implement that was used to dig holes in the field to plant the cane for the next harvest season or grinding season. Stubble cane is sometimes called plant cane or seed cane. The movable property belonging to the estate of John Ermant Leitmeyer also consisted of one lot of machinery, which was appraised at $1,000.00. This lot of machinery consisted of a three roller sugar mill with the engine, boiler, two clarifiers, two pumps, C-1 Garrison Steam pump, and one belt Jena pump, four sugar kettles, one stack, and one steam train foundation. The movable property also consisted of one bench vice, which was appraised at $1.50, seven American mules, appraised at $6.00 a mule, and the total valuation was $42.00, one Texas mare, which was appraised at $35.00, one old crippled Texas horse, which was appraised at $5.00, twelve head herd of cows, which was appraised at $12.00 each cow, and the total valuation is $144.00, one bull, which was appraised at $15.00, nine calves, which were appraised at $5.00 each calf, and the total valuation was $45.00, one saw and bonne, which was appraised at $7.50, one old bookcase, which was appraised at $12.00, one William rocker, which was appraised at $3.00, one cane bottom rocker, which was appraised at $1.00, sox dining room wood chairs, which were appraised at 75 cents a chair, and the total valuation is $4.50, one bedroom set, which included one bed, washstand, bureau, and two chairs, was appraised at $15.60, one small clock was appraised at $1.00, one large cookstove was appraised at $10.00, one dining table was appraised at $1.00, six wooden chairs was appraised at $1.00, one marble top side board was appraised at $10.00, one lot of dishes and table ware, including forks, knives, and spoons, was appraised at $2.00, one bedroom set, which included a bed, marble top dresser, armoir, a marble top sideboard was appraised at $25.00, one Iron Money safe was appraised at $30.00, one Ingrain carpet was appraised at $6.25, one leather trunk with a canvas cover was appraised at $5.00, one bed and marble top dresser was appraised at $10.00, one spring mattress was appraised at $4.00, and one Caliber Winchester Rifle was appraised at $5.00.

The inventory of the movable and immovable property belonging to the ester of John Ermant Leitmeyer was taken by Frederick K. Bauman, a notary public, in the presence of Jamis Sampson and William Edward Satterfield. The appraisers were Warren R. Farmer, one of the partners in the Orange Grove Sugarhouse, and L. L. Bernard.

At the time of the death of John Ermant Leitmeyer, there existed several notes and one judicial judgment that was payable to the estate. The judicial judgement was against Mr. J. Gall, and in favor of John Ermant Leitmeyer, in the amount of $2,679.73. The notes that were still outstanding, and yet to be paid to the estate consisted of a note signed by George W. Dallas, dated September 10, 1892, for $50.00, and payable one year after date to the order of John Ermant Leitmeyer, with 8% interest until paid. Another note signed by George W. Dallas, dated September 28, 1891, for the sum of $1,000.00, payable one year after date to the order of John Ermant Leitmeyer, and this note was conditioned to bear 8% interest until paid in full. Another note signed by William Newman, dated February 26, 1898, for $374.84, payable to the order of John Ermant Leitmeyer, on January 1, 1899, and conditioned to bear interest at the rate of 8% until paid. Another note signed by William Newman, dated March 21, 1898, for the amount of $150.00, payable to John Ermant Leitmeyer on January 1, 1899, and conditioned to bear interest at the rate of 8% until paid. Another note signed by William Newman, dated May 18, 1898, for the amount of $200.00, with 8% interest from the date, and payable to the order of John Ermant Leitmeyer.

There were no debts belonging to the estate of John Ermant Leitmeyer that were unpaid and no leases to execute. The estate property belongs to the community property that existed between John Ermant Leitmeyer and his wife, Esther Heelum Leitmeyer, and the four minor children that were born of this marriage. The widow, Esther H. Leitmeyer had usufruct of the entire property that belongs to the estate of her late husband, John Leitmeyer, in her possession and charge. Alonzo E. Hall was named co-executor without bond of the estate of John Ermant Leitmeyer. Esther Leitmeyer claims the usufruct accorded to her by law of the share of the minors in the entire property belonging to the community. this was Esther H. Leitmeyer's second marriage, her first marriage was to Mr. Newman, and she had one son, William. John Ermant Leitmeyer brought into the marriage with Esther Heelum, his wife, the sum of $20,000.00, which the community of acquets and gains owes to the separate estate of John Ermant Leitmeyer.

The casket for John Ermant Leitmeyer was purchased from Mr. J. Fisher, for the sum of $32.00, and the funeral notices were put on the telephone and light poles, which cost $1.50. The Reverend R. M. Jourdan conducted the funeral services for John Ermant Leitmeyer, and the cost of conducting the funeral services was $15.00. The cemetery lot was purchased for the sum of $25.00 a lot.

The structures that were standing on the Leitmeyer Plantation, at the time of the death of John Ermant Leitmeyer consisted of one dwelling house or the Leitmeyer home, which was situated on Lot 1 on the west side of Bayou Teche, and the Leitmeyer home was valued at $2,000.00. On hay shed, measuring 25 feet by 40 feet, was valued at $30.00, and six double cabins, which were situated on Lot 1, were valued at $60.00 a cabin,a nd the total valuation amounted to $360.00. Also, one sugar house building and shed, which were valued $2,250.00. Also, on Lot 3 was one lot of old 4 strand wire fencing, which was valued at $5.00, and eight arpents of old wire fencing 4 strand, which was valued at $8.00.

The immovable property that belonged to the estate of John Ermant Leitmeyer consisted of eighty eight arpents or seventy four American acres and eight hundredths of an acre of land on the east side of Bayou Teche, below the town of New Iberia, in Iberia Parish. This tract of land was known as Lot 3, and was bounded north by Sonennams & others, south by the property of Bernard, east by the property of George W. Broughton, and west by Duperier. The land that was contained in Lot 3 was valued at $15.00 an arpent, and the total valuation was $1,320.00. Another tract of land, that was known as Lot 1, which consisted of four hundred and nineteen arpents of land, and this land was valued at $25.00 an arpent, bringing the total valuation to $10,475.00. Lot 1 was the main part of the Leitmeyer Plantation, where the Leitmeyer home was situated, the cabins were situated, and the sugar house was situated. Lot 1 was bounded above or north by the lands of J. B. Berard, and below by the Nelson Tract, and in the rear by the lands of Boutte.'

In the year of his death, John Ermant Leitmeyer had advanced to William Newman, his stepson and the son of Esther Heelum Leitmeyer by her first marriage, various sums of money for the purpose of making a crop. That during the years 1900, 1901, and 1902, William Newman incurred indebtedness towards his mother, Esther Heelum Leitmeyer, who gave him advances or money for the purpose of making the crop of sugarcane on the Leitmeyer Plantation, or the land, which she was in possession of as surviving widow and tutrix of her four minor children. William Newman gave her, Esther Leitmeyer, a note for $1,875.00 in 1902, to represent all of his previous obligations or notes that his mother, Esther Leitmeyer consented to his retaining the various amounts of annual balances due. In other words, all the annual balances for the previous notes of William Newman were represented by that one note given to his mother, Esther Leitmeyer, in the year of 1902.

On May 2, 1901, Mrs. John Ermant Leitmeyer, the widow, granted a right of way to the Iberia Parish Police Jury. This right of way consisted of a strip of land, measuring twenty feet wide and running North to South. This land was to be used for a public road. The right of way was on land that was situated on the east side of Bayou Teche, and this property was bounded north by the lands of Duperier, and south by the lands of George Broughton.

On April 22, 1903, Mrs. John Ermant Leitmeyer, the widow, Honore Romero, Garcia Romero, and Miss Marie Boutte', all met to discuss the setting of boundary lines between their respective properties. They all appeared before Leonce de la Houssaye, a notary public, and they all in his office, which was situated in the town of New Iberia, Louisiana. Honore Romero, Garcia Romero, and Miss Marie Boutte' were all owners of the lands of Therese' Boutte', and their lands adjoined the lands of Leitmeyer. The boundary line that divided the lands of Therese' Boutte' started at a point on the west side of the line from point D to point A.
The boundary line that divided the Leitmeyer property from the lands of Therese' Boutte' started at a point marked D on the south side of the Leitmeyer tract, at the depth of forty arpents or thirty four acres. Point D is designated by a post set by a surveyor in 1892. Point D runs to a point marked E at the rear lower end of the Camille Berard tract of land. The boundary line of the Leitmeyer property from the lands of Therese' Boutte'

started at a point on the east side of points D to A. All the land on the east side of these points belonged to the Leitmeyer family.

In November of 1911, Esther Leitmeyer, the widow of John Ermant Leitmeyer, granted a right of way to the Iberia, & St. Mary, & Eastern Railroad Company. The Iberia, & St. Mary, & Eastern Railroad Company was a corporation organized under the laws of the State of Louisiana, and it was domiciled in the Parish of Iberia. The Iberia, & St. Mary, & Eastern Railroad Company had as its representative, its chief engineer, J. W. Troxal.

Mrs. Leitmeyer granted a right of way, which measured sixty feet wide over and across that certain property, situated in the sixth ward of Iberia Parish, between the public road leading from New Iberia to Jeanerette, which road was known as the Old Spanish Trail, and the right of way of Morgan's Louisiana & Texas Railroad & Steamship Company. This right of way was bounded north by the public road, now the Old Spanish Trail, east by the public road leading along the Nelson Canal, south by Morgan's Louisiana & Texas Railroad & Steamship Company right of way, and west by Landry. The amount of land taken for this right of way measured two acres and ninety nine hundredths of an acre of land. The Iberia & St. Mary & Eastern Railroad Company paid Mrs. Leitmeyer a sum of $200.00 per acre of land.

On January 12, 1912, Frederick M. Leitmeyer and other members of the Leitmeyer family granted a right of way to the Southwestern Traction and Power Company. This right of way was twenty five feet wide and ran along the front of that tract of land that was situated below the City of New Iberia, Louisiana. This right of way was on that portion of the Leitmeyer Plantation that was situated between the public road, leading from New Iberia to Jeanerette, or the Old Spanish Trail, and the Iberia & St. Mary & Eastern Railroad. This right of way extended from the northern to the southern limits of the tract of land. The right of way was to start five feet from the line of trees that were planted along the front of the public road, or the Old Spanish Trail. The grantee, Southwestern Traction and Power Company, obligated themselves not to injure any of the trees. These trees that were planted along the roadside were intended for shade and ornamental purposes. The Southwestern and Traction Power Company may do some trimming of the trees, as the trimming will be required to avoid interference with the operation of the road. Also, the Southwestern Traction and Power Company will grade a sidewalk along the western side of the right of way. This gave a wall of six feet to establish a proper drain between the right of way and the sidewalk. The grade of the sidewalk is to be such as to make it effective, and the same grade as that of the roadbed, if required by the Leitmeyer family.

The Southwestern and Traction Power Company will construct

and maintain the area across the sidewalk, and the right of way of no less than five crossings. The grantee, the Southwest Traction and Power Company, will do the maintenance when required by the Leitmeyer family to do the maintenance at places indicated by the Leitmeyer family. The crossings are to be the width for wagons and carriages to be driven over the crossing.

It is required that the Southwestern Traction and Power Company will not interfere with any drainage on the Leitmeyer property. Also, the Southwestern Traction and Power Company will provide for failure of rice plume, whenever it is required to do so.

It is understood that the grantee, Southwestern Traction and Power Company, will constantly maintain and operate a line of interurban railway, which will be operated by electricity. The electricity will be the motion that will power the railroad cars. The rail line of the Southwestern Traction and Power Company will extend from the City of New Iberia to the town of Jeanerette, Louisiana. Also, the grantee, the Southwestern Traction and Power Company, will construct and maintain a station on its right of way. The construction of the rail line is to be completed by July 1, 1912, and failure to complete the line before the date will result in forfeiture of the right of way. The sale price of the right of way was $10.00, which was paid in a cash payment by the Southwestern Traction and Power Company.

The oak trees that were once situated on the plantation of John Ermant Leitmeyer

THE OAKS PLANTATION

(The Wade Gajan Farm)

The Oaks Plantation, which is situated on the Old Spanish Trail, on the west side of Bayou Teche, was once a part of St. Mary's Plantation, which was owned by the S. O. Nelson & Company of the City of New Orleans. Before the commercial firm of S. O. Nelson & Company had purchased what is now known as the Oaks Plantation or the Wade Gajan Farm, this tract of land formed the plantation of Louis Despanet DeBlanc and Charles DeBlanc.

The site where the present day home of Mr. and Mrs. Clyde Bess is located and the site where the home of the late Wade and Maxie Gajan is located was once the homesite of the plantation home of St. Mary's Plantation, and the DeBlanc plantation homesite. Mrs. Clyde Bess is the former Lois Gajan, who is the only daughter of Joseph Wade Gajan and his wife, Martha "Maxie" Taylor Gajan. In the early 1980's, Lois Gajan Bess and her husband, Clyde Bess, retired here in the New Iberia area, and built their present day home.

The homesite of Lois Bess Gajan and her husband, Clyde Bess, stands in the grove of oak trees, which once shaded the plantation home of the Stephen Owens Nelson Plantation on the Old Spanish Trail. Also, the former home of the late Joseph Wade Gajan and his wife, Martha, "Maxie" Taylor Gajan is situated on tract one of the S. O. Nelson Plantation and the DeBlanc Plantation, and presently this tract of land comprises the Oaks Plantation. In the year of 1842, the original plantation home was situated in this grove of oak trees, and this plantation home was occupied by the family of Charles DeBlanc and his wife, Constance LaBranch DeBlanc. After the DeBlanc plantation was sold to the banking firm of S. O. Nelson & Company of the City of New Orleans, this plantation home was occupied by Stephen Owens Nelson and his family.

Lois Bess Gajan and her husband, Clyde Bess, are the parents of two daughters, Elizabeth "Betty" Bess Walton, and Nancy Bess Beall. Betty Bess is married to Richard L. Walton, and has one son. Nancy is the wife of Dr. J. LeBer Beall, and they are the parents of one son. Nancy and her husband, Dr. J. LeBer Beall, reside in Irving, Texas. Betty and her husband, Richard Walton, reside in Bay City, Texas.

Today, the Oaks Plantation consisted of a certain tract of land, which is situated in Township 12 South, Range 7 East. The Oaks Plantation contains 319.38 acres of land. It is bounded on the north side by property owned by Lois Gajan Bess, which are lots 42 through lot 84 of the Wade Gajan subdivision, on the south side by the Forty Arpent Road or Admiral Doyle Drive, on the east side by the property of Edward N. Simmons, and on the west side by the Nelson Canal or the Nelson Canal Road. On the date of August 16, 1853, tract one was described as measuring thirty eight arpents or thirty two American acres and three hundredths of an acre fronting Bayou Teche on the west side of by bayou with a depth of forty arpents or thirty four acres in depth. At this time, the boundaries of tract one were on the upperside by the lands belonging to the widow and heirs of Baron Benoit Bayard I, and below by the lands of Colonel Charles Napoleon Olivier or Orange Grove. That part of tract one that was situated on the east side of Bayou Teche measured twenty seven arpents or twenty two American acres and ninety five hundredths of an acre fronting Bayou Teche, with a depth of forty arpents or thirty four American acres. This part of tract one that was situated on the east side of Bayou Teche was bounded on the upperside by the lands of Henry Camors, and on the lowerside by lands of Adelaide Olivier, a free man of color.

The slave cabins were located where the site of the present day Iberia General Medical Center, the Diamond Shamrock Gasoline Service Station, the Huval Insurance Agency, the Consolata Nursing Home, and the medical offices of Drs. Dipti and Neil Bavishi are located.

During the plantation days, the present day Nelson Canal was a drainage ditch for the waters of small streams to drain off into Bayou Teche, and this deep drainage ditch was situated on the plantation of S. O. Nelson. Also, there was a brick bridge, which allowed for crossing this deep ditch.

On the date of March 1, 1971, the will and last testament of the late Lizzie Kramer Gajan, who was the wife of the late Emile Gajan of the Parish of Iberia, was read and probated in the courtroom of the Iberia Parish Courthouse. Lizzie Kramer Gajan was the mother of Joseph Wade Gajan, who was the owner of the Oaks Plantation, and Ethel Gajan Dupuy. The two subscribing witnesses to this act of agreement were Benjamin White, and Isadore W. Gajan. Isadore W. Gajan was also a notary public in the town of New Iberia, Louisiana and he was a relative of both Emile Gajan and his children, Joseph Wade Gajan and Ethel Gajan Dupuy.

Form the succession of his late mother, Joseph Wade Gajan inherited a tract of land, which was situated in Section 36, Township 12 South, Range 7 East, and also situated in the second ward of Iberia Parish. This tract of land contained ten arpents of land. Also, this tract of land was bounded on the north and the west sides by the lands of David Baez, Junior, and on the east side by the Forty Arpent Road, which is known as Admiral Doyle Drive, and on the south side by the land where the oilfield service company, Movible Offshore, was formerly located.

On the date of May 16, 1926, the succession of Emile Gajan was opened and probated in open court, in the courtroom of Judge James Simon, who served as judge of the 16th Judicial District Court, which was holding court in the Parish of Iberia.

The heirs to this succession were his widow, the former, Lizzie Kramer of New Iberia, and his two children, a daughter, Ethel Gajan and a son, Joseph Wade Gajan. Ethel Gajan and Joseph Wade Gajan were recognized as sole heirs of Emile Gajan, and they are entitled to ownership of one half of all the property left by their late father, Emile Gajan.

Ethel Gajan and Joseph Wade Gajan inherited a tract of land, which was situated in the second ward of the Parish of Iberia. This tract of land contained ten arpents of land. The boundaries of this tract of land were on the north side by the lands of Cecile Fusilier, on the south side by the lands of Cecile Fusilier, on the east side by the public road, and on the west side by the lands of Mrs. Prosper Olivier.

On the date of October 1, 1925, Joseph Gajan leased a tract of land unto Ulysse Gonsoulin, of the Parish of Iberia. Joseph Gajan was a resident of the town of New Iberia, and he was the husband of the former Maggie Landry. Ulysse Gonsoulin was a resident of Iberia Parish, and he was the husband of Edna Berard.

The tract of land that was leased unto Ulysse Gonsoulin was leased for the purpose of cultivating rice. This tract of land was situated in the second ward of the Parish of Iberia. The boundaries of this tract of land was on the north side by the lands of Joseph Wade Gajan, on the east side by the railroad track of Morgan's Louisiana & Texas Railroad & Steamship Company, which is presently the Southern Pacific Railroad Company railroad track, on the west side by lands of Joseph Wade Gajan, and on the south side by the lands of Mrs. Robert F. Broussard. The amount of acreage that is leased is forty five acres of land. This tract of land is part of the property acquired by Joseph Gajan from Wade Gajan.

This lease of the tract of land between Joseph Gajan and Ulysse Gonsoulin was in effect for a period of three years, beginning on the date of January 1, 1926 and ending on the date of December 31, 1928. In this lease agreement, no ditches are to be destroyed, and no trees are to be cut down. The livestock is not allowed to run loose on the land.

Ulysse Gonsoulin has agreed to pay unto Joseph Gajan as rental for the tract of land the sum of $2,000.00 for the whole rent of the three year period. The amount of the rentals is represented by the three promissory notes of Ulysse Gonsoulin. The first promissory note is for the sum of $76.00, and it is due and payable on the date of August 15, 1926. The second promissory note is for the sum of $665.00, and this note is due and payable on the date of August 15, 1927. The third promissory note is for the sum of $575.00, and this note is due and payable on the date of August 15, 1928. All of these three promissory notes are conditioned to bear interest at the rate of 8% from the date of maturity or until paid in full.

Ulysse Gonsoulin, the lessee, is to give Joseph Gajan, the lessor, a stack of rice straw, which is made up of straw coming from the forty five acres of rice. This rice is to be threshed and placed in the bayou yard of Gajan's property.

Also, on the date of October 1, 1925, Wade Gajan entered into an agricultural lease of the farm of the Oaks Plantation with Ulysse Gonsoulin. The execution of this deed of lease took place in the office of Isadore Gajan, who was a notary public in the town of New Iberia, Louisiana.

Ulysse Gonsoulin was the husband of Edna Berard. Joseph Wade Gajan was the husband of Martha, "Maxie" Taylor Gajan, and they resided at the Oaks Plantation on the Old Spanish Trail East, which is a short distance from the corporate limits of the town of New Iberia, Louisiana.

Wade Gajan was leasing the farm land of the Oaks Plantation to Ulysse Gonsoulin for the purpose of the cultivation of rice. The tract of land that was leased measured fifty acres, and composed a part of the Oaks Plantation or the Wade Gajan Plantation. The Oaks Plantation is bounded on the north side by the Nelson Canal Road, on the south side by the lands of Joseph Gajan, on the east side by Bayou Teche, and on the west side by the Forty Arpent Road or Admiral Doyle Drive. This lease is for the duration of a period of three years. Also, this lease commences on the first day of January of the year of 1926, and ends on the 31st day of December of the year of 1928. Ulysse Gonsoulin is to take possession of the tract of fifty acres of land as soon as the crop of sugarcane is harvested.

These fifty acres included the land situated between the first two drainage ditches, which drainage ditches run west of Morgan's Texas & Louisiana Railroad & Steamship Company railroad tracks. Also, this fifty acres of land extends from the northern boundary of the Oaks Plantation or the Wade Gajan property clear to the line of the property of Joseph Gajan on the south side. This tract of land takes in the land that extends and runs in a westerly direction clear the pasture fences, which pasture fences are situated on the west side. This leased tract of land also includes the property of Joseph Gajan, which is situated on the southside and this property runs in a northerly direction so as to make the land area of fifty acres of land.

Ulysse Gonsoulin is to pay as a rental for the tract of which is fifty acres in area, the sum of $6.80 per acre of ach of the three years, or the sum of $1,020.00 for the whole For the repayment of the rental, Ulysse Gonsoulin has given iree promissory notes of even date, and these three promissory are payable to the order of Ulysse Gonsoulin. Also, these promissory notes are endorsed in the blank by Ulysse lin. Each one of the three promissory notes is for the sum 0.00. The first promissory note is due and payable on the f August 15, 1926, the second promissory note is due and e on the date of August 15, 1927, and the third promissory s due and payable on the date of August 15, 1928. All three sory notes are conditioned to bear 8% interest per annum from y.

seph Wade Gajan is to allow Ulysse Gonsoulin the use engine, which engine is now located in the pumping plant property. Ulysse Gonsoulin is allowed to use this engine e understanding that he is to repair and place the engine r use as an emergency pumping plant. In consideration for of the engine, Ulysse Gonsoulin is to supply Joseph Wade Gajan with free water supply, which will be sufficient to irrigate the fifty acres of land planted in rice.

In the event that the engine is not used by Ulysse Gonsoulin at all, and Joseph Wade Gajan finds a sale for the engine, Wade Gajan will be allowed to sell the engine. If the engine is sold, then Wade Gajan is to repay Ulysse Gonsoulin the cost of the repairs of the engine.

In the event that the engine is sold, Joseph Wade Gajan is to pay Ulysse Gonsoulin water rent on the fifty acres of land that Wade Gajan has leased to Ulysse Gonsoulin. The payment of water will be at a rate of 200 pounds per acre of land, and the water rent dates from the year the engine is sold.

Also, no livestock is permitted to run loose at any time, nor is hunting permitted on the fifty acres of land that has been leased to Ulysse Gonsoulin by Joseph Wade Gajan.

On the date of April 13, 1925, Joseph Wade Gajan granted a right of way to the Iberia, St. Mary, and Eastern Railroad Company for a spur track. This transfer did not include ownership of the land, but this transfer conveyed only the rights to the use of the railroad spur track. This right of way grant did not prevent the use of the property by the owner, J. Wade Gajan.

The execution of this deed of transfer took place in the office of notary public, Isadore Gajan, who was a relative of Joseph Wade Gajan. The office of Isadore W. Gajan was situated in the town of New Iberia, Louisiana. Also, the two witnesses to the execution of this deed of transfer were E. H. Buffington and James Gajan.

Also, Joseph Wade Gajan acknowledged to notary public, Isadore W. Gajan that he was the person, who had signed the deed of transfer to the Iberia, St. Mary, & Eastern Railroad Company. The deed of transfer and right of way was executed and signed as a voluntary act.

This railroad spur track commenced at a point of intersection of the westerly line of a road crossing, and the center line of the main railroad track of the Iberia, St. Mary, & Eastern Railroad Company as constructed at the engineers station number 2729, plus 60, thence, northerly at right angles along the road line a distance of thirty feet to the point of beginning, thence westerly thirty feet from and parallel to the center line of the main track of the Iberia, St. Mary, & Eastern Railroad Company a distance of six hundred feet to a point, thence northerly at right angles a distance of twenty feet to a point, thence easterly fifty feet from and parallel to the center line of the main track of the Iberia, St. Mary, & Eastern Railroad Company a distance of six hundred feet to a point in the westerly line of the road, thence, southerly at right angles along the road line a distance of twenty feet to the point of beginning. The right of way contains 0.23 acres of land area.

On the date of February 7, 1920, Joseph Wade Gajan entered into an agreement with Louis Fontenot to purchase a tract of land. This agreement was executed before William G. Weeks, who was an attorney and a notary public in the City of New Iberia. Also, this agreement was an agreement of the intention of Louis Fontenot to purchase a tract of land, which comprised the Oaks plantation.

Louis Fontenot was a resident of the town of Crowley, which is situated in the Parish of Acadia. Also, Louis Fontenot was married to the former Donatile Keller. At this time, Joseph Wade Gajan was residing in the town of New Iberia, Louisiana.

This tract of land that was to be purchased by Louis Fontenot was a tract of land that contained sixteen arpents of land. This tract of land was bounded on the east side by Bayou Teche, on the west side by the Old Spanish Trail or the public road that leads from the town of New Iberia to the town of Jeanerette, Louisiana, on the northside by the lands of John Ermant Leitmeyer, and on the southside by a tract of land, which is known as the Crosby tract.

This tract of land was reserved for the purpose of erecting a syrup mill on. The syrup mill is to be erected in five years from the date of February 7, 1920. In the event, of failure to erect a syrup mill on these premises by the date of February 7,

1925, the principals, Emile Gajan and his son, Joseph Wade Gajan, obligate themselves to sell the tract of land to Louis Fontenot. By this agreement, Emile Gajan and Joseph Wade Gajan will not sell this tract of land to anyone else but Louis Fontenot.

In the event that this tract of land is sold, the tract will be sold for the purchase price of $130.00 per acre of land. The total purchase price of this tract of land will be the sum of $2,080.00.

By another deed executed on the same day of February 7, 1920, which was a deed of conveyance, Joseph Wade Gajan and Joseph Gajan sold to Louis Fontenot two tracts of land. This act of conveyance was executed in the town of New Iberia, at the same time that the deed of agreement was executed between the parties of Emile Gajan, Joseph Wade Gajan, and Louis Fontenot.

At this time, Joseph Wade Gajan and Joseph Gajan were selling Louis Fontenot two tracts of land. The first tract of land was bounded on the northside by Bayou Teche, in the rear by the Forty Arpent Road which is known as Admiral Doyle Drive, below in part by lands belonging to Mr. Grevemberg and Robert F. Broussard, and above by the public road, which is known as the Nelson Canal Road. The Nelson Canal Road was a parish road, which leads from the Old Spanish Trail to the Forty Arpent Road. The first tract of land was known as the Nelson tract, and it contained four hundred and fifty six arpents of land. The reason that this tract was known as the Nelson tract was that at one time all the land comprising the Oaks Plantation comprised a sugar plantation, which was owned by Stephen Owens Nelson, personally, of the City of New Orleans. Before the Nelson tract was owned by Stephen Owens Nelson, personally, it comprised a part of St. Mary's Plantation, which was owned by the commercial firm of S. O. Nelson & Company of the City of New Orleans, Louisiana. The firm of S. O. Nelson & Company was owned by four businessmen, Stephen Owens Nelson, and Ellison Banksmith, both of the City of New Orleans, Thomas A. Nelson of Athens, Alabama, and Henry C. Walker of the City of Memphis, Tennessee. Originally, the Nelson tract formed a part of the DeBlanc plantation, which was a sugar plantation that was owned by Jean Baptiste Despanet DeBlanc, then Louis Despanet DeBlanc, and then later Louis Despanet DeBlanc and Charles DeBlanc.

The second tract of land that was sold to Louis Fontenot was a tract of land that was adjacent to the first tract of land. The boundaries of the second tract of land were Bayou Teche in front, in the rear by tract one or the tract known as the Nelson tract, and below by the estate of the late Robert F. Broussard. This second tract of land contains one hundred and ten arpents of land. The two tracts of land comprised a land area of five hundred and sixty six arpents of land, but it was sold minus sixteen arpents of land. The total land area was five hundred and fifty arpents of land or four hundred and seventy six American acres and

... escrow account with ... agreement, the sum of $5,000.00, ... forfeited to Emile Gajan should Louis Fontenot fail to comply with his agreement to purchase the land, and in the event that Emile and Joseph Wade Gajan fail to comply with the terms of the agreement.

The sum of $5,000.00, which is placed in the escrow account, is to be a payment of the cash portion of the purchase price of the tract of land, which purchase price is the sum of $15,000.00. Upon the deposit of the amount of $5,000.00 in the escrow account, Louis Fontenot is to have the immediate possession of the tract of land under the conditions of this contract. In the event of failure to the terms of this contract to purchase the property, the sums of money that have been paid by Louis Fontenot to Wade Gajan and Joseph Gajan will become the property of Wade and Joseph Gajan.

Upon the reading of an abstract to the title of this property, which took place at the law office of Weeks & Weeks in the town of New Iberia, Louisiana, it was discovered that Louis Fontenot has to pay the sum of $68,675.00, of which the sum of $15,000.00 is to be in cash. For the balance of the purchase price, $53,675.00, Louis Fontenot has to furnish five promissory notes. Each one of the five promissory notes has to be in the sum of $6,735.00. The payment of these five promissory notes are secured by a vendor's lien and privilege on the two tracts of land, and this vendor's lien and privilege will affect the property until the payment of the notes in full. These five promissory notes are conditioned to bear interest at the rate of 7% per annum from the date of February 7, 1920 or until paid in full. The first note is due on the date of December first of each year or until the notes are paid in full.

There is to be reserved from this sale, all the seed cane on the two tracts of land, all the stack rice straw, the feed for the livestock, the agricultural implements, and all the unused lumber, cross ties, and iron culverts.

Wade Gajan and Joseph Gajan are to insure all the buildings and the improvements on the two tracts of land for an amount of not less than $5,000.00. Also, Wade Gajan and Joseph Gajan are to pay the premium of the insurance policy.

Louis Fontenot has agreed that Joseph Gajan is to be allowed to use that part of the land or the second tract of land from Bayou Teche to the Southern Pacific Railroad tracks for the year of 1920. Also, a part of the land is not to be delivered to Louis Fontenot until the date of December 31, 1920.

Wade Gajan and Joseph Gajan reserve the right of way for a railroad switch tract from the main line of the Southern Pacific Railroad, which was formerly Morgan's Louisiana & Texas Railroad & Steamship Company. This railroad switch track is to lead to the sixteen arpent tract, and this right of way is to be taken along the upper boundary of this tract of land.

On the date of March 5, 1920, Joseph Gajan conveyed to Joseph Wade Gajan at a cash sale two tracts of land. Isadore Gajan was the notary public, who executed this act of cash sale,a nd the execution of this act took place in the town of New Iberia, Louisiana. The two subscribing witnesses to the execution of this act of cash sale were Emile Gajan the father of Joseph Wade Gajan, and Gaston Mestayer. Also, at the execution of this act of cash sale, Joseph Wade Gajan paid to Joseph Gajan the purchase price of the two tracts of land, which was the sum of $19,490.00. The purchase price of the sum of $19,490.00 was paid in cash money.

The first tract of land was situated on both sides of the Old Spanish Trail or the public road, which leads from the town of New Iberia to the town of Jeanerette, Louisiana. Also, this tract of land was situated three miles below the town of New Iberia, Louisiana. This tract of land measured two and one half arpents fronting the west side of Bayou Teche, with a depth of sixteen arpents. The total land area of the first tract of land was forty arpents of land. The boundaries of the first tract of land was on the northside by the lands of Joseph Gajan, on the southside by the lands of Warren R. Farmer or Orange Grove Plantation, on the eastside by Bayou Teche, and on the westside by the right of way of Morgan's Louisiana & Texas Railroad & Steamship Company railroad tracks. Presently, the railroad tracks of Morgan's Louisiana & Texas Railroad & Steamship Company is now known as the Southern Pacific railroad tracks.

The second tract of land was situated on the same side as the first tract of land. The second tract of land measured five arpents front with a forty arpent depth. The total area of the second tract of land was two hundred arpents of land. The depth of this tract of land ran between parallel lines of thirty four arpents clear to the right of way of Morgan's Louisiana & Texas Railroad & Steamship Company railroad track. The boundaries of this second tract of land was on the north side by the lands of Joseph Gajan, on the south by the lands of Warrren R. Farmer or Orange Grove plantation, on the east side by the right of way of Morgan's Louisiana & Texas Railroad & Steamship Company, and on the west side by the Forty Arpent Road, which is known as Admiral Doyle Drive. This is the same property that was acquired by Joseph Gajan from Manette Applegate, who was the wife of Robert F. Broussard, by deed of sale on the date of November 2, 1918.

Before the early 1960's, the Forty Arpent Road, which is now Admiral Doyle Drive, was known to native New Iberians and to natives of the area as the Forty Arpents Road. This meant that this public road was situated forty arpents from Bayou Teche, and this road was the backline of all the plantations along the Old Spanish Trail. When the United States Government decided to locate a United States Navy Base in New Iberia, Louisiana, this road was named for an outstanding Navy Admiral. This United Navy base was situated at the extreme west end of the Forty Arpent Road or Admiral Doyle Drive. In honor of a very prestigious Naval officer, the Forty Arpent Road was renamed Admiral Doyle Drive. Toady, Admiral Doyle Drive is a very busy four land road, which has on each side of the roadway many shopping centers, various stores, Catholic High School, and the New Iberia Senior High School besides a high rise office building.

On the date of November 2, 1918, Manette Applegate Broussard conveyed to Joseph Gajan a tract of land at a sale and mortgage. Manette Applegate was the widow of the late Senator Robert F. Broussard, and a resident of the Parish of Iberia. Joseph Gajan was the husband of Maggie Landry, and he was also a resident of the Parish of Iberia.

This tract of land measured five arpents fronting the Forty Arpent Road, which is presently known as Admiral Doyle Drive, and the depth of this tract of land was fifteen arpents. The total area of this tract of land was seventy five arpents of land. The boundaries of this tract of land was on the north side by land owned by the Suburban Realty Company, on the south side by the lands of Warren R. Farmer or Orange Grove Plantation, on the east side by lands of Joseph Gajan and on the west side by the public road, known as the Forty Arpent Road or Admiral Doyle Drive. This was the same tract of land that Manette Applegate Broussard had acquired from the estate of the late Robert F. Broussard. On the date of January 2, 1917, Senator Robert F. Broussard had acquired this tract of land from Joseph Gajan by act of conveyance.

On the date of January 2, 1917, Emile Gajan conveyed a tract of land to Senator Robert F. Broussard at a sale and mortgage. The deed of conveyance was executed before Alfred Reignited, who was a notary public in the town of New Iberia, Louisiana. The two subscribing witnesses to this deed of conveyance were Louis Indest and N. Delcambre.

This tract of land that was conveyed to Senator Robert F. Broussard measured five arpents on the Forty Arpent Road, or Admiral Doyle Drive with a depth of fifteen arpents. The total land area of this tract of land was seventy five arpents. The boundaries of this tract of land were on the north side by land owned by the Suburban Realty Company, on the south side by the lands of Warren R. Farmer or Orange Grove Plantation, on the east side by the lands of Joseph Gajan, and on the west side by the Forty Arpent Road or Admiral Doyle Drive.

For the payment of the purchase price of the sum of $3,750.00, Senator Robert F. Broussard executed three promissory notes. These three promissory notes were made to the order of Robert F. Broussard, and the notes were endorsed in the blank. Each one of the three promissory notes was for the sum of $1,000.00. The first promissory note was payable on the date of January 2, 1918, the second promissory note was payable on the date of January 2, 1919, and the third promissory note was payable on the date of January 2, 1920.

On the date of July 17, 1905, Emile Gajan conveyed a tract of land to Emar Romero at a cash sale. This deed of conveyance was executed in the office of notary public, Alfred Reignited, which office was situated in the town of New Iberia, Louisiana. The two subscribing witnesses to this deed of conveyance were Louis Indest and Erebe Broussard. At the execution of this act of sale, Emar Romero paid the cash sum of $337.00 to Emile Gajan.

The tract of land that was being sold was situated on the west bank of Bayou Teche with a frontage along Bayou Teche of three fourths of an arpent. Also, this tract of land has a frontage along the Old Spanish Trail or the public road leading from the town of New Iberia to the town of Jeanerette, Louisiana. The depth of this tract of land was three arpents of land. The total land area of this tract of land two arpents and 25/100 of an arpent of land. The boundaries of this tract of land were in the front or on the west side by the public road or the Old Spanish Trail, and on the north side by a parcel of land this day sold to E. Barras. The tract of land that was sold was a portion of the property that Emile Gajan had acquired from Alonzo E. Hall on the date of December 30, 1903.

On the date of March 29, 1905, Emile Gajan conveyed three tracts of land to Clovis Louivere. This deed of conveyance was executed before notary public, Alfred Renoudet.

The first tract of land was situated about two and one half miles below the city limits of New Iberia, Louisiana. This fronting the west side of the public road, or the Old Spanish Trail, which leads from the town of New Iberia to the town of Jeanerette, Louisiana. The depth of this tract of land was to the right of way of Morgan's Louisiana & Texas Railroad & Steamship Company. This tract of land contained five and one half arpents in area. The boundaries of this tract of land were in the front by the Old Spanish Trail, or the public road, which is known as Louisiana State Highway # 182, and in the rear by the right of way of the Southern Pacific Railroad tracks, formerly Morgan's Louisiana & Texas Railroad & Steamship Company, on the upperside by the lands of Erebe Broussard, and below by the lands of Emile Gajan.

The second tract of land measured one and one fourth arpents fronting on the east side of the Old Spanish Trail or the public road, which is known as Louisiana State Highway #182. Also, the second tract of land had a depth of three arpents of land to Bayou Teche. The second tract of land contained three arpents and seventy fifths arpents of land. The boundaries of the second tract of land were in front by the public road or the Old Spanish Trail, in the rear by Bayou Teche, above by the lands of Emile Gajan, and below by a tract of land of Erebe Broussard.

The third tract of land was situated in the same locality as the first two tracts of land. The third tract of land measured one and one half arpents fronting the right of way of the railroad track of Morgan's Louisiana & Texas Railroad & Steamship Company. The depth of the third tract of land was thirty five arpents of land. The total land area was fifty two arpents and five hundredths of an arpent of land to the Forty Arpent Road or Admiral Doyle Drive. The boundaries of the third tract of land were on the north side by the lands of Emile Gajan, on the south side by the lands of Erebe Broussard, on the east side by the right of way of Morgan's Louisiana & Texas & Steamship Company railroad track, and on the west side by the Forty Arpent Road or Admiral Doyle Drive.

The crop of sugarcane that is now growing on the three tracts of land is the property of Emile Gajan and delivery of the three tracts of land to Clovis Louivere will be on the date of January 1, 1906. Existing on the three tracts of land to the benefit of the balance of the lands sold to Erebe Broussard remains a right for the lands to drain through the remaining lands of Emile Gajan. The drainage ditches are to be kept up mutually by Erebe Broussard, Emile Gajan, and Clovis Louivere.

The purchase price of these three tracts of land is the sum of $5,300.00. Clovis Louivere has furnished his four promissory notes, and these notes are made to the order of Clovis Louivere. These four promissory notes are dated March 29, 1908. The first promissory note is for the sum of $2,600.00, and it is payable on the date of January 12, 1906. The other three promissory notes are for the sum of $883.33. The second promissory notes are for the sum of $883.33. The second promissory note is due and payable on the date of January 1, 1907, the third promissory note is due and payable on the date of January 1, 1908, and the fourth promissory note is due and payable on the date of January 1, 1909. All four promissory notes are conditioned to bear interest at the rate of 8% per annum. All the Iberia Parish, Louisiana State and federal taxes on these three tracts of land have been paid. Clovis Louivere has agreed to pay the Iberia Parish, Louisiana State, and federal taxes on these three tracts of land in the year of 1905.

On the date of February 13, 1905, Emile Gajan sold a tract of land to Erebe Broussard at a sale and mortgage. This deed of conveyance was executed in the office of the presiding notary public, Alfred Renoudet, whose office was located in the town of New Iberia, Louisiana. The two subscribing witnesses to the execution of this deed of conveyance were Louis Indest and Errole Romero.

The tract of land that was sold was situated on the west side of Bayou Teche in the Parish of Iberia, and the State of Louisiana. This tract of land had a frontage of three arpents on the Old Spanish Trail or the public road, which leads from the town of New Iberia to the town of Jeanerette, Louisiana. The depth of this tract of land was three arpents. The total area of this tract of land was nine arpents of land. Also, the boundaries of this tract of land were on the west side and in the front by the Old Spanish Trail or the public road, which is known as Louisiana State Highway # 182, and on the north side by the lands of Emile Gajan. This tract of land was a portion of the lands purchased by Emile Gajan from Alonzo E. Hall on the date of December 30, 1903.

The purchase price of this tract of land was the sum of $337.50, and Erebe Broussard has executed a promissory note to pay the purchase price. This promissory note is made to the order of Erebe Broussard, and this promissory note is endorsed in the blank. The promissory note is dated the date of February 17, 1905, and the amount of the promissory note is the sum of $337.50. This promissory note is payable in two years after the date, which is the date of February 13, 1907.

On the date of December 30, 1902, Alonzo E. Hall sold a tract of land to Emile Gajan. At the execution of this act of conveyance, John T. White served as notary public. The two witnesses to this act of conveyance were R. E. Southwell and Thomas E. Lahine.

The tract of land that was being sold was situated three miles below the city limits of New Iberia, Louisiana, which is situated in the Parish of Iberia. This tract of land was situated on the west side of Bayou Teche, and it contains five hundred and sixty six American acres and forty nine hundredths of an acre of by the public road, which is known as the Spanish Trail, and the lands of John Ermant Leitmeyer, on the lowerside by lands of Warren R. Farmer, which forms part of the Orange Grove plantation, and in the rear by the public road, which is known as the Forty Arpent Road. This tract of land was all the land that remained to Alonzo E. Hall from the original purchase of land from the Citizens Bank of Louisiana on the date of April 27, 1889. Also, sold with this act of conveyance were seventeen mules, a lot of harnesses, and a lot of agricultural implements or farming utensils.

There was a mortgage bearing on this tract of land, and this mortgage was in favor of John M. Terry of Cedar Rapids, Iowa. The amount of this mortgage was the sum of $8,000.00. This mortgage was payable in the amount of $6,000.00, which was due on the date of April 8, 1904, and the remaining $2,000.00, which was due on the date of January 1, 1904. This mortgage was extended for a period of three years from the date of the mortgage, which made this mortgage due on the date of January 1, 1907 and April 8, 1907. Emile Gajan has assumed the mortgage bearing against the property of the Oaks Plantation, and this mortgage is in favor of John M. Terry of Cedar Rapids, Iowa.

At the execution of this act of conveyance, Emile Gajan paid the sum of ten thousand dollars ($10,000.00). This left a balance of $10,000.00, which was to be paid in four promissory notes. These four promissory notes were endorsed in the blank and bore the date of December 30, 1903. The first promissory note was for the sum of $2,000.00, and it was due in ten days from the date of December 30, 1903, and the second promissory note was for the sum of $2,000.00, and this second promissory note was due in thirty days or January 30, 1904. The third promissory note was for the sum of $2,000.00, and this third promissory note was due in sixty days from the date of December 30, 1903, and the fourth promissory note was for the sum of $4,000.00, and this fourth promissory note was due in twelve months or on the date of December 30, 1904.

On the date of December 27, 1901, John T. White acting as notary public executed a mortgage, which was in favor of John M. Terry of Cedar Rapids, Iowa. This mortgage was to bear on the Oaks Plantation, which was owned by Alonzo E. Hall, of the Parish of Iberia, and the State of Louisiana. Also, this mortgage was for the sum of $2,000.00 with interest at 8% per annum from the date of January 1, 1902. The two witnesses to the execution of this act of conveyance were A. LeBlanc and H. Mactur.

Alonzo E. Hall executed one promissory note, which promissory note was made payable to the order of John M. Terry of Cedar Rapids, Iowa. The sum of this mortgage was the amount of $2,000.00 in principal, and this promissory note was conditioned to bear interest at the rate of 8% per annum from the date of January 1, 1902. The interest due on this mortgage was payable annually. Also, this promissory note was due and payable on the date of July

1, 1904.

Alonzo E. Hall, a resident of Iberia Parish and the State of Louisiana, mortgaged unto John M. Terry, a resident of Cedar Rapids, which is situated in the State of Iowa, the Oaks Plantation, which is situated in the Parish of Iberia and the State of Louisiana. The Oaks Plantation is situated about thee miles below the town of New Iberia, Louisiana, on the west bank of Bayou Teche. This tract of land contained five hundred and sixty six arpents and forty ninths hundredths of an arpent of land. The boundaries of the Oaks Plantation were in the front by the Bayou Teche, in the rear by the public road, which is known as the Old Spanish Trail, and above by lands belonging to the estate of John Ermant Leitmeyer, and below by lands belonging to Warren R. Farmer, which was designated as Orange Grove Plantation.

On the date of December 24, 1903, Alonzo E. Hall, a resident of the Parish of Iberia, declared that under Article 244 of the Constitution of the State of Louisiana, which is relative to homesteads, that he has a homestead bearing on the Oaks Plantation, which is owned and occupied by him. The property is appraised at the valuation of the sum of $10,000.00. Also, Alonzo E. Hall declared that it is his desire to waive and abandon the homestead right and exemption from seizure, which is reserved to him as set forth by Article 244 of the Constitution of the State of Louisiana. Also, Effie L. Ortte Hall, who was the wife of Alonzo E. Hall, declared that she has taken full cognizance of the general waiver of homestead, which was made by her husband, Alonzo E. Hall. Also, Effie L. Ortte Hall consents and signs her name on her own free will. This declaration made by Effie L. Ortte Hall was done to protect her legal matrimonial, dotal, and paraphernal rights, which is granted to a wife by the laws of the State of Louisiana.

On the date of April 16, 1900, Alonzo E. Hall sold a tract of land to John C. Davis at a sale and mortgage. The deed of conveyance was executed before William G. Weeks, who was a lawyer and a notary public in the town of New Iberia, Louisiana. The two subscribing witnesses to this deed of conveyance were Donaldson Caffery, Junior and G. Winter. This deed of conveyance was executed in the town of New Iberia, Louisiana.

At the execution of the deed of conveyance, Clifford Hall appeared and acted on the behalf and in the interest of Alonzo E. Hall by virtue of an act of power of attorney. Clifford Hall was selling a tract of land to John C. Davis of the Parish of St. Mary.

The tract of land that John C. Davis was purchasing was situated on the east side of Bayou Teche, and it contained five hundred and nineteen arpents of land. The boundaries of this tract of land were in the front by the lands of Delis Lucky, in the rear by the lands of Walsh, below by lands belonging to James Clifton Hall, and on the west side by Bayou Teche. This was a portion of the tract of land that Alonzo E. Hall had acquired from the Citizens Bank of New Orleans on the date of April 27, 1891. Originally, this land formed a portion of the lands on the east side of Bayou Teche of the plantation of the commercial firm of S. O. Nelson & Company of the City of New Orleans, and then the plantation of Stephen Owens Nelson of the City of New Orleans, Louisiana, personally. This tract of land was sold by Alonzo E. Hall to James Clifton Hall by act of conveyance on the date of January 27, 1891.

The purchase price of this tract of land was the sum of $13,029.43, John C. Davis has executed six promissory notes, which are made to the order of himself and endorsed by him in the blank. These six promissory notes are dated the date of April 28, 1900. Three promissory notes are each for the sum of $2,333.33. The first promissory note is due and payable on the date of April 16, 1901, the second promissory note is due and payable on the date of April 16, 1902, and the third promissory note is due and payable on the date of April 16, 1903. The next three notes are due in the years of 1904, 1905, and the last and final note is due and payable on the date of April 16, 1906. All six promissory notes are conditioned to bear interest at the rate of 8% per annum from the date of April 16, 1900 or until paid. The interest is payable annually. The property will remain specially mortgaged with a vendor's lien and privilege existing until the full and the final payment.

On the date of March 12, 1900, Alonzo E. Hall entered into a right of way agreement with Morgan's Louisiana & Texas Railroad & Steamship Company. This deed of agreement was executed before notary public, William G. Weeks, of the City of New Iberia, Louisiana. The two subscribing witnesses to this deed of agreement was Henry A. King and Achille Patin. At the execution of this deed of agreement, Morgan's Louisiana & Texas Railroad & Steamship Company was represented by its representative, William F. Owens, who served the company as its superintendent and agent.

Morgan's Louisiana & Texas Railroad & Steamship Company was a corporation, which was created under the corporate laws of the State of Louisiana. The company of Morgan's Louisiana & Texas Railroad & Steamship Company was domiciled in the City of New Orleans, Louisiana. This company was in the business of laying down railroad tracks, running train lines, and running a steamship line, which operated on the bayous and rivers of the State of Louisiana.

Alonzo E. Hall owned a tract of land in the Parish of Iberia, and this tract of land contained five hundred and sixty six American acres and forty ninths hundredths of an acre of land. This tract of land was bounded in front by Bayou Teche, in the rear by the public road, known as Louisiana State Highway # 182 or the Old Spanish Trail, above or on the west side by the lands of John

Old Spanish Trail, above or on the west side by the lands of John Ermant Leitmeyer, and below or on the east side by the lands of Warren R. Farmer or Orange Grove Plantation.

The strip of land that Alonzo E. Hall was granting this right of way to Morgan's Louisiana & Texas Railroad & Steamship Company was a strip of land forty two feet wide. This strip of land ran parallel and contiguous to the existing right of way of Morgan's Louisiana & Texas Railroad & Steamship Company on the south side of the railroad track, and through the tract of land belonging to Alonzo E. Hall. This strip of land is between engineers station 6442I97 and 6469I65. This strip of land is bounded on the east side by the lands of Warren R. Farmer or the Orange Grove plantation, on the west by the lands of John Ermant Leitmeyer, and the public road or the Old Spanish Trail.

Also, this strip of land is of sufficient width to make a total width of right of way of the southside of the entire line of the main railroad track of Morgan's Louisiana & Texas Railroad & Steamship Company. This strip of land begins at a point where the east line of the tract of land of Alonzo E. Hall intersects the center line of the main railroad track of Morgan's Louisiana & Texas Railroad & Steamship Company at engineers station 644I92. Then this strip of land runs in a southerly direction from the east line of the tract of land of Alonzo E. Hall thirty three feet to the northeast corner of the strip of land, and continues in a straight line in a southerly direction forty two feet to the southern corner of the forty two foot strip of land. Then, the right of way continues two thousand six hundred and sixty eight feet in a westerly direction on a line parallel to seventy five feet distant from the entire line of the railroad track of Morgan's Louisiana & Texas Railroad & Steamship Company to the southwest corner of the forty two foot strip of land, then in a northerly direction on the west line of the lands of Alonzo E. Hall, forty two feet to the northwest corner of the forty two foot strip of land. The right of way continues in an easterly direction two thousand six hundred and sixty eight feet along the southern boundary to the right of way of Morgan's Louisiana & Texas Railroad & Steamship Company to the point of beginning. The strip of land contains 2.57 acres of land.

The purchase price of this forty two foot strip of land is the sum of $257.00, which was paid in cash money at the execution of this act of sale by the representative and superintendent, William F. Owen, of Morgan's Louisiana & Texas Railroad & Steamship Company. Also, all the Iberia Parish, Louisiana State and federal taxes are paid by Alonzo E. Hall.

On the date of April 8, 1899, Alonzo E. Hall declared under Article 244 of the Constitution of the State of Louisiana, that he has a homestead on the Oaks Plantation, which is owned and occupied by him. Article 244 of the Constitution of the State of Louisiana were h-

...at Alonzo E. Hall declared that he has a ... bearing on the Oaks Plantation was due to the fact that he was waiving and abandoning the homestead right and exemption from seizure reserved to him by the Articles of the Constitution of the State of Louisiana. This declaration makes the waiver and abandonment in general for all interests and purposes. The homestead right and exemption is granted under the laws of the State of Louisiana.

Also, on the date of April 8, 1899, Alonzo E. Hall, of the Parish of Iberia, was legally indebted to John W. Terry, who was a resident of the City of Cedar Rapids, Iowa, for the sum of $6,000.00. At this execution of the act of mortgage, John W. Terry was legally represented by John R. Davis, who was acting as the special attorney on the behalf and in the interests of John W. Terry of the City of Cedar Rapids, Iowa. The authority of John R. Davis to represent John W. Terry was recognized by Alonzo E. Hall.

The promissory note was executed by Alonzo E. Hall, and it was endorsed in the blank by Alonzo E. Hall. Also, this promissory note was dated the date of April 8, 1899, and it was due on the date of April 8, 1904, or in five years from the date of the execution of the act of mortgage. Also, this promissory note was conditioned to bear interest at the rate of 8% from the date of April 8, 1899 or until paid in full. The interest on this promissory note was to be paid annually. This mortgage was to bear on the property of the Oaks Plantation, which was owned by Alonzo E. Hall.

On the date of April 10, 1899, Alonzo E. Hall, of the Parish of Iberia, borrowed money in the form of a promissory note from James Clifton Hall, who was a resident of the County of Cherokee, Iowa. The amount of this promissory note was the sum of $9,000.00, and it was represented by two promissory notes. This mortgage was bearing on the property of the Oaks Plantation, which was owned and occupied by Alonzo E. Hall.

Alonzo E. Hall executed two promissory notes, and each note was for the sum of $4,500.00, which made a total of $9,000.00. This promissory note was made to the order of Alonzo E. Hall, and was endorsed by Alonzo E. Hall. These two promissory notes fell due on the date of April 10, 1901, which was two years after the date of this mortgage. Also, these two promissory notes were conditioned to bear interest at the rate of 8%, and this interest was payable annually. These two promissory notes were delivered to James Clifton Hall by Alonzo E. Hall.

James Clifton Hall by Alonzo E. Hall.

On the date of June 1, 1889, the Citizens Bank conveyed a tract of land to Alonzo E. Hall and William L. Bailey. The execution of this act of conveyance was passed by notary public, Edgar Grima, in the City of New Orleans, Louisiana. At the execution of this act of conveyance, Stanley O. Thomas, who was President of the Citizens Bank, was acting on behalf and in the interests of the Citizens Bank. The witnesses to the execution of this act of conveyance were Charles M. Audry and Clement C. Belaire.

The Citizens Bank was a corporation, which was domiciled in the City of New Orleans, Louisiana. Also, the Citizens Bank operated under a corporate charter, which was written under the corporate laws of the State of Louisiana. The President of the Citizens Bank was authorized to act on behalf and in the interests of the Citizens Bank in all legal notarial acts, and other business interests of the bank.

The President of the Citizens Bank, Stanley O. Thomas, declared by an act, which was executed before notary public, P. Laurent Renoudet of the Parish of Iberia, on the date of May 25, 1889, Alonzo E. Hall by his attorney in fact, William G. Hall, has sold to William L. Bailey a tract of land. This tract of land was located about three miles below the town of New Iberia, Louisiana, which is situated in the Parish of Iberia. Also, this tract of land is situated on the west bank of Bayou Teche,a nd contains about ten arpents of land. The boundaries of this tract of land were on the front by Bayou Teche, in the rear by the railroad track of Morgan's Louisiana & Texas Railroad & Steamship Company, above by lands belonging to the Citizens Bank, and below by lands belonging to the Citizens Bank. This tract of land is two hundred and eighty feet wide on Bayou Teche with a depth running between parallel lines to Morgan's Louisiana & Texas Railroad & Steamship Company. This tract of land forms part of lots fourteen and fifteen.

The purchase price of this tract of land is the sum of $100.00, which was paid in cash to the Citizens Bank at the execution of this act of conveyance. This cash payment was made by William G. Hall, who was acting in his capacity as attorney in fact for Alonzo E. Hall. The President of the Citizens Bank, Stanley O. Thomas, acknowledged the receipt for the purchase price on the tract of land. Also, granted to Alonzo E. Hall by Stanley O. Thomas, who was serving as President of the Citizens Bank, was a release of the mortgage and vendor's privilege, which was bearing on the tract of land. This vendor's privilege existed on the property, which was purchased by Alonzo E. Hall, on the date of April 29, 1889, and this vendor's privilege was in favor of the Citizens Bank of Louisiana. This tract of land formed a part of the plantation, which was purchased by Alonzo E. Hall. This vendor's privilege was a way of securing the four promissory notes, which were drawn and executed by Alonzo E. Hall, for the principal and interest, which amounted to the sum of $4,715.34. The first promissory note was payable on the date of January 1, 1890, the second promissory note was payable on the date of January 1, 1891, the third promissory note was payable on the date of January 1, 1892, and the fourth promissory note was payable on the date of January 1, 1893. All four promissory notes were conditioned to bear interest at the rate of 7% per annum from the date of January 1, 1889 or until paid in full. These four promissory notes were paraphed by notary public, P. Laurent Renoudet, "Ne Varietur," to identify these four promissory notes with this particular act of sale. At the execution of this act of conveyance, the President of the Citizens Bank of Louisiana, Stanley O. Thomas, authorized the clerk of court of the Parish of Iberia to erase and cancel these four promissory notes from the mortgage records of the Parish of Iberia.

On the date of May 19, 1898, Alonzo E. Hall declared that he was legally indebted to Mr. J. Hechisrau of the City of New Orleans for the sum of $5,000.00. This mortgage was conditioned to bear interest at the rate of 8% per annum from the date of May 19, 1898 or until paid in full. This mortgage was due on the date of January 1, 1899. Also, this mortgage was considered a farm mortgage, which means that it was given for the raising of the crops of sugarcane and corn on the Oaks Plantation. This mortgage was bearing on the property of the Oaks Plantation.

On the date of December 23, 1896, Alonzo E. Hall was legally indebted to Felix Patout for the sum of $3,500.00. This mortgage is represented by a promissory note, which Alonzo E. Hall has furnished. This promissory note is drawn and subscribed to the order of Alonzo E. Hall, and it is also endorsed by Alonzo E. hall in the blank. This promissory note bears the date of December 23, 1896, and it is payable on the date of January 1, 1898. Also, this promissory note is conditioned to bear interest at the rate of 8% per annum. This mortgage bears on the property of the Oaks Plantation, which is owned by Alonzo E. Hall.

The Oaks Plantation consisted of a tract of land, which was located about three miles below the town of New Iberia, on both the east and the west sides of Bayou Teche, in the Parish of Iberia. The Oaks Plantation was the balance of the lands, which was purchased by Alonzo E. Hall from the Citizens Bank of Louisiana. The site of the Oaks Plantation was also the site of the plantation residences of both the DeBlanc Plantation, and then the Nelson Place or St. Mary's Plantation, which was owned by the commercial firm of S. O. Nelson & Company, and then personally by Stephen Owens Nelson of the City of New Orleans, Louisiana. The boundaries of the Oaks Plantation was on the upperside by lands belonging to John Ermant Leitmeyer and Delis

Verret, on the west side by the public road, known as the Forty Arpent Road, and on the lowerisde by lands belonging to James Clifton Hall. The Oaks Plantation contained one thousand and ninety five arpents of land. Also, the Forty Arpent Road was situated just below the Grand Marais or the Great Swamp.

The reason that Felix Patout was making this loan to Alonzo Hall was to enable Alonzo E. Hall to plant, cultivate, and save his crops of sugarcane and corn, which was being raised in the fields of the Oaks Plantation. By being the grantor of this mortgage, Felix Patout has the privilege granted by the laws of the State of Louisiana to furnishers of money and supplies, which are necessary for the cultivation of the crops. Alonzo E. Hall pledges and pawns the crops of corn and sugarcane to Felix Patout as a means of further security for the payment of the amount of money loaned to him. The amount of money, which is loaned to Alonzo E. Hall is represented by a promissory note. This promissory note has been executed in accordance with the provisions of Act Number 66 of the Louisiana legislature on the date of April 30, 1874.

Also, on the date of March 16, 1896, Alonzo E. Hall declared that he is legally indebted to Felix Patout for the sum of $2,000.00. For the repayment of this mortgage, Alonzo E. Hall executed a promissory note in favor of Felix Patout. This promissory note was drawn and subscribed by Alonzo E. Hall, and this promissory note was made to the order of Alonzo E. Hall . Also, this promissory note was for the sum of $2,000.00, and it was conditioned to bear interest at the rate of 8% per annum from the date of maturity until paid in full. This promissory note was payable on the date of January 1, 1897. For the repayment of this loan, Alonzo E. Hall mortgaged the Oaks Plantation to Felix Patout.

The reason that Felix Patout made this loan of the sum of $2,000.00 to Alonzo E. Hall was to enable Alonzo E. Hall to plant, cultivate, and harvest his crops of sugarcane and corn, which were growing in the fields of the Oaks Plantation. The crops of corn and sugarcane are pawned and pledged in favor of Felix Patout, and the property of the Oaks Plantation is to remain mortgaged until the full and the final payment of this loan. Felix Patout was a furnisher of money and supplies to farmers, which would enable these farmers to be able to plant, cultivate, and harvest their crops.

On the date of August 11, 1894, Alonzo E. Hall and John Ermant Leitmeyer conveyed a tract of land to the Police Jury of the Parish of Iberia to make a road, which road would lead from the Old Spanish Trail to the Forty Arpent Road. This road would bear the name of the Nelson Canal Road. The name of this road was taken from the name of the former owners of this tract of land. At one time, where the Nelson Canal Road is presently located, this tract of land formed a part of tract one of the Nelson Place or St. Mary's Plantation. The Nelson Place or the Nelson Plantation, or St. Mary's Plantation was owned by the commercial firm of S. O. Nelson & Company of the City of New Orleans, Louisiana, and then by Stephen Owens Nelson, personally.

At this time, in the year of 1894, the Iberia Parish Police Jury consisted of Ovide Landry, Robert A. Olivier, who was an owner of Orange Grove Plantation, Adolphe B. Romero, Alfred Romero, Frank Gauthraux, and Honra Romero. All the members of the Iberia Parish Police Jury had taken an oath and sworn that they will lay out a road, which would be named the Nelson Canal Road. Also, this road is directed to be laid out by the Iberia Police Jury to the greatest ease and advantage of the inhabitants, and with as little prejudice to enclosures as may be without any favoritism be shown.

Robert A. Olivier was the son of Eugene Barthlemy Olivier and his wife, Therese Laura Dalcour Olivier, who were at one time the owners and proprietors of the Orange Grove Plantation. Also, Robert A. Olivier was the grandson of the late Colonel Charles Napoleon Olivier and his second wife, Felicite Chretin Olivier.

On the date of August 10, 1894, Alonzo E. Hall and John Ermant Leitmeyer were served propositions of this proposal of the public road, which was to be laid out through their lands. Both, Alonzo E. Hall and John Ermant Leitmeyer were served notices of the intentions of the Iberia Parish Police Jury to build this public road.

Also, on the date of August 10, 1894, Alonzo E. Hall and John E. Leitmeyer met at the brick bridge, which was situated across the Nelson Canal, which was the place designated in the notices. Then the members of the Iberia Parish Police Jury traced out the proposed site of the future Nelson Canal Road in the presence of Alonzo E. Hall and John Ermant Leitmeyer.

On the date of August 11, 1984, the Jury of Freeholders of the Parish of Iberia was to trace and lay out a public road, which was to be eventually called the Nelson Canal Road. This road was to lead from the main road, which leads from the town of New Iberia to the town of Jeanerette, Louisiana, and this road is known as the Old Spanish Trail. Also, the Nelson Canal Road was to be laid out through the lands of John Ermant Leitmeyer and Alonzo E. Hall to the Forty Arpent Road. This road was to be along the Nelson Canal, which originally had been a drainage ditch on the plantation of S. O. Nelson.

The Nelson Canal Road begins at a point just above the Nelson Canal at the brick bridge. Actually, the Nelson Canal Road beginning point is ten to fifteen feet from the brick bridge, and this road runs through the lands of John Ermant arpent of land. The Nelson Canal Road takes out of the lands of John Ermant Leitmeyer, a triangle area of land, which triangle contains .90 of an arpent of land or 90/100 of an arpent of land. Also, the Nelson Canal Road runs through the land of Alonzo E. Hall for a distance of thirty arpents, and this road takes out of the land of Alonzo E. Hall six arpents of land. The strip of land on which the Nelson Canal Road is laid out is a strip of land forty five feet wide, which strip of land runs along the Nelson Canal. This is the least injurious place on both the lands of John Ermant Leitmeyer and Alonzo E. Hall to lay out a public roadway.

Alonzo E. Hall and John Ermant Leitmeyer mutually agreed on the boundaries, and they showed the Iberia Parish Police Jury where the boundaries for this roadway was to be. Then, they designated the boundaries of the roadway by putting stakes and plowing furrows so as to be easily recognized to the termination of the future Nelson Canal Road. The Nelson Canal Road is forty five feet wide throughout its entire length, and was so traced and staked out as to be plainly visible.

John Ermant Leitmeyer received $25.00 an arpent of land from the Iberia Parish Police Jury for the 90/100 of an arpent of land, which was prorated from John Ermant Leitmeyer. Alonzo E. Hall received the sum of $225.00, which was the price of the land prorated from him. This price of $225.00 was fixed due to the fact that Alonzo E. hall has already worked the road and drained and hedged it in field.

On the date of January 27, 1891, Alonzo E. Hall conveyed a tract of land to James C. Hall, Robert Hall, Edmund D. Huxford, and Ernest Henricks, who were all residents of Cherokee County, which is situated in the State of Iowa. At the execution of this act of conveyance, Edmund D. Huxford and Ernest Henrick, were represented by James C. Hall, who was acting in their behalf and in their interests by virtue of an act of power of attorney. The execution of this act of conveyance took place in the office of P. Laurent Renoudet, which office was situated in the town of New Iberia, Louisiana. The two witnesses to the execution of this act of conveyance were John Lamprez and Frank M. Welch, who was an attorney in the town of Jeanerette, Louisiana.

Alonzo E. Hall was selling to James C. Hall, Robert Hall, Edmund D. Huxford, and Ernest Henrick a tract of land, which was situated in the Parish of Iberia, on the west and the east banks of Bayou Teche. This tract of land was situated about three miles below the town of New Iberia, Louisiana. This tract of land contains one thousand arpents of land, and it is bounded on the upperside by land belonging to Alonzo E. Hall, on the lowerside by land belonging in part to Warren R. Farmer, which was Orange Grove Plantation, and in part by land belonging to William L. Bailey, and on the west side by the public road or the road known as the Old Spanish Trail.

on the west side by the public road or the road known as the Old Spanish Trail.

Also, this tract of land consumes the southeast side of the lands of Alonzo E. Hall and William L. Bailey. This tract of land measures in the front twelve arpents of land and eighty eight hundredths of an arpent of land with thirty eight arpents and ninety hundredths of an arpent of land, which is the tract measured on a line perpendicular to the side lines and extending back between parallel lines to a depth of forty arpents. This tract of land is situated on both the east and the west sides of Bayou Teche, and this land is situated in Iberia Parish. Prior to October 1869, this land was considered as being situated in the upper part of St. Mary Parish, because the Parish of Iberia had not been created. This tract of land had a frontage on the west side of Bayou Teche of ten arpents of land with a width in front of 13.77 chains or 908.82 feet. This tract of land extended from Bayou Teche to the railroad of the Southern Pacific Railroad, which was formerly known as Morgan's Louisiana & Texas Railroad & Steamship Company.

The purchase price of this tract of land was the sum of $20,000.00. At the execution of this act of conveyance, James C. Hall, Robert Hall, Ernest Henricks, and Edmund Huxford paid in part payment or the cash downpayment the sum of $8,000.00. the balance of the purchase price was the sum of $12,000.00, and for the repayment of the balance of the purchase price, three promissory notes were executed. These three promissory notes were drawn and subscribed to the order of James C. Hall, Robert Hall, Ernest Henricks, and Edmund Huxford. These three promissory notes were endorsed in the blank, and these notes were dated the date of January 27, 1891. Also, these three promissory notes were payable and due in one, two, and three years. The first promissory note was due in the year of 1892, the second promissory note was due in the year of 1893, and the third promissory note was due in the year of 1894. Each one of the three promissory notes was for the sum of $4,000.00, and these three promissory notes were conditioned to bear interest at the rate of 7% per annum from the date of January 10, 1891 or until paid in full.

Before the signing of this act of conveyance, it was well and distinctly understood that the purchasers, James C. Hall, Robert Hall, Ernest Henricks, and Edmund Huxford deliver unto themselves a tract of land containing exactly one thousand arpents of land, which tract of land was exchanged for ten arpents of land belonging to William L. Bailey. Also, the railroad track of the Southern Pacific Railroad Company, formerly Morgan's Louisiana & Texas Railroad & Steamship Company ran through this tract of land, and this railroad track measured sixty six feet in width.

On the date of June 1, 1889, the Citizens Bank sold a tract of land to Alonzo E. Hall and William L. Bailey. The

Thomas, who was serving as President of the Citizens Bank of Louisiana, represented the interests of the Citizens Bank of Louisiana. The execution of this deed of conveyance took place in the office of notary public, Edgar Grima, in the City of New Orleans, Louisiana. The two subscribing witnesses to this deed of conveyance were Charles G. Audry and Clement C. Belaire.

The President of the Citizens Bank, Stanley O. Thomas, declared by an act, which was passed before notary public, P. Laurent Renoudet, of the Parish of Iberia, on the date of May 25, 1889, Alonzo E. Hall, by his attorney in fact, William Hall, has sold to William L. Bailey a tract of land. This tract of land is situated three miles below the city limits of the town of New Iberia, Louisiana, and this tract of land is situated in the Parish of Iberia. Also, this tract of land is situated on the west bank of Bayou Teche. The boundaries of this tract of land are in front by Bayou Teche, in the rear by the railroad track of Morgan's Louisiana & Texas Railroad & Steamship Company, above by the lands of the Citizens Bank, and below by the lands of the Citizens Bank. This tract of land has a width of two hundred and eighty feet along Bayou Teche with a depth that runs between parallel lines to Morgan's Louisiana & Texas Railroad & Steamship Company. This tract of land forms part of lots fourteen and fifteen of the lands of Alonzo E. Hall.

The purchase price of this tract of land was the sum of $100.00 in cash, which was paid to the Citizens Bank at the execution of this act of conveyance by William G. Hall, the attorney in fact of Alonzo E. Hall. This amount of $100.00 was paid on the mortgage debt. The receipt was acknowledged by Stanley O. Thomas, who was President of the Citizens Bank, and acquittal was granted to Alonzo E. Hall. A vendor's privilege existed on the land, and this vendor's privilege was in favor of the Citizens Bank of Louisiana per act dated April 29, 1889. At this sale, the tract of land was sold to William L. Bailey, and Alonzo E. Hall was released from the mortgage and the vendor's privilege.

To secure the payment of the principal and the interest of four promissory notes of Alonzo E. Hall for the sum of $4,715.34. The first promissory note is payable on the date of January 1, 1890, the second note is payable on the date of January 1, 1891, the third note was payable on the date of January 1, 1892, and the fourth note was payable on the date of January 1,1893. All four promissory notes were conditioned to bear interest at the rate of 7% per annum from the date of January 1, 1889. These four promissory notes are identified with an act of sale, which was passed on the date of April 27, 1889. This mortgage is identified with this particular act of sale by the paraphed, "Ne Varietur," which identifies another notarial act or deed with another notarial act or deed. Stanley O. Thomas, the President of the Citizens Bank, authorized the clerk of court of the Parish of Iberia to cancel and erase the mortgage from the mortgage records.

Alonzo E. Hall was a resident of the Parish of Iberia, and he resided on the Oaks Plantation in the plantation home for a while. Alonzo E. Hall was married to the former Olive Ortte of New Iberia, Louisiana. Alonzo E. Hall lived in the plantation home that originally was the plantation home of the DeBlanc family. Also, Alonzo E. Hall died in the City of New Iberia on the date of April 19, 1954. Alonzo E. Hall and his wife, Olive Ortte Hall, had no children, but they had numerous nieces and nephews.

On the date of May 7, 1889, Alonzo E. Hall had his attorney, P. Laurent Renoudet, execute a mortgage, which was in favor of Rufus Smith. This mortgage was to bear on the property comprising the Oaks Plantation.

Alonzo E. Hall executed one promissory note, which was drawn and subscribed to the order of Alonzo E. Hall. Also, this promissory note was endorsed by Alonzo E. Hall, and dated the date of May 7, 1889. The sum of this promissory note was the amount of $4,000.00, and this note was conditioned to bear interest at the rate of 8% per annum. Also, this note was payable on the date of January 15, 1895. At the execution of this act of mortgage, which occurred in the office of P. Laurent Reignited, which was situated in the town of New Iberia, Louisiana. Alonzo E. Hall delivered this promissory note to Rufus Smith, who was the holder of the promissory note.

This sum of $4,000.00 was borrowed by Alonzo E. Hall to enable him to plant, cultivate, and save his crops of sugarcane and corn, now being made in the fields of the Oaks Plantation. Alonzo E. Hall was to use this money to pay the wages of the laborers, buy fertilizers, for repairs to machinery and equipment, and to buy new machinery and equipment to replace the old, worn out machinery and equipment.

It was also agreed that the buildings and improvements, which were situated on the lands of the Oaks Plantation were to be insured and kept insured against the loss by fire. The amount of the fire insurance policy or policies was to be at least in the amount of $6,000.00.

Alonzo E. Hall recognizes in favor of Rufus Smith a lien and privilege, which is to bear on the crops of corn and sugarcane, which is being raised in the fields of the Oaks Plantation. This lien and privilege is granted to Rufus Smith under the laws of the State of Louisiana to furnishers of money and supplies, which are necessary for the cultivation, making, and saving of the crops. Also, Alonzo E. Hall pledged and pawned the crops of corn and sugarcane to Rufus Smith.

As a measure of further security for the payment of the promissory note, in accordance with the provisions of Act 66 of the legislature of the State of Louisiana, which is granted as

As a measure of further security for the payment of the promissory note, in accordance with the provisions of Act 66 of the legislature of the State of Louisiana, which is granted as protection to furnishers of money and supplies for the cultivation, making, and saving of crops, the property of the Oaks Plantation is to remain specially mortgaged until the final payment of the promissory note.

Live oak, which were planted by Stephen Owens Nelson, the owner St. Mary's Plantation.

A view of the cane fields of the former St. Mary's plantation

KYLE ACRES

Kyle Acres, the residence home and property of Mrs. Virginia Kyle Hine and Miss Barbara Kyle, once formed the lower part of tract one of the plantation proper of St. Mary's Plantation, which was owned by S. O. Nelson & Company, of New Orleans, Louisiana.

On June 4, 1842, this plantation and the slaves that were residing on the plantation were conveyed by Peltier de la Houssaye, who was Sheriff of St. Mary Parish, to Messrs. Louis Despanet DeBlanc and Charles B. DeBlanc, at a sheriff sale. Peltier de la Houssaye had in his possession two writs of seizure and sale, which were issued out of the Parish Court of the Parish of Orleans. The reason these two writs of seizure and sale were issued was in connection with a civil lawsuit against the goods, chattels, lands, and tenements of Jean Baptiste Despanet DeBlanc (Charles DeBlanc), deceased).

All the rights, titles, and interest of Jean Baptiste Despanet DeBlanc were seized by Sheriff Peltier de la Houssaye, who was serving as Sheriff of St. Mary Parish, to pay for the debts contracted by Jean Baptiste Despanet DeBlanc with the Union Bank of the City of New Orleans, Louisiana. Originally, the lands of the plantation consisted of tract one, tract two, and tract three. In the year 1842, this plantation was cultivated as a sugar plantation.

The slaves that were living on this plantation, in the year of 1842, were Zenon, a mulatto man of 24 years of age, Celine, a slave woman, age 23 years old, Charles, a black man slave, age 21 years of age, Onezime, a man slave, age 23 years, Elio, a griffe man slave, age 21 years old, Edmund, a teenage slave, age 18 years of age, Joseph, a black man slave, age 31 years of age, Polidore, a man slave, age 32 years of age, Eugene, a griffe man slave, age 23 years old, Baptiste, a man slave, age 21 years of age, Narcisse, a black man slave, age 33 years of age, Daniel, a man slave, age 20 years of age, Alexander, a man slave, age 20 years of age, Ephraise, a man slave, age 39 years of age, Sandy, a man slave of 38 years of age, Abraham, a man slave, age 35 years of age, Page, a griffe man slave, age 35 years of age, Henry, a man slave, age 28 years of age, Hayden, a black man slave, age 28 years of age, Jesse, a slave man, age 29 years of age, Jordan, a slave man, age 40 years of age, Thomas, a slave man age 22 years of age, Watt or Matt, a slave man, age 33 years of age, Joe, a slave man, age 26 years of age, Sam, a slave man, age 30 years of age, Big Sam, a slave man, age 35 years old, Hamsted, a slave man, age 25 years of age, John, a slave man, age 21 years of age, Jack, a slave man, age 30 years of age, Fredrick, a yellow man or a mulatto salve, age 35 years of age, William, a black man, age 20 years old, Helary, a slave man, age 30 years of age, Janis, a slave, age 33 years of age, William Malus, a slave of 30 years of age, Scpiou, a teenage slave, age 14 years of age, Jean Baptiste, a child slave, age 9 years old, Zelie, a child slave, age 9 years old, Emilie, a slave child, age 7 years old, Florence, a child slave, age 3 years of age, Melissaine, a woman slave, age 33 years old, and the four children of the slave woman, Melissaine, and these four slave children were Josephine, a slave girl child, age 11 years of age, Pauline, a girl slave child age 3 years of age, Leau, a baby slave, age 9 months old, and Leontine, a slave child of 7 years of age, Evariste, a slave boy child age 4 years of age, Maia, a black woman slave, age 40 years of age, Delphine, a slave woman, age 38 years of age, Therese, a slave teenage girl, age 16 years of age, Cecile, a teenage girl slave, age 13 years old, Charlotte, a slave girl child, age 11 years old, Agaz, a slave woman, age 45 years old, Angelle, a girl teenage slave, age 16 years of age, Mary, a slave woman, age 27 years old, and her two slave children, Philimone, a slave child, age 4 years old, and Celestin, a slave child, age 1 years old, Caroline, a slave woman, age 26 years old, and her two children, Josephine, a girl slave child age 4 years old, and St. Ville, a toddler slave child, age 15 months, Victoria, a slave woman age 26 years old, and her five children, Michael, a slave boy, age 11 years old, Lisette, a girl slave child age 7 years old, Francisque, a slave child age 5 years old, Victorin, a slave child, age 3 years old, and Victores, a baby slave, age 11 months old, Mary Lombard, a slave woman, age 30 years of age, and her child, Baptiste, a slave child of 2 years of age, Anna, a slave woman, age 28 years of age, and her two children, Silvie, a slave child age 4 years of age, and Toby, a slave baby, age 3 months old, Mehela, a griffe woman slave, age 30 years old, and her child, Julie, a girl slave child of age 4 years old, Marianne, a black woman slave, age 22 years of age, and her three children, Ben, a slave boy, age 3 years of age, Clementine, a slave toddler, age 18 months old, and Justin, a girl slave child, age 3 months old, Rose, a slave woman, age 20 years of age, and her two slave children, Roseine, a slave girl child of age 2 years, and Jelesno, a baby slave of 2 months old, Nice, a yellow woman or a mulatto woman slave, age 35 years of age, and her child, Louis, a boy toddler slave child of age 1 years old, Helene, a black woman slave, age 32 years of age, and her four children, Phebe, a girl slave of age 12 years of age, Marguerite, a girl slave child of age 7 years old, Sarah, a girl slave child age 5 years of age, and Tochine, a toddler slave child of 1 year of age, Chalimette, a woman slave of 36 years of age, and her six children, Feleicite, a girl slave age 18 years old, Pierre a slave boy, age 15 years old, Sam's Fur, a slave boy age 13 years of age, Zelie, a slave child of 9 years of age, Emilie, a slave child of 7 years of age, and Florence, a girl slave child of 3 years of age. These slaves were slaves for life, and all these slaves were residing in the slave cabins on the DeBlanc Plantation. The slave cabins were situated on tract one, where today the St. Peter Street extension exists.

Also, eight hundred shares of capital stock of the Union Bank of Louisiana, which were secured by a mortgage on the three tracts of land, which comprised the DeBlanc plantation, and the slaves that were also residing on the DeBlanc plantation. In the antebellum period, the slaves were considered property, and could be mortgaged, as well as sold. This public sheriff sale was held on June 4, 1842, between the hours of eleven o'clock in the morning and four o'clock in the afternoon.

The mortgages that were bearing on the DeBlanc plantation were considered a special mortgage in favor of the Union Bank of Louisiana, for two hundred and ninety nine shares of capital Union Bank stock, and the amount of the value of these shares of stock were worth amounted to was $13,200.00. This mortgage originated from the purchase of the property, or the main part of the plantation of Charles DeBlanc from Louis DeBlanc as per act of sale, which was passed before the Honorable Judge John Moore, who was the Parish Judge and Ex-Officio Notary of St. Mary Parish, Louisiana, on the date of December 26, 1835.

There was also a special mortgage in favor of the Union Bank to secure the three hundred shares of capital stock of the Union Bank of Louisiana, and this mortgage was passed on January 1, 1833. Also, a special mortgage in favor of the Union Bank to secure thirteen shares of capital stock in the Union Bank, and this mortgage was dated August 26, 1834. These were the only two mortgages that were shown by the books of record of the clerk of court in St. Mary Parish, but on examining a copy of the act of sale made by Louis DeBlanc to Charles DeBlanc, that copy of the act was sent to New Orleans, and it appeared that the certificate showed another mortgage in favor of the Union Bank to secure eighty shares of capital stock in the Union Bank. This act of mortgage was executed on July 11, 1833. The last mortgage having apparently been left out by the recorder of the act of conveyances at the office of the clerk of court in St. Mary Parish, Louisiana.

There was also a special mortgage in favor of the Union Bank of Louisiana, and against the DeBlanc plantation for five hundred and one shares of capital stock in the Union Bank of Louisiana, and this act was dated April 16, 1839. Also, another special mortgage in favor of the Union Bank of Louisiana and against the DeBlanc plantation for the sum of $12,000.00, and this act was dated April 17, 1839. This note was the amount of the stock note that was due by Charles DeBlanc. Another special mortgage bearing on the DeBlanc plantation, and in favor of Messrs. Peter Rapp, Evariste Mauris, and Auguste Reggio for the sum of $42,000.00, and this note was dated June 15, 1839. There was a judicial mortgage in favor of Jean Baptiste Despanet DeBlanc and against the DeBlanc plantation in the amount of $14,000.00 with interest at 10% from June 8, 1841, which resulted from the civil law suit of Jean Baptiste Despanet DeBlanc versus Louis DeBlanc. This judgement was rendered on February 26, 1842 in favor of the plaintiff. There was another judicial mortgage in favor of Louis Despanet DeBlanc in the amount of $21,000.00 with interest at 10% from June 8, 1844, resulted from the same civil law suit entitled, "Jean Baptiste DeBlanc versus Louis D. DeBlanc." There was another judicial mortgage in favor of the Union Bank of Louisiana and against Messrs Edouard Durrive and Charles DeBlanc, and the firm of Durrive and DeBlanc in solido for the sum of $14,100.00 with interest until paid. The terms of paying this judicial mortgage off were as follows, the first payment of $5,500 was due on January 11, 1842, the second mortgage payment $2,400.00 was due on January 18, 1842, the third mortgage payment of $2,000.00 was due on January 28, 1842, and the fourth mortgage payment of $2,500.00 was due on February 14, 1842, and the fifth mortgage payment of $2,000.00 was due on February 28, 1842. Also, $12.50 cost was added for protest and the cost of the civil suit. Another judicial mortgage in favor of the Union Bank of Louisiana and against the firm of Durrive & DeBlanc in solido for the amount of $2,500.00 with interest at 6% from February 10, 1842. Also, a $2.50 charge for protest and the cost of the civil suit was added. Another judicial mortgage in favor of the Union Bank of Louisiana and against W. G. Herrs & Company in solido for the sum of $2,500.00 with interest at 5% from February 10, 1842. Also, a $2.50 charge was added for protest and the cost of the civil law suit. Another judicial mortgage in favor of the Union Bank of Louisiana and against the firm of Layet & Amelin & Durrive & DeBlanc in solido in the amount of the sum of $1,714.00 with interest at 6% per annum from January 24, 1842. A $2.50 charge was added for protest and the cost of the civil law suit.

The two certificates of mortgage in favor of the Union Bank of Louisiana and against W. G. Herrs & Company were recorded in the recorder's office or the office of the clerk of court on April 29, 1842. The recorder or the clerk of court does not know what commerce or business there might be between Charles DeBlanc and the firm of W. G. Herrs & Company, when the judgment in the suit was rendered on April 15, 1842. Whatever influence this suit may exercise in the property of the plantation may be ascertained by the nature of the judgment of the civil lawsuit, and the nature of the firm of W. G. Herrs & Company.

Christopher Adams, Junior of the Parish and the City of New Orleans, and President of the Union Bank of Louisiana, showed a certificate that was written and executed before A. Mazoreau, a notary public, on April 4, 1842. From this certificate of mortgage, it appears that Christopher Adams, Junior, in his capacity as President of the Union Bank of Louisiana, granted to Louis Despanet DeBlanc, the vendor, a release to the total extent of ninety four shares of capital stock in the Union Bank of Louisiana. The amount of these ninety four shares of capital stock had been the part of the President of the Union Bank, Christopher Adams, Junior, on the deduction in the appointment made among the subscribers of the stock of the Union Bank of Louisiana after their original subscription. Therefore, the ninety four shares of capital stock may be deducted from the mortgage for three hundred shares of capital stock, and this was stated in the act of conveyance certificate that was produced at the sale of Louis DeBlanc to Charles DeBlanc.

On June 4, 1842, Louis D.DeBlanc became the purchaser of all the rights, titles, and interest of Charles DeBlanc in and to the DeBlanc Plantation, for the price and sum of $72,000.00. Louis D. DeBlanc was granted all the rights, titles, and interest in the property and slaves of the plantation which Charles DeBlanc had before February 20, 1842. This conveyance was a sale and mortgage of the land of the plantation and the slaves that were on the plantation, and it was passed before Barthlemy d'Augueseua de la Houssaye, a notary public in St. Mary Parish. The Union Bank of Louisiana was again the holder of the mortgage on the plantation that was recently purchased by Louis Despanet DeBlanc.

On December 26, 1835, at a sale and mortgage, Louis Despanet DeBlanc, a sugar cane planter of St. Mary Parish, Louisiana, declared that he intended to sell his sugar plantation, which was located on the Old Spanish Trail, in upper St. Mary Parish, to Charles DeBlanc, of the City of New Orleans, Louisiana. Louis Despanet DeBlanc was residing in 1835 at this plantation residence house or the main plantation home of the DeBlanc plantation. The main plantation home of the DeBlanc plantation was situated on the Old Spanish Trail, on the tract of land where the Clyde Bess home and the Joseph Wade Gajan home are situated.

The DeBlanc plantation had seventeen arpents or fourteen American acres and forty five hundredths of an acre fronting Bayou Teche, with a depth of forty arpents or thirty four American acres in depth, on the west side of Bayou Teche. The tract of land situated on the west side of Bayou Teche was bounded above or on the upperside by lands belonging to Charles DeBlanc, and on the lowerside or below by land belonging to Colonel Charles Napoleon Olivier. The tract of land of the plantation that was located on the eastern side of Bayou Teche measured thirteen arpents or eleven American acres and five hundredths of an acre fronting Bayou Teche, with a depth of forty arpents or thirty four American acres. The tract situated on the eastern side of Bayou Teche was bounded on the upperside by the lands of Maxmilliam (Max) Decuir, and on the lowerside by land belonging to Louis Dautreuil. The total land area of the DeBlanc plantation was six hundred and eighty arpents of land on the west side of Bayou Teche and five hundred and twenty arpents of land on the east side of Bayou Teche. The total land area of the DeBlanc plantation was one thousand and two hundred arpents of land.

There were buildings and improvements on the plantation, in the year of 1835, as well as apparatus for the making of brown sugar and molasses. The apparatus for making sugar included a sugar mill with kettles. Also, there were forty six slaves residing on the DeBlanc plantation at the time of the sale in the year of 1835.

These slaves were Leon, a slave man, age 31 years of age, Jarvis, a black slave man, age 30 years of age, Sandy, a colored slave man, age 28 years of age, Onezime, a black slave man, age 18 years of age, Julien, a black slave boy, age 10 years old, Ephiraree, a black slave man, age 30 years of age, Fnlory, a black slave man, age 32 years of age, Hayaud, a black slave man, age 22 years of age, Peter, a black slave man, age 24 years of age, Little Sam, a black slave man, age 22 years of age, Matt, a black slave man, age 28 years old, Jordan, a black slave man, age 31 years old, Jack, a black slave man, age 33 years old, Big Sam, a black slave man, age 28 years old, Joe, a black slave man, age 19 years old, Abram, a black slave man, age 30 years old, Charles, a black teenage slave, age 16 years old, Agag, a black slave woman, age 40 years of age, with her three children, Eloi, a teenage slave child, age 15 years old, Edmund, a slave boy, age 12 years of age, and Augelle, a slave child, age 10 years old, Peggy, a slave woman, age 45 years of age, Chalinette, a slave woman, age 30 years old, with her four children, Felicie, a slave child, age 10 years of age, Pierre, a slave boy, age 6 years of age, St. Ford, a slave child of 4 years of age, and Aseleiz, a slave child of 10 years of age, Melepaire, a mulatto slave woman, age 28 years of age, with her two children, Sophia, a girl slave child, age 4 years of age, and Leon, a slave boy, age 2 years of age, Elenore, a mulattress or a mulatto slave woman, age 21 years old, with her two children, Phebe, a slave child of 5 years of age, and Julie, a slave girl child of 2 years of age, Victorine, a slave woman, age 22 years of age, and her two children, Michel, a slave boy age 13 years of age, and Lesitte, a girl slave child age 1 years old, Mary, a black slave woman, age 20 years old, with her son, Davis, a slave boy, age 2 years of age, Marianne, a teenage slave girl, age 16 years of age, Rose, a slave girl of 12 years of age, Therese, a slave girl of 10 years of age, Cecile, a slave teenager, age 14 years of age, Charlotte, a slave girl of 4 years of age, and John, a black slave boy, age 8 years of old.

Also, sold with the lands of the DeBlanc plantation, and the forty six slaves were all the tools and implements of husbandry, such as hoes, scythes, rakes, and plows. The livestock on the plantation consisted of a herd of twenty eight American and Creole horses, and twenty five pairs or a herd of fifty work oxen, and a herd of two hundred cattle.

Louis DeBlanc sold his plantation to Charles DeBlanc for the price and sum of $50,000.00. At the act of conveyance, which occurred on the plantation, at the residence of Louis Despanet DeBlanc, the purchase price of $36,800.00 was paid in cash to Louis D. DeBlanc by Charles B. DeBlanc, for the debts of Louis Despanet DeBlanc. The receipt was acknowledged by Louis Despanet Deblanc as having received $36,800.00 cash as a down payment for the plantation. Also, two hundred and ninety nine shares of capital stock in the Union Bank of Louisiana were transferred that day. The residence of Louis Despanet DeBlanc was the plantation home of the DeBlanc plantation, which was situated on the Old Spanish Trail, on the tract of land that today partly is located the home of Mr. and Mrs. Clyde Bess, and the home of the late Mr. and Mrs. Joseph Wade Gajan. Mrs. Clyde Bess is the daughter of Mr. and Mrs. Joseph Wade Gajan, and her maiden name is Lois Gajan.

Tract four of the DeBlanc plantation was purchased on May 4, 1844, by an act of conveyance from Henry Penn or Paine, Senior to Charles DeBlanc. This sale was passed at the residence of Charles Napoleon Olivier. This tract of land was situated on both the east and the west sides of Bayou Teche, in the Parish of St. Mary. Also, this tract of land measured five arpents or four American acres and twenty five hundredths of an acre fronting Bayou Teche on both the east and the west sides of Bayou Teche, with a depth of eighty arpents or sixty eight American acres on the tract of land on the east side of Bayou Teche.

On June 22, 1847, at a sale and mortgage, tract two of the DeBlanc plantation property was acquired by Louis Despanet DeBlanc from Casmir Olivier, a free man of color. Barthlemy d' Auguesua de la Houssaye was the notary public for the sale. This sale was for the price and sum of $800.00, for which the purchaser, Louis Despanet DeBlanc, has made and executed three promissory notes. These three promissory notes were signed and delivered to Casmir Olivier, a free man of color, who was the vendor or the seller of the property by Louis D. DeBlanc, and acquittal for the purchase price was given by Casmir Olivier. These three promissory notes were conditioned to bear 8% interest from specified maturity until paid.

Each of the undivided half of this certain parcel of land, was acquried by Charles and Louis DeBlanc. This tract of land, which was situated in the Parish of St. Mary, on the east side of Bayou Teche, had a frontage of two arpents by the depth of forty arpents. The total land area of this tract of land was eighty arpents. This tract of land was bounded above or on the upperside by the lands of Louis Despanet DeBlanc and below or on the lowerside by the lands of Adelaide Dubriel, a free woman of color.

Tract two was purchased on July 28, 1835 at a cash sale by Casmir Olivier, a free man of color, from his mother, Madame LaCoste, a free woman of color. After bargaining and delivering the parcel of land to her son, Casmir Olivier, Madame LaCoste, a resident of St. Mary Parish, Louisiana, accepted $300.00 in cash money for this tract of land. Casmir Olivier, a free man of color, reserved from this sale, the use of wood, which woods were situated on the upper part of the tract of land. This tract of land contained twenty arpents or seventeen American acres in depth and one arpent or eight five hundredths of an American acre frontage on Bayou Teche running from Bayou Teche to the prairie. The total land area of this tract of land was twenty arpents or fourteen arpents and forty five hundredths of an arpent.

On August 16, 1853, the S. O. Nelson & Company, through one of its representatives, Stephen Owens Nelson, purchased the sugar plantation from Louis Despanet DeBlanc and his brother, Charles B. DeBlanc. This sugar plantation originally consisted of four separate tracts of land, which were situated on both the east and the west sides of Bayou Teche. On this 16th day of August of the year of 1853, at the residence home or the plantation home, which was situated on the tract of land, where the present day home of Mr. and Mrs. Clyde Bess, and the home of the late Mr. and Mrs. Joseph Wade Gajan are presently located. Charles and Louis DeBlanc met with Stephen Owens Nelson, the representative and one of the owners of the commercial firm of S. O. Nelson & Company of New Orleans, Louisiana, to pass an act of conveyance, which would transfer their title, rights, and privileges to the DeBlanc sugar plantation to the commercial firm of S. O. Nelson & Company of the City of New Orleans, Louisiana. Also, present at this sale were Eugene Barthlemy Olivier, and his brother, Alexander Olivier, who would serve as witnesses to this act of conveyance, and the Honorable John A. Dumartrait, who would serve in his capacity as recorder of the act of conveyance. This sugar plantation was located on the Old Spanish Trail, in upper St. Mary Parish. At the time this plantation was sold, it consisted of one thousand and two hundred and eighty arpents of land or one thousand and eighty eight acres. Today, these tracts of land are located in present day Iberia Parish.

Iberia Parish was created on October 29, 1869 by taking a portion of St. Martin Parish and a portion of St. Mary Parish to form the present day Iberia Parish. All the property that was located below present day Evangeline Street, in the town of New Iberia, was a part of St. Mary Parish, and all the property that was located above Evangeline Street in the present day town of New Iberia was a part of St. Martin Parish.

Besides the four tracts of land purchased by the firm of S. O. Nelson & Company, also purchased were the residence house or the plantation home, the slave houses or the slave cabins, the barns, the stables, the fences, and the out houses. The brick sugar house with its steam engine, and the sugar kettles were also purchased. The sugar kettles were used in the processing of the cane juice into molasses on the plantation. The brown sugar was processed in the brick sugar house. For the slaves to haul the green stalks of sugar cane from the fields of the plantation, to the brick sugar house also on the plantation, the cane carts were also purchased. At this time, in the State of Louisiana, the sugar was processed from the cane juice into brown sugar and molasses on the plantation in the plantation sugar house, which was usually located on the plantation. Every large plantation had its own sugar mill and grinned its own sugar cane. Sometimes, plantation sugar mills took in the sugar cane of their neighbors to grind, but this was only after all the sugar cane on the plantation had been grinned and harvested.

The livestock that were living on the plantation, at the time of the sale in the year of 1853, consisted of a herd of twenty one creole mares, a herd of forty mules, a herd of ten creole horses, sheep, oxen, calves, cows, mares, and colts.

Most of the men and the women slaves were originally on the plantation, at the time of the civil lawsuit, of the plaintiff, Jean Baptiste Despanet DeBlanc versus Charles DeBlanc, in the year of 1842. This civil law suit bore the docket number of 14484, and was filed and tried in the Parish court for the City of New Orleans, and the Parish of Orleans.

Also with the act of conveyance of St. Mary's or the DeBlanc plantation to the commercial firm of S. O. Nelson & Company of the City of New Orleans, Louisiana, a greater portion of the slaves, which were originally acquired by Louis Despanet DeBlanc from a purchase at a sheriff's sale, dated June 4, 1842, were conveyed or sold to the commercial firm of S. O. Nelson & Company. These slaves were Zenon, age 46 years old, Paze, age 46 years of age, Julien, age 33 years old, Henry Miller, age 38 years old, Charles, age 31 years old, Hayden, age 38 years of age, Onezime, age 33 years old, Thomas, age 32 years of age, Eloi, age 31 years of age, Watt, age 45 years of age, Edmund, age 28 years of age, Joe, age 38 years of age, Joseph, age 41 years of age, Jurden, age 50 years of age, Eugene, age 35 years of age, Sam, age 40 years of age, Baptiste, age 33 years of age, Big Sam, age 45 years of age, Narcisse, age 43 years of age, John, age 31 years of age, Daniel, age 30 years of age, Jack, age 49 years of age, Alexander, age 54 years of age, Frederick, age 45 years of age, Sandy, age 46 years of age, Helay, age 48 years of age, James Jones, age 43 years of age, Melirs, age 40 years of age, Pierre, age 17 years of age, Lac'on, age 19 years of age, Michael, age 19 years of age, Rolly, age 33 years of age, Corline, age 35 years of age, William, age 22 years of age, James Brown, age 40 years of age, Chamis, age 56 years of age, Henry, age 23 years of age, Philomeon, age 16 years of age, Petit James, age 21 years of age, and Linden, age 40 years of age. All these men slaves were all residing on the plantation in the month of August of the year of 1853, and they were all slaves for life.

These were the women slaves and the slave children that were residing on the plantation, at the time of the sale to the commercial firm of S. O. Nelson & Company. Leontine, age 17 years of age, Victorine, age 36 years of age, Caroline, age 36 years of age, Lesite, age 17 years of age, Josephine, age 16 years of age, Anna, age 36 years of age, Mary Lombard, age 40 years of age,

Mahley, age 40 years of age, Rose, age 29 years of age, Marianne, age 31 years of age, Babe', age 38 years of age, Rosalie, age 16 years of age, Old Maid, age 50 years of age, Marguerite, age 17 years of age, Mary Creole, age 35 years of age, Noie, age 46 years of age, Phoebe, age 22 years of age, Violet, age 27 years of age, and Helena, age 40 years old.

These were the slave children, who were born on the plantation, and their parents were the slave men and the slave women, who worked on the plantation. Jean Baptiste, age 12 years old, Noel, age 12 years of age, Joseph, age 9 years of age, Tecimesise, age 6 years of age, Adran, age 7 years of age, Zelia, age 14 years of age, Josephine, age 11 years of age, Annette, age 8 years old, Victorine, age 13 years old, Jacques, age 4 years of age, Leandre, age 8 years of age, Da is age 8 years of age, Led, age 6 years of age, Rosina, age 12 years of age, Baptiste, age 12 years of age, Christine, age 10 years of age, William, age 4 years of age, Louise, age 10 years of age, Elizabeth, age 9 years of age, Magdelin, age 2 years of age, Toby, age 10 years old, Maffe', an infant slave age 8 months of age, Jenna, age 14 years of age, Isidore, age 11 years of age, Alfred, age 6 years of age, Onezeme, age 2 years old, Jesmetter, age 4 years of age, Rosemona, an infant of one month old, an unnamed infant slave, Sarah, a slave girl of 14 years of age, and Thomas, a slave boy, who was 8 years old.

On July 29, 1854, Mrs. Coralie de Reggio, the legal wife of Charles DeBlanc, declared in the presence of the Honorable John A. Dumartrait, who served as the recorder of this act of conveyance of the sale of the DeBlanc plantation or St. Mary's Plantation, that she was aware and well acquainted with the nature and the terms of the sale of St. Mary's Plantation to the commercial firm of S. O. Nelson & Company of New Orleans, Louisiana. The reason that Mrs. Coralie DeBlanc was required to make this statement in front of the notary public, John A. Dumartrait, was to protect her dotal rights and privileges.

At this time, by the laws of the State of Louisiana, Mrs. Coralie de Reggio DeBlanc, as other married woman in the State of Louisiana, had certain dotal rights and privileges on the property owned by their husbands. The reason for this legal mortgage was for the restitution of a wife's dowry for the renunciation of her dotal property, acquired by the wife since marriage, whether acquired by possession of a succession and from the date the succession was opened, or by donation and the date the donation was perfected, and thirdly for the nuptial presents. Fourthly, for the debts contracted by a wife or her husband, or the debts contracted by both a wife and her husband, and fifth for the paraphernal property alienated by the wife or the husband, or disposed of for a husband's individual interests. For these five reasons, a lawyer or a notary public, in the State of Louisiana, in the year of 1853, had to take a married woman out of the hearing and physical presence of her husband, and inform the wife of the upcoming act of conveyance or sale of any property a husband may wish to dispose of or to sell. A wife had a right to renounce her dotal rights and privileges to the property, or to keep her dotal rights and privileges. At this time, a notary public or a lawyer showed a wife an authentic copy of the act of conveyance of the property, a husband was contracting with another party for the purpose of conveying the property. At this sale, Charles DeBlanc exhibited or showed a certificate, which had been executed in the City of New Orleans, Louisiana, bearing the date of August 5, 1853, showing that Charles DeBlanc had accounted to Mrs. Coralie de Reggio DeBlanc, his lawful wife, for all the claims and demands of her (Mrs. Coralie de Reggio DeBlanc) against him (Charles DeBlanc). A judgment of separation of property between Charles and Coralie de Reggio DeBlanc was rendered by the Parish Court of the Parish of Orleans, on the date of May 4, 1842. So therefore, Charles DeBlanc waived the neplity of the intervention and renunciation of Mrs. Coralie de Reggio DeBlanc, his wife, in these premises.

After a due reading of the foregoing act of conveyance between the parties of the commercial firm of S. O. Nelson & Company and Messrs. Louis Despanet DeBlanc and his brother, Charles DeBlanc, Mrs. Constance LeBranch, the lawful wife of Louis Despanet DeBlanc, took cognizance of the act of conveyance. Also, Mrs. Constance LeBranch DeBlanc stated that it is her wish and intention to release in favor of the commercial firm of S. O. Nelson & Company and its representative, Stephen Owens Nelson, the real property and the slaves that were conveyed or sold from the dotal, paraphernal, and other rights, and from any claims, mortgages, or privileges, which she may be entitled to, and by virtue of her marriage to Louis Despanet DeBlanc. This was known in the State of Louisiana as the act of renunciation.

The four tracts of land comprising this plantation or St. Mary's plantation at this time in 1853, were described as tract one measuring thirty eight arpents or thirty two American acres and three hundredths of an American acre fronting Bayou Teche, on the west side of the bayou, with a depth of forty arpents or thirty four American acres in depth. Tract one was bounded above or on the upperside by lands belonging to the widow and the heirs of Baron Benoit Bayard, and below or on the lowerside by the lands of Colonel Charles Napoleon Olivier. That part of tract one that was situated on the east side of Bayou Teche measured twenty seven arpents or twenty two American acres and ninety five hundredths of an American acre fronting Bayou Teche, with a depth of forty arpents or thirty four American acres. This part of tract one which was situated on the east side or the east bank of Bayou Teche was bounded on the upperside by tract two of St. Mary's plantation, and by the lands of Henry Camors, and on the lowerside by the lands of Adelaide Olivier, a free man of color.

Tract two of St. Mary's plantation measured eleven arpents or nine American acres and thirty five hundredths of an American acre fronting Bayou Teche, with a depth of four arpents or three American acres and forty hundredths of an American acre. This tract was bounded on the upperside or above by the lands belonging to Mrs. Baron Benoit Bayard, and on the lowerside by lands of St. Mary's plantation or tract one.

Tract three was located in St. Martin Parish, Louisiana, and this tract had a frontage along Bayou Teche of three arpents and one half of an arpent, or two American acres and two hundred and ninety five hundredths of an American acre, with a depth of fifteen or eighteen arpents or twelve American acres and seventy five hundredths of an American acre and thirty hundredths of an American acre. This tract was bounded on the upperside by lands of Charles Hebert, and on the lowerside by land of Agricole LeBlanc.

Tract four was situated in St. Mary Parish, Louisiana, with a frontage on Bayou Teche of five arpents or four American acres and twenty five hundredths of an American acre on both the east and the west sides of Bayou Teche. Also, this tract was forty arpents or thirty four American acres in depth on the western side of Bayou Teche, and eighty arpents or sixty eight American acres in depth on the eastern side of Bayou Teche. Tract four was bounded above or on the upperside on both the east and the west sides of Bayou Teche by lands belonging to Celestin Penn, the wife of Ursin Provost, and on the lowerside by lands formerly belonging of Henry Penn or Paine.

The purchase of St. Mary's Plantation on August 16, 1853 by the commercial firm of S. O. Nelson & Company was made for the sum of $140,000.00. The firm of S. O. Nelson & Company paid $20,000.00 in cash money at the time of the act or sale or conveyance was being passed, and the balance of $120,000.00 was to be paid in credit installments. The receipt of $20,000.00 cash down payment was acknowledged by the DeBlanc brothers. For the balance of the purchase money of $120,000.00, the firm of S. O. Nelson & Company, with Stephen Owens Nelson duly authorized to execute his drafts of even date for the sum of $34,000.00. These notes were drawn on the counting house of S. O. Nelson & Company of the City of New Orleans. This was accepted as ready money by Charles and Louis Despanet DeBlanc.

For the balance of the purchase price, which amounted to $90,000.00, Stephen Owens Nelson has executed and signed ten promissory notes. These promissory notes are to the amount of $9,000.00 each promissory note. Five of these ten promissory notes or $45,000.00 are made payable to the order of S. O. Nelson and endorsed by S. O. Nelson are for the use of Charles DeBlanc at the counting house of S. O. Nelson & Company in the City of New Orleans. These promissory notes were to fall due on February 4, 1855, and every year until February 4, 1859. Also, these promissory notes were conditioned to bear six percent interest per annum from August 4, 1853, or until paid in full. The date of

August 4, 1853 was chosen because it was the date the sale or the conveyance of St. Mary's Plantation was agreed on by the contracting parties, the DeBlanc brothers and the representatives of the commercial firm of S. O. Nelson & Company. The last five of the promissory notes were made payable to the order of S. O. Nelson and endorsed by S. O. Nelson for the use of Louis Despanet DeBlanc at the counting house of S. O. Nelson & Company in the City of New Orleans, Louisiana. These promissory notes for Louis Despanet DeBlanc's use were conditioned to bear 8% interest per annum from August 4, 1853 or until paid in full.

On January 5, 1854, Stephen Owens Nelson of the City of New Orleans, Louisiana, purchased several slaves from the firm of Slade & Barett for the cash price of $13,275.60.

Also, on January 5, 1854, Stephen Owens Nelson purchased from Rufus Lusk, a slave named, William, who was 25 years of age, for the sum of $1,250.00.

On February 23, 1854, Slade & Barett conveyed back to Stephen Owens Nelson of the City of New Orleans, Louisiana, a negress slave, named Fanny, who was 17 years of age. The reason Slade & Barett did this way to replace the slave woman, Hannah, who was purchased by Stephen Owens Nelson on the date of January 5, 1854. The slave woman, Hannah, died shortly after she was purchased by Stephen Owens Nelson.

On June 21, 1854, three of the partners of the commercial firm of S. O. Nelson & Company agreed to convey or to sell their one fourth interest, rights, titles, and privileges in and to St. Mary's plantation to Stephen Owens Nelson. Therefore, Stephen Owens Nelson would become the sole proprietor of St. Mary's plantation, and he would be legally responsible and liable for all the debts of the plantation. Also, Stephen Owens Nelson would enjoy all the proceeds from the gains or the profits of the plantation.

Also, on June 21, 1854, Ellison Banksmith, one of the partners of the commercial firm of S. O. Nelson & Company, declared that it was agreed that he convey or sell his one fourth interest in St. Mary's plantation to Stephen Owens Nelson of the City of New Orleans, Louisiana. Therefore, Ellison Banksmith of the City of New Orleans, Louisiana, conveyed his rights, titles, and interest to the real property claims of every kind and nature whatsoever unto Stephen Owens Nelson. Also, on June 25, 1854, Thomas A. Nelson, a resident of Athens, Alabama, which is situated in the County of Limestone, Alabama, followed Ellison Banksmith, and sold his one fourth share and interest in the rights, titles, privileges, and claims to the property, and the slaves of St. Mary's plantation to Stephen Owens Nelson of the City of New Orleans, Louisiana. Henry C. Walker, of Memphis, which is situated in the County of Shelby, Tennessee, also sold his one fourth share and interest in St. Mary's plantation to Stephen Owens Nelson. After his three business partners, Thomas A. Nelson, Henry C. Walker, and Ellison Banksmith sold their share to Stephen Owens Nelson, he became the sole proprietor of the plantation and the slaves on the plantation. When the four members of the commercial firm of S. O. Nelson & Company owned the plantation and the slaves, it was considered a partnership investment. The reason that the three partners of Stephen Owens Nelson sold their one fourth share of ownership of the plantation and the slaves was because the plantation was not making money to pay off the outstanding debts.

Stephen Owens Nelson did purchase the plantation and the slaves for the investment of the commercial firm of S. O. Nelson & Company of New Orleans, Louisiana. The debits and credits of the plantation, and the cost of operating the plantation was carried on the books of the firm of S. O. Nelson & Company in the account titled, St. Mary's Plantation. This included all the entries that were contained, the cash payments, and all the installments of the balance of the purchase price of $90,000.00. Also, the acceptances of the firm of S. O. Nelson & Company of $30,000.00 was a also paid to the account of St. Mary's Plantation. The previous installments of the note have been paid with funds of the firm of S. O. Nelson & Company.

Now, that the three business partners had sold their one fourth share of ownership to the plantation to Stephen Owens Nelson, the plantation became a sold proprietorship. Stephen Owens Nelson was now legally responsible for the debts and credits of the plantation, and the cost of operating the plantation. Also, Stephen Owens Nelson had to make all the payments maturing on the installment as his obligation. Stephen Owens Nelson assumed all the past payments on the installments and notes that were made by the firm of S. O. Nelson & Company before the sellout.

Because of the Louisiana law of 1835, that a legal and lawful wife held a legal mortgage on the property owned by her husband, the wives of the three business partners had to make a renunciation of their dotal, matrimonial, and paraphernal rights, privileges, and interests in St. Mary's Plantation. Even though the wife of Henry C. Walker of Memphis, Tennessee, and the wife of Thomas A. Nelson of Limestone County, Alabama were not residing in the State of Louisiana, they came under this Louisiana law because their husbands owned property situated in the State of Louisiana. Therefore, Mrs. Lizzie Walker, the wife of Henry C. Walker of Memphis, Tennessee, had to make a renunciation of her dotal, matrimonial, and paraphernal rights, titles, and privileges to her husband's one fourth interest in St. Mary's Plantation. Also, Mrs. Marianne W. Nelson, the wife of Thomas A. Nelson, of Limestone County, Alabama, renounced her dotal, matrimonial, and paraphernal rights, titles, and interest to her husband's one fourth interest in St. Mary's Plantation.

Instead of the plantation account being carried on the books of the firm of S. O. Nelson & Company in the account of St. Mary's Plantation, it was now carried on the books in the personal account of Stephen Owens Nelson. As per June 1, 1854, the personal account of Stephen Owens Nelson was charged with thee amounts of payments made for the plantation. Also, on this date the balance that was transferred from the account of St. Mary's Plantation to the personal account of Stephen Owens Nelson was $73,006.23 with interest at 8%.

Stephen Owens Nelson became the sole proprietor of the plantation property and slaves. At this time, that the plantation investment became a sold proprietorship, the accounting records of the commercial firm of S. O. Nelson & Company showed a balance in the account of St. Mary's Plantation of $73,006.23, bearing interest at eight per cent per annum. Therefore, Stephen O. Nelson, personally, became legally responsible for all the payments of the original mortgage as these mortgage payments matured. Now, the books or the accounting records of the commercial firm of S. O. Nelson & Company charged Stephen Owens Nelson, himself, for the amounts due for the payments of St. Mary's Plantation, instead of charging the account of St. Mary's Plantation with the mortgage payments and the expenses of the plantation.

After Stephen Owens Nelson became the sole proprietor of St. Mary's Plantation, situated in upper St. Mary Parish, the plantation had these attachments to it, such as sugar, molasses, corn, fodder, hay, sheep, horses, mules, and other livestock.

On June 26, 1858, an inventory of the community property that existed between Stephen Owens Nelson and his late wife, Martha Ann Tanner Nelson, was taken by Eugene B. Olivier and Alfred Weeks. Martha Ann Tanner Nelson died on August 27, 1855. The reason for this inventory was to show what immovable property and movable property belonged to the estate of the deceased Martha Ann Tanner Nelson or Mrs. Stephen Owens Nelson. Alfred Weeks and Eugene B. Olivier were sworn to faithfully appraise the property of the plantation, and the effects that were situated on St. Mary's Plantation. The plantation property still consisted of the original four tracts of land, and this property was valued at $77,140.00. Tract one which was located between the land of Mrs. Baron Benoit Bayard on the upperside or north, and by Colonel Charles Napoleon Olivier on the lowerside or the south was valued at $60,000.00. Tract two, which was a small tract of land, with an eleven arpent frontage on Bayou Teche and a depth of forty arpents was located in the same vicinity. Tract two was bounded above by Mrs. Bayard's lands and below by tract one or the plantation proper. This property of tract two was valued at $880.00. Tract three consisted of a frontage along Bayou Teche of three and one half arpents, with a depth of fifteen or eighteen arpents. Tract three was lying in St. Martin Parish, but adjoining tract two.

Also, tract three was bounded on the upperside or the northside by the lands of Charles Hebert, and on the lowerside or the southside by lands of Agricole LeBlanc. The lands of tract three was valued at $20.00 per arpent, making the total sum of the valuation of tract three $1,260.00. The fourth tract of land belonging to St. Mary's Plantation was located in St. Mary Parish, Louisiana. Tract four had a frontage along Bayou Teche of five arpents, with a depth of the usual forty arpents or thirty four American acres, which was situated on the west side of Bayou Teche. On the east side of Bayou Teche, tract four had a frontage of five arpents along Bayou Teche,with a depth of eighty arpents or sixty eight American acres. Tract four was bounded on both the east and the west sides of Bayou Teche by lands formerly belonging to Celestin Penn, wife of Ursin Provost, and below by land formerly belonging to Henry Penn or Paine.

By July 18, 1868, Stephen Owens Nelson had died, and the personal estate and the commercial firm of S. O. Nelson & Company was in the state of bankruptcy. As a result of the bankrupt conditions of the estate Stephen Owens Nelson, the Crescent City Bank and the Citizens Bank of Louisiana were the owners of all the lands belonging to the estate of Stephen Owens Nelson. S. O. Nelson & Company had purchased the DeBlanc plantation for an investment opportunity,but that investment opportunity had turned sour. The firm of S. O. Nelson & Company was in the sugar factoring business, and their customers were many sugar planters from this area. The reason of the purchase is because the owners of the firm thought that the proceeds from a sugar plantation would be an investment opportunity for the company. Evidently, St. Mary's Plantation did not make the money that the owners of the firm of S. O. Nelson & Company thought it would, and that was the reason why three of the partners, Mr. Banksmith, Mr. Walker, and Mr. T. A. Nelson, all sold their share of interest to Stephen Owens Nelson. Stephen Owens Nelson did not want to sell St. Mary's Plantation, so he assumed legal responsibility for the plantation, personally. The troubles for the South that were caused by the Civil War brought on the bankruptcy of Stephen Owens Nelson, both personally and as a member of the firm of S. O. Nelson & Company. Also, by the year 1876, the Crescent City Bank of New Orleans was able to sell their one third interest in the bankrupt estate of Stephen Owens Nelson to the Citizens Bank of Louisiana. Now, the Citizens Bank of Louisiana had full ownership to the lands formerly owned by Stephen Owens Nelson,both individually and as a member of the commercial firm of S. O. Nelson & Company. The Citizens Bank of Louisiana, acting through its Board of Directors and its President, decided to put up for sale St. Mary's Plantation or the lands formerly owned by Stephen Owens Nelson.

On April 27, 1889, Edgar Grima, a notary public, in the City of New Orleans, Louisiana, executed an act of conveyance of a portion of the property formerly owned by Stephen Owens Nelson to Alonzo E. Hall, who was a resident of Fort Plains, in the County of

Montgomery, which is situated in the State of New York. At the execution of the act of conveyance, Alonzo E. Hall was not present, but he was represented by William Hall, a resident of Kentwood, which is situated in the Parish of Tangipahoa, which is situated in the State of Louisiana. Also, Stanley O. Thomas, who was then President of the Citizens Bank of Louisiana, and was representing the Citizens Bank of Louisiana, was present at the execution of the act of conveyance.

The property that was being sold was a tract of land, formerly a part of tract one of the land that comprised the Nelson Plantation in Iberia Parish, formerly upper St. Mary Parish. This tract of land consisted of tracts of land, which were situated on both the east and the west sides of Bayou Teche. This tract of land was subdivided into thirty five lots, which were situated on both the east and the west sides of Bayou Teche. These lots were designated by lot numbers, which numbered lots one through eighteen were situated on the right bank of Bayou Teche, and lots numbering twenty eight through forty four were situated on the left bank of Bayou Teche. This plan of the lots was in conformity with a plan made by W. A. Heis, a surveyor, and the survey was dated August 29, 1876. The original survey map was deposited with the Citizens Bank of Louisiana in the City of New Orleans. But a survey map on a smaller scale was certified by Louis H. Peltier, a deputy surveyor, and was this copy that was annexed to the bill of sale that was executed in the act of conveyance to Alonzo E. Hall.

Lots numbering one through eleven, each consisted of fifty eight arpents and eighty one hundredths of an arpent of land. The total arpentage of land comprising lots number one through eleven was six hundred and forty six arpents and ninety two hundredths of an arpent of land. A lot is a portion of land. Lot twelve through lot eighteen were all located on the east side of Bayou Teche. Lot twelve consisted of seventy four arpents and ninety five hundredths of an arpent of land. Lot thirteen consisted of fifty eight arpents and eighty six hundredths of an arpent of land. Lot fourteen consisted of fifty eight arpents and eighty four hundredths of an arpent of land. Lot fifteen consisted of fifty eight arpents and sixty five hundredths of an arpent of land. Lot sixteen consisted of fifty eight arpents and fifty five hundredths of an arpent of land. Lot seventeen consisted of fifty eight arpents and fifty five hundredths of an arpent of land. Lot eighteen consisted of fifty eight arpents and thirty seven hundredths of an arpent of land. The total arpentage of land comprising lots twelve through lot eighteen consisted of one thousand and seventy three arpents and sixty eights hundredths of an arpent of land.

Lot twenty eight through lot forty four were all located on the west side of Bayou Teche. Lot twenty eight consisted of sixty arpents and ninety one hundredths of an arpent of land.

Lot twenty nine consisted of sixty arpents and forty hundredths of an arpent of land. Lot thirty consisted of sixty arpents and forty hundredths of an arpent of land. Lot thirty one consisted of fifty nine arpents and fifty four hundredths of an arpent of land. Lot thirty two consisted of fifty nine arpents and thirteen hundredths of an arpent of land. Lot thirty three consisted of seventy four arpents and ninety five hundredths of an arpent of land. Lots numbers thirty four through forty, each consisted of fifty eight arpents and eighty one hundredths of an arpent of land.

The whole tract of land has a total of two thousand and ninety five arpents and fifty seven hundredths of an arpent of land. The property is bounded below or on the lowerside by the Olivier Plantation, now Orange Grove. Lot forty four is bounded below by a tract of land separating the property from the Olivier Plantation. Lot eighteen is bounded above by lot nineteen. Lot nineteen was sold by the Citizens Bank to John Ermant Leitmeyer. Lot twenty eight is bounded above by lot twenty seven, which was sold by the Citizens Bank to Dennis Lucky. These lots are all that remain of the Nelson Plantation or St. Mary's Plantation that have not been sold by the Citizens Bank of Louisiana.

The Citizens Bank acquired this property by an act passed before Felix Grima, a notary public, on July 18, 1868. The Citizens Bank acquired one third interest of the Nelson Plantation by the Crescent City Bank by a purchase from Emory E. Norton, who served as the assignee in bankruptcy of the estate of Stephen Owens Nelson and Thomas A. Nelson, bankrupt individually and as members of the counting house of S. O. Nelson & Company. By a virtue of two orders issued out of the Honorable District Court of the United States for the District of Louisiana, settling in bankrupt cases, issued in the matter of the commercial firm of S. O.Nelson & Company, No. 45 of the Docket of the court, and dated the 11th and the 15th of July of the year of 1868. The Crescent City Bank sold its one third undivided interest in the Nelson Plantation to Charles Dupuy by an act of conveyance, which was passed before Edward Barnett, a notary public, in the City of New Orleans, on February 28, 1872. Therefore, the Citizens Bank of Louisiana acquired this one third undivided interest in the Nelson Plantation from Charles Dupuy by an act of conveyance, which was passed before John Bindernagh, a notary public, in the City of New Orleans, on January 24, 1876.

On July 18, 1868, the Citizens Bank of Louisiana sold to Emory E. Norton, the assignee in the bankruptcy case of the estate of S. O. Nelson and T. A. Nelson, its shares and interest to the lands formerly comprising the Nelson Place or St. Mary's Plantation. The Nelson Place consisted of four tracts of land. Tract one measured thirty eight arpents fronting Bayou Teche on the west side of Bayou Teche by the depth of forty arpents, and it was bounded in 1868, on the upperside or above by the lands of the widow Bayard and the heirs of Bayard, and on the lowerside or below by the lands of Charles Olivier, now known as Orange Grove and Olivier. On the east side of Bayou Teche, tract one had a twenty seven arpent frontage on Bayou Teche, by the depth of forty arpents, and the portion of tract one, which was situated on the east side of Bayou Teche was bounded above or on the upperside by the portion of tract one that was located on the west side of Bayou Teche, and the lands of Henry Carrier or assigns, formerly the land of Francois Duplesis, and below by the lands of Adelaide Olivier.

Tract two that comprised the Nelson Place measured eleven arpents fronting the east side of Bayou Teche, with a depth of four arpents. This tract was bounded above or on the upperside by the lands of the widow and heirs of Baron Bayard, and below by tract one of the Nelson Place or St. Mary's Plantation.

There was another tract of land that contains three and one half arpents fronting Bayou Teche, with a depth of fifteen or eighteen arpents. Tract three was bounded on August 3, 1856, on the upperside or above by the lands of Charles Hebert, and below or on the lowerside by the lands of Agricole LeBlanc.

Tract four was located in the Parish of St. Mary, and it measured five arpents fronting Bayou Teche, on both the east and the west sides of Bayou Teche, with a depth of forty arpents on the west side of Bayou Teche, and a depth of eight arpents on the east side of Bayou Teche. Tract four was bounded on both the east and the west sides of Bayou Teche by the lands formerly belonging to Celestin Penn, who was the wife of Ursin Provost, and below or on the lowerside by the lands formerly belonging to Henry Penn, Senior. Tract four was situated on both the east and the west sides of Bayou Teche. Today, tract four is located in Iberia Parish, but in the year of 1856 tract four was located in St. Mary Parish, Louisiana.

At this conveyance to Emory E. Norton, the Citizens Bank of Louisiana sold besides all the real property, all the buildings and improvements, such as drainage ditches, and bridges. Also, the dwelling house or the plantation house, the out houses, the slave houses or cabins, the brick sugar house with the machinery and the steam engine, the sugar kettles, barns, stables, and the fences were all sold.

The purchase price of this conveyance was $10.00, and a further sum for the charges, costs, and expenses, which amounted to $6,000.00, which was the sum of the purchase price. The purchaser, the Citizens Bank of Louisiana paid the total purchase sum in treasury notes of the United States government. Each treasury note was in proportion of the respective interest in the purchase of the property, which had formerly belonged to the bankrupt estate of Stephen Owens Nelson. The assignees, Emroy E. Norton, acknowledged the receipt and gave the Citizens Bank of Louisiana full acquittal and discharge for the purchase price of the property.

The purchaser, the Citizens Bank of Louisiana, received this sale and transfer subject to the mortgage held by the Citizens Bank of Louisiana,a nd the mortgage set forth in the court order by the Honorable Court in the Parish of Orleans, and the City of New Orleans, on the date of July 11, 1868.

Also, the parties, the Citizens Bank of Louisiana, the purchaser, and the assignee, Emory E. Norton, dispense with the production certificate. A production certificate is a certificate, which shows the mortgages that are bearing on a piece of property, or against the property. A production certificate is recorded in the Parish where the property is located. For this conveyance, the production certificate would be recorded in the Parishes of St. Mary and St. Martin, where the property belonging to the bankrupt estate was located. A production certificate is required by law in the State of Louisiana, at the execution of a conveyance, but the production certificate may be dispensed of at the conveyance in agreement with the contracting parties.

On January 24, 1872, the Crescent City Bank of New Orleans sold their one third interest to this property to Charles Dupuy, a resident of the City of New Orleans, Louisiana. This sale was executed at the office of the notary public in the City of New Orleans, and the two witnesses were Messrs. Edward Smith and Martin L. Ainsworth.

This sale was made and accepted by the Crescent City Bank for the three promissory notes, that were drawn by Louis E. Dupuy, a single man of the City of New Orleans, to the order of Louis E. Dupuy or to his own order, and by himself (Louis E. Dupuy) endorsed. These three promissory notes were due on February 28, 1872, and made payable to the Crescent City Bank of New Orleans. Each promissory note was for the sum of $11,000.00 with 8% interest. The first note was payable in twelve months from the date of the sale, the second note was payable in eighteen months from the date of sale, and the third note was payable in twenty four months for the date of sale.

At the execution of this conveyance, the Crescent City Bank of New Orleans sold their one third interest to the bankrupt estate of Stephen Owens Nelson.

On April 27, 1889, Alonzo E. Hall, a resident of Fort Plains, which is situated in the County of Montgomery, which is also situated in the State of New York, purchased at a sale and mortgage the property of which a portion would become the present day, Kyle Acres.

This act of conveyance was executed in the City of New Orleans, Louisiana, with Edgar Grima, a notary public in New

Orleans, passing the act of conveyance. Stanley O. Thomas, who was then the President of the Citizens Bank of Louisiana, and was representing in his capacity as President and William G. Hall, who was representing Alonzo E. Hall, the purchaser, by the power of a special power of attorney. William G. Hall was a resident of Kentwood, Louisiana.

The property that was being purchased was described as a certain plantation, which was situated in Iberia Parish, Louisiana. This tract of land was composed of thirty five lots or portions of land, which were designated as Lots Number 1 - 18 on the right bank of Bayou Teche, and Lots Number 28 - 44 on the left bank of Bayou Teche. This was in conformity with a plan made by W. A. Hine, a surveyor, and this plan was dated August 29, 1876. This plan of survey was deposited in the Citizens Bank of Louisiana. The bank had the survey made in order to divide a portion of the former tract of the Nelson Plantation or the lands formerly belonging to S. O. Nelson up into lots. The bank thought the lots would be easier to sell, instead of a huge tract of land. A copy of this survey on a smaller scale than the original survey was annexed to this act of conveyance.

The lots or the portions of land contained the following arpents of land. Lots Number 1 through 11 had 58.81 arpents in each lot. These lot number 1 through 11 were located on the right bank of Bayou Teche or the East side of Bayou Teche. Altogether, the arpentage of lots 1 through 11 was 646.92 arpents of land.

Lot 12 through Lot 18 were all located on the east side of Bayou Teche. Lot 12 contained 74.95 arpents of land. Lot 13 contained 58.86 arpents of land. Lot 14 contained 58.84 arpents of land. Lot 15 contained 58.65 arpents of land. Lot 16 contained 58.55 arpents of land. Lot 17 contained 58.55 arpents of land. Lot 18 contained 58.37 arpents of land. Together lots number 12 through lot number 18 contained 1,073.68 arpents of land.

Lot 28 through Lot 44 were located on the west side of Bayou Teche. Lot 28 contained 60.91 arpents of land. Lot 29 contained 60.40 arpents of land. Lot 30 contained 60.40 arpents of land. Lot 31 contained 59.54 arpents of land. Lot 32 contained 59.13 arpents of land. Lot 33 contained 74.95 arpents of land. Lots 34 thorough 44 contained 58.81 arpents of land. Together lot 28 through 44 contained 642.92 arpents of land.

The whole arpentage of the land being conveyed is 2,095.57 arpents of land. The total arpentage of land on the east side of Bayou Teche is 1,021.89 arpents of land. Also, the total arpentage on the west side of Bayou Teche is 1,073.68 arpents of land.

The boundaries of the plantation being conveyed are lot one is bounded below by the Olivier Plantation, the present day Orange Grove. The Olivier Plantation in the year of 1889, belonged to the Citizens Bank of Louisiana. Lot 44 is bounded below by a tract of land separating the property from the Olivier plantation. Lot 18 is bounded above by Lot 19. Lot 19 was sold by the Citizens Bank of Louisiana to John Ermant Leitmeyer. Lot 28 is bounded above by Lot 27. Lot 27 is sold by the Citizens Bank to Dennis Lucky.

This was all that was unsold by the Citizens Bank of Louisiana of a certain property, known and designated as the Nelson Plantation. The Nelson Plantation was formerly St. Mary's Plantation, and the lands that were formerly owned by Stephen Owens Nelson. The Citizens Bank of Louisiana acquired the Nelson Plantation in the proportions of two undivided thirds of the Citizens Bank of Louisiana from the bankrupt estate of S. O. Nelson and T. A. Nelson, and one undivided third from the Crescent City Bank by purchase from Emory E. Norton, the assignee in the bankruptcy of the estate of S. O. Nelson and T. A. Nelson.

On May 11, 1889, Fred Duperier, then serving as Clerk of Court of Iberia Parish, in the year of 1889, received a copy of the act of conveyance or the act of sale of the remaining property of the former Nelson Plantation by the Citizens Bank of Louisiana to Alonzo E. Hall. The sale of the property to Alonzo E. Hall was for the purchase price of $25,146.84. William G. Hall, who was acting by a power of special attorney for Alonzo E. Hall, paid $6,286.71 at the execution of the act of conveyance. This left a balance of $18,886.13 to be paid by notes, and the property purchased was secured by a vendor's privilege. William G. Hall, who was acting for Alonzo E. Hall, furnished four promissory notes, which notes were subscribed by William G. Hall. Each one of the four promissory notes was in the amount of $4,175.00. These four promissory notes were conditioned to bear interest at 7% from January 1, 1889 until paid. Also, these four promissory notes were payable at the Citizens Bank of Louisiana in the City of New Orleans, Louisiana. The first promissory note in the amount of $4,175.00 was due on January 1, 1890. The second promissory note in the amount of $4,175.00 was due on January 1, 1891. The third promissory note, in the amount of $4,175.00, was due on January 1, 1892. The fourth promissory note in the amount of $4,175.00 was due on January 1, 1893. All these notes were to be paid to Stanley O. Thomas, then the President of the Citizens Bank of Louisiana, and the representative of the Citizens Bank in this act of conveyance.

On May 25, 1889, William G. Hall, acting on behalf of Alonzo E. Hall of Fort Plains, New York, sold or conveyed to William Bailey, of Iberia Parish, a tract of land. This tract of land was a portion of tract one of the former Nelson Plantation, and a portion of the lands conveyed to Alonzo Hall by the Citizens

Bank of Louisiana.

The tract of land that was being conveyed by William G. Hall to William L. Bailey was described as being situated three miles below the town of New Iberia, on the west bank of Bayou Teche. This tract of land contained ten arpents, and it was bounded on the front by Bayou Teche, in the rear by Morgan's Railroad Company, above by lands of Alonzo E. Hall, which were other portions of the former Nelson Plantation, and below by the Olivier Plantation, the present day Orange Grove, which was owned in the year of 1889 by the Citizens Bank of Louisiana. This tract measured 260 feet wide on Bayou Teche by a depth of parallel lines to Morgan's railroad track. This tract of land comprised lot 14 and lot 15.

The price for this tract of land was $250.00, which was paid in cash money at the execution of the act of sale. The execution of the act of sale took place in the office of P. Laurent Renoudet in the City of New Iberia.

Stanley O. Thomas, who was President of the Citizens Bank of Louisiana, stated that the tract of land forms lot 14 and lot 15 of the former Nelson Plantation. Stanley O. Thomas granted full acquittal and a release of the mortgage and vendor's privilege, which was granted to Alonzo E. Hall, on the date of April 27, 1889. This tract of land that was conveyed forms a part of the tract of land that was sold to Alonzo E. Hall and mortgaged in favor of the Citizens Bank of Louisiana.

On April 2, 1895, William L. Bailey, of New Iberia, conveyed at a cash sale this tract of land to the Iberia Building Association. The Iberia Building Association was represented at the execution of this act of conveyance by its President, E. F. Millard. The Iberia Building Association was a corporation that was erected under the laws of the State of Louisiana. Also, the Iberia Building Association was domiciled in the City of New Iberia, Louisiana, which is situated in Iberia Parish, Louisiana. Today, the Iberia Building Association is presently known as the Iberia Savings Bank. The purchase of this tract of land was pursuant to the charter and by laws of the Iberia Building Association for the use and benefit of Joseph A. Hertrinx, who holds and owns eight shares of stock. These eight shares of stock are represented by book 604 and book 639. Also, on March 20, 1895, Joseph A. Hertrinx contracted a loan of $1,600.00 on the shares of stock with 6% interest per annum or a year from the date of March 20, 1895.

The act of conveyance of this property was executed on April 2, 1895 by Leonce de la Houssaye, a notary public in the City of New Iberia, Louisiana. The President, E. F. Millard, of the Iberia Building Association was the purchaser of this tract of land. The Iberia Building Association paid $1,600.00 in cash money, which was the amount of the loan to Joseph A. DeBluix or Hertrinx. Also, Joseph A. Hertrinx had conformed himself to all conditions contained in the charter and by laws of the Iberia Building Association, and he does consent and agree that the title of this property shall remain vested in the Iberia Building Association until full and complete payment of $1,600.00 with interest to Joseph A. DeBluix or Hertrinx. Whenever, Joseph A. DeBluix or Hertrinx is repaid this property shall be reconveyed to him free from all liens, mortgages, and other encumbrances.

The eight shares of stock in the Iberia Building Association were represented by book 604 and book 639. The payments and accumulated of both present and future interest are to remain so pledged until the full and entire payment of the purchase price or the mortgage amount of $1,600.00 with interest. The method of payment is specified in the charter and by laws of the Iberia Building Association.

Also, Joseph A. DeBluix or Hertrinx had to insure and keep insured all the buildings on the tract of land from the loss by fire until the final repayment of the loan of $1,600.00 with interest. The fire insurance company is to be selected and designated by the Board of Directors of the Iberia Building Association for the amounts to be insured. The Board of Directors of the Iberia Building Association will fix the amounts that the building and improvements, which are situated on the tract of land will be insured for. This amount will probably be the amount of money or the cost it would take to fix or replace these buildings and improvements, if these buildings and improvements should be destroyed by fire. The policy of the fire insurance that will be issued on these buildings and improvements will be delivered to the President of the Iberia Building Association, E. F. Millard, when the policy is taken out and paid for.

The forfeiture of the eight shares of stock could result from non payment of the installments on the mortgage of the property, or non payment of the interest, or other duties. Also, it carries with it the right of redemption.

On February 3, 1897, William G. Weeks, a notary public in the City of New Iberia, executed and passed an act of conveyance for the Iberia Building Association, through its President E. F. Millard, of this property to Andrew S. Auld, a resident of New Iberia, Louisiana. E. F. Millard was representing the Iberia Building Association and acting on a virtue of a resolution, which was adopted by the Board of Directors of the Iberia Building Association at their regular meeting, which was held on February 3, 1897. The reason that this property, which was situated three miles below the town of New Iberia was being sold to someone else other than Joseph A. DeBluix was because Joseph A. DeBluix had forfeiture the right of redemption of this property. Therefore, the Iberia Building Association only warrants the title to this property against its (the Iberia Building Association) acts while the Iberia Building Association was in possession of the property and no further.

Andrew S. Auld, the purchaser, binds himself to keep the buildings and improvements now existing on the property or any other buildings and improvements, which may be erected during the ownership of the property by Andrew S. Auld. Also, Andrew S. Auld is to keep these buildings and improvements constantly insured against loss by fire. This fire insurance policy should be kept in a good and solvent insurance company that is doing business in the State of Louisiana. The least amount that these buildings and improvements should be insured against from loss by fire is the amount of $400.00, until the full and final payment of notes to the Iberia Building Association is paid in full. Also, Andrew S. Auld is to deliver and transfer the fire insurance policy or the fire insurance policies to the Iberia Building Association. Also, the Iberia Building Association is empowered to cause the fire insurance to be made and affected at the cost of Andrew S. Auld, but this shall not be constricted as an obligation or binding upon the Iberia Building Association or any other future holder of this note, or to hold Andrew S. Auld responsible for any damages which may result from the non insurance of the building and improvements, which are situated on the property.

This act of conveyance was a sale and mortgage. The cash down payment was $500.00 with the balance of $500.00 being held in notes by the Iberia Building Association. Andrew S. Auld gave three promissory notes that were dated February 3, 1897, and these three notes were bearing interest at 8% per annum from the date of February 3, 1897 until paid in full. Notes one and two were each in the amount of $166.75. Note one was due on February 3, 1898, and note two was due on February 3, 1899. Note three was in the amount of $166.50, and it was due on February 3, 1900. Also, all the taxes assessed on this piece of property were paid.

Jules Gervais Arnandez and Hanan H. Shadell served in their capacity as competent witnesses to this act of conveyance.

On September 16, 1898, Andrew S. Auld conveyed this property, which was situated three miles below the town of New Iberia, Louisiana to Martial E. Sorrel. This act of conveyance was a sale and mortgage. The purchase price of this piece of property was $832.25. Martial Sorrel paid $500.00 as a cash down payment at the execution of the act of conveyance. There were the three original mortgage notes that existed at the conveyance of Andrew S. Auld purchasing the property from the Iberia Building Association. These mortgage notes were drawn and endorsed by Andrew S. Auld. Also, these three mortgage notes, which two mortgage notes were dated February 3, 1897, and the third mortgage note was due on February 3, 1900 were payable at the office of the Iberia Building

Association, which office was located in the town of New Iberia, Louisiana.

A view of the Huval Insurance Company building that is situated on land that once formed a part of the former plantations of John Ermant Leitmeyer, Stephen Owens Nelson, Louis and Charles DeBlanc, and Jean Baptiste Despanets DeBlanc

The Nelson Canal Bridge is situated on land that comprised the plantation of St. Mary's, which was owned by Stephen Owens Nelson

ORANGE GROVE PLANTATION

On the date of July 10, 1935, the Whitney National Bank of the City of New Orleans conveyed at a sale and mortgage to the Teche Sugar Company, Incorporated, a certain sugar plantation, known as the Orange Grove Plantation. At the execution of the act of conveyance, Charles W. Fox, who was serving as Vice President of the Whitney National Bank, was acting in his capacity for the bank. Also, the executive committee of the Board of Directors of the Whitney National Bank issued a resolution for the conveyance of the Orange Grove sugar plantation. This resolution was paraphed, "Ne Varietur," by the notary public, to identify this resolution, which was issued by the executive committee of the Board of Directors, with this act of conveyance.

Also, involved in this act of conveyance was Branches, Incorporated, which was a corporation organized under the laws of the State of Louisiana. Branches, Incorporated had its registered office and domicile in the City of New Orleans. H. G. Thompson, who was serving as Vice President of the Branches, Incorporated, was acting on behalf of Branches, Incorporated, at the execution of this act of conveyance. Also, a certified copy of the resolution of the Board of Directors of Branches, Incorporated was paraphed, "Ne Varietur," to identify this resolution with the act of conveyance. Branches, Incorporated was the corporation that was organized for the operation of branch banks in the Parish of Orleans of the Whitney National Bank.

The Teche Sugar Company, Incorporated was a corporation that was organized under the laws of the State of Louisiana, and was operating in the Parish of Iberia. Also, the Teche Sugar Company had its registered office and domicile in the town of Jeanerette, Louisiana. At the execution of the act of conveyance, Robert Lee Roane, who was serving as Vice President of the Teche Sugar Company, acted in his capacity in representing the interest of the Teche Sugar Company. Also, the Board of Directors had issued a resolution, and a certified copy of this resolution was paraphed,"Ne Varietur," to identify this resolution with this act of conveyance. Through, Robert Lee Roane, the Teche Sugar Company, was purchasing and accepting the Orange Grove sugar plantation from the Whitney National Bank, and Branches, Incorporated.

The Teche Sugar Company purchased the Orange Grove sugar plantation consisted of these three tracts of land. The first tract of land was situated in the second ward of the Parish of Iberia, Louisiana, and it contained nine hundred and fifty acres of land. The first tract of land was bounded on the north side by Bayou Teche, and the Southern Pacific Railroad, on the east side by Bolivar, and on the south side by the public road, which road is commonly known as the Forty Arpent Road, or Admiral Doyle Drive, and on the west side by the Southern Pacific Railroad and Broussard. The second tract of land consisted of land situated in the third ward of the Parish of Iberia, and this second tract of land contained one hundred and sixty acres of land. This second tract of land was bounded on the north side by the property of Bertrand, on the east side by land of Sorrel, on the south side by Bayou Teche, and on the west side by Bouriagu, which was commonly known as designated as the Douglas and DeBlanc Place. The third tract of land was situated in the third ward of the Parish of Iberia, and it contained one hundred and twenty acres of land. This tract of land was bounded on the north side by Decuir, on the east side by the road, on the south side by the road and Bayou Teche, and on the west side by Dencause. This third tract of land was known and designated as the Franklin Place. The fourth tract of land was situated in the second ward of Iberia Parish, and it contained forty acres of land. This fourth tract of land was bounded on the north side by the public road, on the east side by Romero, on the south side by Romero. and on the west side by O. Romero. This fourth tract of land was known as designated as the P. Romero Place.

Besides the land sold, the Teche Sugar Company, through its Vice President, Robert Lee Roane, also purchased the Orange Grove sugarhouse and sugar refinery. The right of ways for roads, the train road, and canals through the property were also purchased for use by the Teche Sugar Company.

This sale was made and accepted for the purchase price and consideration of $85,000.00. At the execution of this act of conveyance, the sum of $25,00.00 was paid to the Whitney National Bank of New Orleans, and Branches, Incorporated by the Vice President of the Teche Sugar Company, Incorporated, Robert Lee Roane. The Vice President of the Whitney National Bank of New Orleans, Charles W. Fox, and H. G. Thompson, who was serving as Vice President of Branches, Incorporated, acknowledged the receipt of the sum of $25,000.00, and granted acquittal to the Teche Sugar Company. Now, the balance of the purchase price was $60,000.00, which was payable in promissory notes.

In representation of the credit portion of the purchase price, of the sum of $60,000.00, the Teche Sugar Company, through its Vice President, Robert Lee Roane, made, drew, and signed four promissory notes. Each promissory note was for the sum of $15,000.00. The first promissory note was due in one year from the date of the execution of the act of conveyance, which was July 10, 1936, the second promissory note was due on July 10, 1937, or two years from the date of the execution of the act of conveyance, the third promissory note was due on July 10, 1938, or three years after the date of the execution of the act of conveyance, and the fourth promissory note was due on the date of July 10, 1939, or four years after the date of the execution of the act of conveyance. These four promissory notes were payable at the Whitney National Bank of the City of New Orleans, Louisiana. Each promissory note was conditioned to bear interest at the rate of 6% per annum from the date until paid. The interest on the balance of the purchase price of the sum of $60,000.00 was payable annually. Also, the four promissory notes were paraphed, "Ne Varietur," to identify the four promissory notes with the original act of conveyance by the notary public, Eugene D. Sanders.

The Teche Sugar Company, which was the mortgagor, binded itself to keep the sugar factory and the machinery in the sugar factory on the Orange Grove Plantation constantly insured against the loss by fire. This fire insurance policy had to be kept in a good and solvent insurance company, which was operating in the State of Louisiana, and the sum of the insurance policy was at least $60,000.00, or a rate of premium not to exceed 3% per annum until the full and the final payment of the promissory notes. Any loss that is incurred is payable to the mortgagee, the Whitney National Bank and Branches, Incorporated, or the holder or the holders of the promissory notes. The fire insurance policy would be in the name of any future holder or holders of the promissory notes.

On the date of March 30, 1906, the planting firm of J. N. Pharr & Sons, Ltd., represented through Henry Newton Pharr, conveyed a right of way to Charles C. Henshaw and others. This right of way was seventy five feet wide, and it ran over and across a certain tract of land, which was situated on the east side of Bayou Teche. This tract of land contained seventy arpents, and it was bounded on the north side by the lands of J. N. Pharr & Sons, Ltd., on the south side by the lands of Frilot, and on the west side by Bayou Teche. Also, this act of conveyance would result in the probable enhancement in the value of their adjacent land in care of building of the railroad through the land of J. N. Pharr & Sons, Ltd.

It was understood that this conveyance was made upon the assurance of Charles C. Henshaw that he will not assign or transfer the right of way to any railroad company, now operating in the State of Louisiana, and forming part of the Southern Pacific System. Also, Charles C. Henshaw and his partners will be a company altogether independent of any railroad company, that was then operating in the State of Louisiana, and forming part of the Southern Pacific system. This transfer or conveyance of the right of way is to lapse and be of no effect in the event that Charles C. Henshaw sells the same to any railroad company, that was operating in the State of Louisiana, and forming part of the Southern Pacific system.

The consideration of the transfer or the purchase price, is the amount of one dollar, which amount is paid in cash or current money at the execution of this act of conveyance. Henry

Newton Pharr, the Vice President of the firm of J. N. Pharr & Sons, Ltd, acknowledged the receipt of the payment of the purchase price.

Charles C. Henshaw accepts the conveyance and agreed to construct a line of railway over and across along the right of way. This line of railway will commence with a through line and will have to be completed on or before the first of July in the year of 1909. Also, Charles, Henshaw agreed to construct and maintain proper railroad crossings over the railway watch at grange for the convenience of the planting firm of J. N. Pharr & Sons, Ltd. Charles Henshaw also agreed to leave the proper springs in all drains and ditches, which are now existing on the right of way. Failure by Charles C. Henshaw to construct a line of railway over the right of way on or before the first day of the month of July in the year of 1909 shall cause a forfeiture by Charles Henshaw of all rights in and to the right of way. But if the railway line is completed on or before the first day of July of the year of 1909, the right of way shall be the property of Charles C. Henshaw and others.

J. N. Pharr & Sons, Ltd., a planting company, conveyed a right of way to Morgan's Louisiana & Texas Railroad & Steamship Company, and the act of conveyance was executed on the date of May 31st, 1906. Henry N. Pharr, who was Vice President of J. N. Pharr & Sons, Ltd, was given power of attorney to grant to Morgan's Louisiana & Texas Railroad & Steamship Company the right to build a switch from the main line of the railroad to the Orange Grove sugarhouse.

On June 15, 1906, J. N. Pharr & Sons, Ltd., a planting company, was represented by Henry N. Pharr, who was authorized to sign the act of conveyance with Morgan's Louisiana & Texas Railroad & Steamship Company. Morgan's Louisiana & Texas Railroad & Steamship Company was now locating a railroad spur track starting at Olivier. The spur track connected with the home track of Morgan's Louisiana & Texas Railroad & Steamship Company. The spur track started at the Olivier train station, and ran in a northerly direction, crossing Orange Grove sugar refinery near Bayou Teche. The length of the spur track was one thousand and seven hundred and ten feet, and came through to its full length through the land of J. N. Pharr & Sons, Ltd.

In consideration of the measured value, which will be given to the lands and the advantages, benefits, and conveyances resulting from the building of the spur track, J. N. Pharr & Sons, Ltd., have conveyed a right of way to Morgan's Louisiana & Texas Railroad & Steamship Company. This right of way consisted of a strip of land, which strip of land contained one acre of land. This right of way was to be used for all other railroad purposes across and through the lands of J. N. Pharr & Sons, Ltd.

This right of way begins at a point in the northern line of the station grounds of Morgan's Louisiana & Texas Railroad & Steamship Company at Olivier, Louisiana. At the railroad station at Olivier, Louisiana, the point of beginning was at the intersection of the center line of the railroad spur tracks with the north line of the station grounds, which begins at a point from the beginning corner in a northeasterly direction. The uniform width of ten feet on each side to a point, which point is the beginning of a seven degree 30 minute curve to the left spur railroad track. In a northeasterly direction curving to the left with a radius of 764.5 feet following the center of the spur railroad track with a uniform width of ten feet on each side of 7,955 feet to make a seven degree 30 minute curve in the railroad spur track.

Henry Newton Pharr, who was serving the planting company of J. N. Pharr & Sons, Ltd., in the capacity of Vice President, was authorized to purchase for the firm of J. N. Pharr & Sons, Ltd., all right of way privileges in the second ward of Iberia Parish, Louisiana that are necessary for the exploitation of twenty or more miles of the narrow gauge railroad on the Orange Grove plantation.

Also, Henry Newton Pharr was authorized to purchase for the planting firm of J. N. Pharr & Sons, Ltd., a certain privilege for hauling sugar cane over the property of any parties he may deem it necessary to purchase from in the second ward of Iberia Parish, Louisiana. This right of way conveyance was filed in the conveyance records of the clerk of court in the Parish of Iberia on the date of August 31, 1906.

On the date of April 26, 1907, the firm of J. N. Pharr & Sons, Ltd., with Henry Pharr serving as its Vice President, conveyed to the Bayou Teche Railway & Light Company a right of way. This right of way was twenty feet wide and thirty feet in length. This right of way ran in a north and southeasterly direction over and across that certain tract of land belonging to J. N. Pharr & Sons, Ltd. This right of way was bounded on the east side by Bayou Teche, on the west side by the public road, known as the Forty Arpent Road or the present day Admiral Doyle Drive, on the north side by the land of Smith, and on the south side by Olivier. This right of way was situated near the public road. The benefits and advantages that will accrue to the planting firm of J. N. Pharr & Sons, Ltd. will enhance the value of the adjacent lands in the building of the railroad through the land.

The consideration for this right of way was the purchase price of $1.00. This one dollar was paid in cash money at the execution of this act of conveyance. Also, at this time, Henry Pharr, the Vice President of the firm of J. N. Pharr & Sons, Ltd., acknowledged the receipt.

Bayou Teche Railway & Light Company agreed to construct a line of railway over and across the right of way connecting with a through line. This construction of the line of railway is to be completed on or before July 1, 1904, and the Bayou Teche Railway & Light Company further agreed to construct and maintain proper railroad crossings over the railway track at the Orange Grove plantation for the convenience of the planting firm of J.N. Pharr & Sons, Ltd. Also, the Bayou Teche Railway & Light Company agreed to leave proper drainage openings in all drains and ditches that were in existence on the right of way. If the Bayou Teche Railway & Light Company does not construct a line of railway in over and across the right of way, then the permission to do so will be forfeited. Also, the right of way is nontransferable until the operation of electric cars on the railroad track situated on the right of way.

On May 20, 1904, Heneritta Pharr conveyed to John Newton Pharr & Sons, Ltd. a tract of land with all the buildings and improvements on the tract of land. This tract of land measured about eighteen arpents and five hundredths of an arpent of land fronting on the west side of Bayou Teche, in Iberia Parish, with a forty arpent depth in the tract of land. This tract of land was bounded above or on the upperside by the property formerly belonging to S. O. Nelson, and below or on the lowerside by lot #1, which lot was sold to Victor Vautrin, and comprising lot #2 through lot # 19. These lots were shown on a map, which map was made on August 25, 1887. This property is presently bounded in the year of 1904, in front by Bayou Teche, above or on the upperside in part by the property of the Iberia Building Association, and in part by the property of Alphonse Tertrou or Orange Grove, and below by the property of Pierre LeBron.

Also, there was another tract of land, which was also conveyed, and this tract of land was situated in the Parish of Iberia. This tract of land measured nine arpents fronting Bayou Teche, with a forty arpent depth to the property. The lines closing in the rear, so as to make the width of five or six arpents. This tract of land was bounded below by the property belonging to Baptiste Collette, and above by the property being conveyed, which is tract number three, and in the rear by lands formerly belonging to the succession of Charles Napoleon Olivier.

The third tract of land was also situated between Bayou Teche and the public road, which lead from the town of New Iberia to the town of Jeanerette, known as the Old Spanish Trail. This tract of land measured about six arpents fronting on Bayou Teche, with an irregular depth of about forty five arpents. The boundaries of this tract are on the upperside or above by lands formerly belonging to the succession of Charles Napoleon Olivier, and below or on the lowerside by tract two, which is also being conveyed.

The fourth tract of land was also situated on the public road, which leads from the town of New Iberia to the town of Jeanerette, Louisiana, and known as the Old Spanish Trail. This tract has a frontage along Bayou Teche of six arpents, with a depth of thirty six arpents. The total land area of tract number four was two hundred and sixteen arpents of land. The fourth tract of land is bounded above or on the upperside by lands belonging to the succession of Charles Napoleon Olivier, and below by lands, which comprise tract three of this act of conveyance. Also, sold with the property is all the buildings and the improvements.

Also, included in this act of conveyance was an undivided two thirds of a certain tract of land, and all the buildings and the improvements to the land were also included. This tract of land was situated on the east side of Bayou Teche in Iberia Parish. This tract of land contained one hundred and forty arpents. The boundaries of this tract of land were on the north side by the lands of Philesphore Dupuy, on the south side by Bayou Teche, on the east side by lands of Etienne Frilot, and on the west side by lands belonging to the Citizens Bank of Louisiana. This was the same property that was acquired from Pierre LeBron from Henri Cogneheim on the date of January 24, 1903.

On June 3, 1904, Eugene A. Pharr conveyed to the E. A. Pharr & Company, Ltd., a certain tract of land together with all the buildings and improvements. This tract of land was situated at Olivier station, which was a train station in Iberia Parish. This tract of land measured two hundred and three feet long by five sixth arpents in depth. This tract of land is bounded on the north side by the lands of Charles Decuir, on the south side by the public road, on the east side by the lands of Carbrand, and on the west side by the public road and property of the vendor, Eugene A. Pharr. The purchase price of this tract of land is $1,500.00.

It is well understood and distinctly equal that the vendor, Eugene A. Pharr, reserves and the E. A. Pharr & Company, Ltd, the purchaser, grants to him (Eugene A. Pharr) the right to carry out any and all contracts made by Eugene A. Pharr for the land and especially, the Leon Godchaux Company, Ltd., which was a sugar refining company. The Leon Godchaux Company, Ltd., has the right to use this tract of land for the boarding and shipping of sugar cane. It is further understood that the vendor, Eugene A. Pharr, reserves from this conveyance or sale of this tract of land that he may have the right to resume at any time that he sees fit to use the land for his own benefit.

On May 2, 1897, Frank J. Mestayer, who was serving Iberia Parish as a Deputy Sheriff, seized the Orange Grove sugar plantation and sugar refinery and mill. The reason that this sugar planation and sugar refinery and mill were seized was because of a civil lawsuit, entitled, "Erman & Cahn versus Robert A. Olivier," of the civil docket of the Parish of Iberia, and this write of seizure and sale (or writ of Fieri Facias) was issued against the property of Robert A. Olivier and Eugene Abraham. The Sheriff of the Parish of Iberia did seize the Orange Grove sugar plantation and sugar refinery and mill, and this property was taken into the possession of the Sheriff of the Parish of Iberia.

The property that consisted of the Orange Grove sugar plantation was described in the year of 1897, as a tract of land measuring nine arpents fronting Bayou Teche, with a forty arpentage depth. The total land area of the Orange Grove sugar plantation was three hundred and sixty arpents of land. The lines of the tract of land were closing in the rear of the tract of land, so as to make two or six arpents. This tract of land was bounded above or on the upperside by the lands of the heirs of Baptiste Collette, and below by tract two of the Orange Grove plantation, and in the rear by the lands formerly belonging to the succession of Charles Napoleon Olivier.

There was another tract of land, which was situated between Bayou Teche and the public road. This public road is known as the Old Spanish Trail, and it leads from the town of New Iberia to the town of Jeanerette. This tract of land measured six arpents fronting Bayou Teche by an irregular depth of five arpents. The total area of this tract of land was thirty arpents of land. The tract of land is bounded above or on the upperside by the lands formerly belonging to the succession of Charles Napoleon Olivier, and below by tract one of the Orange Grove plantation.

The third tract of land that comprised the Orange Grove plantation was situated on the public road. This tract of land had a frontage along Bayou Teche of six arpents with a depth of thirty six arpents. The land area of the third tract of land was two hundred and sixteen arpents. This third tract of land was bounded above or on the upperside by the lands belonging to the succession of Charles Napoleon Olivier, and tract two of the Orange Grove plantation. Also, all the buildings and the improvements, which were situated on the Orange Grove sugar plantation were included in the sale.

After the Sheriff of the Parish of Iberia had completed all the legal formalities concerning the seizure of the Orange Grove sugar plantation, then the property was advertised for sale. The property that comprised the Orange Grove sugar plantation was advertised in the weekly newspaper, called "The Weekly Iberian," on May 22, 1897. The Weekly Iberian newspaper was published in the town of New Iberia, Louisiana. The advertisement stated that the Sheriff sale of the Orange Grove plantation would take place at the Iberia Parish Courthouse, which was situated on Main Street in the town of New Iberia, Louisiana, on the date of June 26, 1897, between the legal sale hours for cash.

In the morning of June 26, 1897, a Deputy Sheriff of Iberia Parish, Louisiana, did offer the property, which comprised the Orange Grove sugar plantation for sale. The Deputy Sheriff read in a loud, clear, and audible voice, first the order of the seizure and sale of the property comprising the Orange Grove plantation, secondly the advertisement of the property for sale, and thirdly the certificate of mortgage, which was obtained from the office of the clerk of court for the purpose of the sheriff sale.

After the crying of the property for a considerable period of time, the Deputy Sheriff sold the property of the Orange Grove sugar plantation to John Newton Pharr. The Orange Grove sugar plantation was sold for the price and sum of $21,535.00. John Newton Pharr had the same rights, titles, and interest to the property of the Orange Grove sugar plantation as did the former owners, Robert A. Olivier, and Eugene Abraham.

John Newton Pharr paid the cash amount of $9,092.38 in principal, $379.00 in interest, $473.00 in attorney's fees, and $192.00 in court costs. The balance of the purchase price was $11,397.32, was going to make up the amount of the bid that John Newton Pharr had retained in his hands to meet the bid for the second and the third mortgages bearing on the property of the Orange Grove sugar plantation.

Actually, John Newton Pharr was not present at the sheriff sale of the Orange Grove plantation, but Walter J. Burke acted as agent for him. The reason that Walter J. Burke was able to act as agent was by virtue of a letter authorization, which empowered Walter J. Burke to act on the behalf of John Newton Pharr.

On May 1, 1897, Erman & Cahn, a commercial firm in the City of New Orleans, Louisiana, and composed of Abraham Erman and Leon Cahn filed a civil lawsuit against Robert A. Olivier, and Eugene Abraham of the partnership of Olivier and Abraham. Robert A. Olivier was a resident of the Parish of Iberia, and Eugene Abraham was a resident of the City of New Orleans, Louisiana. Walter J. Burke served as counsel or attorney for the firm of Erman & Cahn or the plaintiff in this case.

The basis of this civil lawsuit was the unpaid note, which note was made by Eugene Abraham of the City of New Orleans, and this note was dated July 7, 1892. This note was payable one year after date or on July 7, 1893, and the note was conditioned to bear interest at the rate of 8% from the date of the note until paid in full. This note was payable to the order of the People's Bank in the City of New Orleans. The amount of this note was $12,000.00. On April 1, 1893, a payment was made in which this note was reduced to the amount of $10,078.94, and all the interest was paid. Then on December 20, 1893, another payment was made in which the note was reduced to $9,092.38.

On January 7, 1892, this note was paraphed, "Ne Varietur," by F. J. Dreyfus to identify this note with an act of mortgage. The act of mortgage bore upon the property of the Orange Grove plantation, which was owned by Robert A. Olivier and Eugene Abraham.

Robert A. Olivier was a resident of the Parish of Iberia, and he was in charge of the operations of Olivier and Abraham, which were being carried on at the Orange Grove plantation. Eugene Abraham was a member of the commercial firm of Charles Gest & Company of Rapides Parish, Louisiana. Also, the commercial firm of Charles Gest & Company had gone into insolvency, and Walter C. Johnson was duly appointed as assignee of insolvency in bankruptcy. There was a certified copy of the bankruptcy proceedings, which set forth the appointment of Walter C. Johnson as assignee of insolvency in bankruptcy. This certificate was filed with the clerk of court as well as the qualifications, oath, and bond of Walter C. Johnson.

Walter C. Johnson, the assignee in bankruptcy, seized the property, which comprised the Orange Grove plantation. The reason that this property was seized was to satisfy the debt set forth in the mortgage, which was executed on January 7, 1892. Also, included in the debt was the amount of the principal owed, the interest bearing on the principal, the cost of attorney's fees, and the court costs. The only costs that were excluded from the costs were the cost of coal and sugar that were located on the property of the Orange Grove plantation. The act of mortgage bore on the property of the Orange Grove plantation.

On January 7, 1892, Robert A. Olivier, of Iberia Parish, Louisiana, and Eugene Abraham, of the City of New Orleans, Louisiana, each purchased one undivided half of Warren R. Farmer's rights, title, and interest to the Orange Grove sugar plantation. The real property that was purchased consisted of a tract of land that measured nine arpents fronting Bayou Teche with a depth of forty arpents. The lines closing in the rear so as to make the width of five or six arpents. This tract of land was bounded below by the lands of the heirs of Baptiste Colette, and above by tracts two and three of the Orange Grove plantation, in the rear by lands formerly belonging to the succession of Charles Napoleon Olivier.

The second tract of land of the Orange Grove plantation, which was also purchased was situated between Bayou Teche and the public road. This public road leads from the town of New Iberia to the town of Jeanerette, and is known as the Old Spanish Trail. This second tract of land measures six arpents fronting on Bayou Teche by an irregular depth of five arpents. This second tract is bounded above by the lands formerly belonging to the succession of Charles Napoleon Olivier, and tract one of the Orange Grove plantation.

A third tract of land was also purchased, and this third tract of land was situated on the public road. This third tract of land had a frontage along Bayou Teche of six arpents with a depth of thirty six arpents. This third tract of land was bounded above by the lands formerly belonging to the succession of Charles Napoleon Olivier, and below by the land, which comprises tract two of the Orange Grove plantation.

This sale was made and accepted for the purchase price of $12,000.00. The purchasers, Robert A. Olivier, and Eugene Abraham have furnished their promissory notes, which notes are subscribed jointly by Robert A. Olivier and Eugene Abraham. Also, these promissory notes were made to the order of Robert A. Olivier and Eugene Abraham. These promissory notes were payable one year later or on January 7, 1893 at the People's Bank in the City of New Orleans. Also, these promissory notes were conditioned to bear interest at the rate of 8% until the promissory note was paid in full. In order to secure the principal and interest, together with the 5% attorney's fees on the amount sued for, and any possible vendor reserves, such as the vendor's privilege and right on the property now sold. The whole of the property mortgaged is in the vendor's favor and in favor of any subsequent holder or holders of the promissory notes. Therefore, the purchasers, Robert A. Olivier and Eugene Abraham binded themselves not to sell or alienate, encumber, or deteriorate the property, which comprises the Orange Grove plantation.

On the date of June 7, 1895, the inscription of the mortgage in the name of Eugene Abraham, Robert A. Olivier, and Warren R. Farmer was erased and cancelled in relation to two mortgage notes, which were dated January 7, 1892. These mortgage notes was for the amount of $300.00. The first mortgage note was due on December 15, 1891, and the second mortgage note was due on January 1, 1892.

The commercial firm of Erman & Cahn filed a civil lawsuit, which set forth that the firm of Erman & Cahn were the holders or the owners of a certain promissory note. This promissory note was dated January 7, 1893, and it was made payable to the order of Robert A. Olivier and Eugene Abraham. The amount of the promissory note was $12,000.00, which was subject to credits and reduced by various payments to the sum of $9,092.36. Also, this promissory note was conditioned to bear interest at the rate of 8% from the date of December 29, 1896.

An order of seizure and sale of the Orange Grove sugar plantation and sugar refinery and mill was granted by the 19th Judicial District Court on the date of May 1, 1897. The only way that this order of seizure and sale could be reversed or rescended was if Robert A. Olivier and Eugene Abraham paid or caused the notes to be satisfied. These promissory notes were held by the commercial firm of Erman & Cahn of the City of New Orleans,

Louisiana. The notice of the delinquent payments of the promissory notes was held by Erman & Cahn. In case of judicial proceedings to collect the monies owed, there was levied a 5% charge for attorney's fees and costs within three days from the service of the notice. One additional day was allowed for every twenty miles from the residence of the person being served to the residence of the judge granting the order of seizure and sale.

On May 1, 1897, the Sheriff of the Parish of Iberia, George Henderson, authorized by a writ of seizure and sale, took into his possession the property comprising the Orange Grove sugar plantation and sugar refinery and mill. Also, Sheriff George Henderson proceeded to seize the property of the Orange Grove sugar plantation according to the laws of the State of Louisiana, and then to eventually sell the property of the Orange Grove sugar plantation for cash at a sheriff's sale.

Robert A. Olivier was served with an order of seizure and sale of the Orange Grove sugar plantation on the date of May 22, 1897. After being served to Robert A. Olivier, this notice was returned to the Iberia Parish Sheriff's office.

On May 27, 1897, Eugene Abraham was ordered to make a choice of an appraiser to appraise or to estimate the value of the property of the Orange Grove sugar plantation. J. D. Roane and George Simon were appointed appraisers for the property of the Orange Grove sugar plantation, and they made an appraisement of the property. Also, Antoine Moresi was appointed as a third person to do the appraisement of the Orange Grove sugar plantain, and he was paid the appraisement fee of $21.50.

On the date of June 22, 1897, J. D. Roane and George Simon made an appraisement of the property, which comprised the Orange Grove sugar plantation, which was situated at Olivier, Louisiana. Tract one of the Orange Grove plantation, which consisted of a tract of land measuring nine arpents fronting Bayou Teche, with a depth of forty arpents, and this tract of land was appraised at the amount of $5,333.33. The first tract of land consisted of three hundred and sixty arpents of land. The second tract of land was situated between Bayou Teche and the public road, which road is known as the Old Spanish Trail. This public road leads from the town of New Iberia to the town of Jeanerette, Louisiana. This second tract of land measures six arpents fronting Bayou Teche by an irregular depth of five arpents, and this second tract of land was appraised at the amount of $4,866.66. The third tract of land, which comprised the Orange Grove sugar plantation was situated on the public road, known as the Old Spanish Trail. This third tract of land had a frontage along Bayou Teche of six arpents of land with a depth of thirty six arpents, and this tract of land was appraised at the amount of $11,333.33. The third tract of land consisted of two hundred and sixteen arpents of land. Also, all the buildings and the improvements which were situated on the three tracts of land were also sold.

John Newton Pharr, of St. Mary Parish, Louisiana, was the bidder for the three tracts of land, which comprised the Orange Grove sugar plantation. John Newton Pharr bid the amount of $5,334.00 for tract one of the Orange Grove sugar plantation, and he bid the amount of $4,867.00 for the second tract of land. Also, John Newton Pharr bid the amount of $11,334.00 for the third tract of land. John Newton Pharr was the highest bidder for the property, which comprised the Orange Grove sugar plantation.

Also, Antoine Moresi was named in the civil lawsuit as a third opponent. On the date of June 5, 1897, Antoine Moresi had filed a petition, which stated that he had a sufficient interest in the civil proceedings regarding the seizure and sale of the Orange Grove sugar plantation. The reason that Antoine Moresi had an interest was in the year of 1889, he had sold the partnership of Olivier and Abraham, two turn plates for the Orange Grove sugar refinery and mill. These two turn plates were manufactured at the Moresi Foundry, which was situated in the town of Jeanerette, Louisiana. These two turn plates were used and attached to the machinery, which was used in the Orange Grove sugar mill. The value of these two turn plates was $84.51, and this amount had never been paid by the partnership of Olivier and Abraham. These two turn plates were under a vendor's lien and privilege meaning that Antoine Moresi acting on behalf of the Moresi Foundry had a right to try to recover these two turn plates. Also, these two turn plates were situated in the Orange Grove sugar mill, which sugar mill had been seized by the Sheriff of the Parish of Iberia. As a result of these two turn plates being situated in the sugar house on the Orange Grove plantation, these turn plates were included in the advertisement of the sheriff sale of the Orange Grove plantation, which advertisement appeared in the Weekly Iberian, a newspaper, which was published in the town of New Iberia, Louisiana.

Antoine Moresi wanted the two turn plates to be appraised separately and sold separately. Also, Antoine Moresi felt that he should be paid by preference of having sold Olivier and Abraham the two turn plates for the Orange Grove sugar house. The partnership of Olivier and Abraham had never paid Antoine Moresi or the Moresi Foundry, the purchase price of the two turn plates. C. T. Cade, who was serving as Sheriff of the Parish of Iberia, recognized Antoine Moresi as the prime and privilege creditor upon the proceeds of the sale of the two turn plates. These two turn plates were appraised at the value of $54.00. On August 31, 1897, a trial was held with Judge Felix Voorhies serving as the presiding judge. A judgment was rendered in the favor of Antoine Moresi and against the partnership of Olivier and Abraham.

In the month of February of the year of 1896, Robert A. Olivier, of the Parish of Iberia, conveyed to the firm of J. & L. Dreyfus, which was a commercial business establishment, various types of movable property. The firm of J. & L. Dreyfus was composed of Jules and Leon Dreyfus, who were acting in their capacity by purchasing this movable property. The movable property that was being sold consisted of one dark bay horse mule, which was seven years old, three light bay horse mules, which were seven years old, one iron gray mare mule, which was seven years old, one iron gray horse mule, which was six years old, three Avery plows with harnesses, size DO, one Avery cultivator, two Avery plows, two cane carts, and one undivided interest in the Avery stubble digger. All this movable property was sold for the purchase price of $800.00.

Robert A. Olivier, of the Parish of Iberia, reserves the right to redeem the movable property in one year from the commercial firm of J. & L. Dreyfus. The firm of J. & L. Dreyfus was to repair the machinery as repairs are needed to be made. Also, the commercial firm of J. & L. Dreyfus agreed and consented to the possibility of redemption of the movable property by Robert A. Olivier.

On the date of February 6, 1896, the commercial firm of J. & L. Dreyfus conveyed to Robert A. Olivier the same movable property. Leonce de la Houssaye acted in his capacity as notary public for the act of conveyance. The movable property that was sold consisted of one gray bay horse mule, age seven years old, three light bay mare mules, age seven years old, one iron gray mare mule, age seven years old, one iron gray horse mule, age six years old, three Avery plows and harnesses, two cane carts, one undivided one third interest in one Avery stubble cane digger. The sale price of all of this movable property was the purchase price of $1,230.00.

On the date of February 13, 1895, Robert A. Olivier, of the Parish of Iberia, conveyed to the commercial business firm of J. & L. Dreyfus movable property. This act of conveyance was executed at the office of Leonce de la Houssaye, a notary public, in the town of New Iberia, Louisiana. The commercial firm of J. & L. Dreyfus was domiciled in the town of New Iberia, Louisiana. Jules and Leon Dreyfus, the owners of the commercial business firm, were present and accepting delivery and possession of the movable property for the firm of J. & L. Dreyfus.

The movable property consisted of one dark bay horse mule, age seven years old, three light bay mare mules, each mule was seven years old, one iron gray mare mule, age seven years old, one iron gray horse mule, age six years old, two cane carts, and one undivided third interest in the Avery stubble digger. The purchase price of this movable property was $1,230.00. At the execution of this act of conveyance, the purchase price of the movable property was paid in cash. Robert A. Olivier acknowledged the receipt of the purchase price, and he granted acquittal to the commercial business firm of J. & L. Dreyfus.

On January 7, 1892, Warren R. Farmer and Robert A. Olivier, both of the Parish of Iberia, and Eugene Abraham, of the City of New Orleans, Louisiana, all met to execute an act of conveyance. At this time, all three partners had agreed to rescind and annual the former act of conveyance, in which Eugene Abraham and Warren Farmer, were the vendors, and Robert A. Olivier was the vendee, and this act of conveyance was executed before Leonce de la Houssaye.

Now, Warren R. Farmer agrees to sell or convey his one undivided half share of interest, rights, and title to the Orange Grove plantation, and the purchasers were Robert A. Olivier, of the Parish of Iberia, and Eugene Abraham, of the City of New Orleans, Louisiana. Also, all the sugar, molasses, and mash that was manufactured on the Orange Grove sugar plantation during the season of 1891 was also sold.

This conveyance was made and accepted for the purchase price of $12,000.00. The purchasers, Robert A. Olivier and Eugene Abraham have furnished their promissory notes, which notes are dated this day of January 7, 1892. The promissory notes had been subscribed by them, Robert A. Olivier, and Eugene Abraham, jointly, and made to their own order. Also, the promissory notes are made payable to the People's Bank of the City of New Orleans, Louisiana, and is due in one year from the date of the act of conveyance or January 7, 1893. These promissory notes are conditioned to bear interest at the rate of 8% per annum.

In order to secure the payment of the mortgage note and the interest, together with all attorney fees at 5% on the amount sued for in case of judicial proceedings being instituted to enforce the payment or any part of the payment, the vendor, Warren R. Farmer, reserves the vendor's right and privilege on the property. Therefore, the property comprising the Orange Grove sugar plantation is mortgaged in the favor of Warren R. Farmer or any future holder or holders of the promissory notes. Also, the purchasers, Robert A. Olivier and Eugene Abraham, bind themselves not to sell or alienate, encumber, or deteriorate the whole of the property comprising the Orange Grove sugar plantation to the prejudice of the mortgage.

Also, this sale is made for the further consideration of the obligation taken by the purchasers, Robert A. Olivier and Eugene Abraham, to pay and then grant complete discharge of the vendor's, Warren R. Farmer's, share and portion due by Warren R. Farmer on a promissory note. This promissory note was dated October 5, 1886, and the amount of the promissory note was $2,375.00. Also, this promissory note was conditioned to bear interest at the rate of 8%

from the date until the promissory note was paid in full. This promissory note was secured by a vendor's lien and privilege on the property comprising the Orange Grove sugar plantation. P. Laurent Renoudet, a notary public in the town of New Iberia, Louisiana, had paraphed this promissory notes, "Ne Varietur," to identify this promissory note with an act passed on the date of October 8, 1888.

Also, there were two other promissory notes, and these two promissory notes were dated February 13, 1891. These two promissory notes were subscribed by Farmer, Olivier & Company to their own order, and the note was endorsed by them. One promissory note was for the amount of $3,000.00, which was payable on December 15, 1892, and the other promissory note was for the amount of $4,000.00, which note was payable on January 1, 1893. Both of these promissory notes were payable at the People's Bank in the City of New Orleans, and both promissory notes were conditioned to bear interest at the rate of 8%.

On January 7, 1892, Robert A. Olivier, of Iberia Parish, Louisiana, and Eugene Abraham, of the City of New Orleans, Louisiana, purchased the one undivided share, rights, title, and interest of Warren R. Farmer. This act of conveyance was passed before Felix Dreyfus, a notary public in the Parish of Orleans, and a resident of the City of New Orleans, Louisiana. Warren Farmer, Robert Olivier, and Eugene Abraham declared that in review of a the modifications made by them in their former agreement, they do, receive, annul, and abrogate the act of sale made by Warren R. Farmer, Eugene Abraham, and Robert A. Olivier. This last mentioned act of sale was passed before Leonce de la Houssaye, a notary public in the town of New Iberia, Louisiana, on the date of January 5, 1892. Warren Farmer sold to Robert A. Olivier and Eugene Abraham one undivided half of his (Farmer's) rights, titles, share, and interest in the Orange Grove sugar plantation. Also, the sugar and molasses that was made on the Orange Grove plantation during the season of 1891 and 1892 was also sold. The promissory note was presented to Warren Farmer, who acknowledged the receipt.

This sale is made for the further consideration of the obligation taken by the purchasers, Robert A. Olivier and Eugene Abraham, to pay the complete discharge of the vendor, Warren R. Farmer, share and the portion due by Eugene Abraham and Robert A. Olivier on a certain promissory note. This promissory note was secured by a mortgage and a vendor's privilege for the sum of $2,375.00, and this promissory note was payable in two years. Also the promissory note was conditioned to bear interest at the rate of 8% until paid.

There were two promissory notes made to the order of Farmer, Olivier, & Company, and both of these promissory notes were dated February 13, 1891. One promissory note was payable on December 15, 1891. The second promissory note was for the sum of $4,000.00, and it was payable on January 1, 1892. Both of these promissory notes were payable at the People's Bank in the City of New Orleans, Louisiana. Also, both promissory notes were conditioned to bear interest at the rate of 8% per annum from maturity until paid. Both promissory notes were paraphed, "Ne Varietur," to identify these two promissory notes with the act of mortgage, which act of mortgage was dated February 13, 1891.

At this time, in the year of 1892, the Orange Grove sugar plantation consisted of three separate tracts of land. Tract one measured nine arpents fronting on Bayou Teche, with a depth of forty arpents. The lines of tract one closed in the rear of the tract so as to give a width of five or six arpents. Tract one was bounded below or on the lowerside by the lands of the heirs of Baptiste Collette, and above or on the upperside by the lands of tract two of the Orange Grove plantation, and in the rear by the lands formerly belonging to the succession of Charles Napoleon Olivier.

Tract two of the Orange Grove plantation was situated between Bayou Teche and the public road, which road is known as the Old Spanish Trail. This public road leads from the town of New Iberia to the town of Jeanerette. Tract two measured six arpents fronting Bayou Teche with an irregular depth of five arpents. Tract two was bounded above by the lands formerly belonging to the succession of Charles Napoleon Olivier, and below by tract one of the Orange Grove plantation, and in the rear by the public road or the Old Spanish Trail.

Tract three of the Orange Grove sugar plantation was situated on the public road or the Old Spanish Trail. Tract three had a frontage of six arpents along Bayou Teche with a depth of thirty six arpents. Tract three was bounded above by the lands belonging to the succession of Charles Napoleon Olivier, and below by lands comprising tract two of the Orange Grove sugar plantation. Also, sold were all the buildings and the improvements that were situated on the land.

On the day of the 20th of the month of January of the year of 1890, Warren R. Farmer, Eugene Abraham, and Robert A. Olivier entered into a contract of partnership. Leonce de la Houssaye, who was a notary public in the Parish of Iberia, presided at the execution of the contract of partnership. All three, Robert A. Olivier, Warren R. Farmer, both of Iberia Parish, and Eugene Abraham, of the Parish of Lafourche, were all owners in portion of the Orange Grove sugar plantation. Warren R. Farmer was the owner of a one half portion of the Orange Grove sugar plantation, Robert A. Olivier was the owner of a one fourth share of the Orange Grove sugar plantation, and Eugene Abraham was the owner of a one fourth share of the Orange Grove sugar plantation.

Also, on the 20th day of the month of January of the year

of 1890, all three men, Warren R. Farmer, Robert A. Olivier, and Eugene Abraham, desired to have executed a contact known as the Articles of Partnership. This partnership will be known and operated under the name of Farmer, Olivier, and Abraham. The object of this partnership will be for the purchase of sugar and the manufacture and refining of sugar. Also, the partnership of Farmer, Olivier, and Abraham will have a life span of two years, which will commence on the date of January 1, 1890 and will extinguish on the date of January 1, 1892. The domicile of the partnership of Farmer, Olivier, and Abraham will be in Iberia Parish, Louisiana.

At the end of the partnership on the date of January 1, 1892, Eugene Abraham and Warren R. Farmer will obligated themselves to give the refusal of the partnership purchase. Also, Robert A. Olivier, who on his part, obligated himself in case he should not buy the share and interest of either or both Warren Farmer and Eugene Abraham in the partnership of Farmer, Olivier, and Abraham or the Orange Grove sugar plantation. Robert A. Olivier can sell his share and interest in the partnership of Farmer, Olivier, and Abraham or the Orange Grove plantation to either or both Warren Farmer or Eugene Abraham at the prorated price.

It was understood by Warren R. Farmer, Robert A. Olivier, and Eugene Abraham, that none of the products manufactured on the Orange Grove plantation, such as raw sugar, brown sugar, molasses, and syrup are to be appropriate to the individual use or benefit of any of the three partners. The products that are produced and manufactured on the Orange Grove sugar plantation are to be shipped to the merchant with whom the partnership of Farmer, Olivier, and Abraham may be dealing with and the products are to be sold for the benefit or the profit of the partnership of Farmer, Olivier, and Abraham.

Warren R. Farmer, Robert A. Olivier, and Eugene Abraham agreed that in the event of the death of either partner, the partnership of Farmer, Olivier, and Abraham will continue to exist for the term for which the partnership contract was made, being two years.

On April 15, 1889, Robert A. Olivier purchased at a sheriff sale the tracts of land, which comprised the Orange Grove plantation. The former owner of the Orange Grove sugar plantation was his mother, Laura Therese Dalcourt, who was the lawful and legal wife of Eugene Barthlemy Olivier.

The property of the Orange Grove sugar plantation had been seized by Sheriff A. G. Baruard, of the Parish of Iberia. The reason that the sheriff had a right to seize the property was by virtue of the power vested by the law in accordance with Article Number 80 of the General Assembly of the State of Louisiana of the year of 1888. Also, Article Number 80 was approved on July 12, year of 1888. Also, Article Number 80 was approved on July 12, 1888. On March 6, 1888, the property comprising the Orange Grove sugar plantation was advertised for sale in the New Iberia Enterprise, which was a newspaper published in the town of New Iberia, Louisiana. After having complied with all the legal formalities required by the laws of the State of Louisiana, and specified in the act, the Sheriff of Iberia Parish did seize the Orange Grove sugar plantation, and cause the Orange Grove sugar plantation to be sold at public auction. The public auction of the property of the Orange Grove sugar plantation was held on April 15, 1889 at the courthouse doors of the Iberia Parish Courthouse, which was situated on Main Street in the town of New Iberia, Louisiana.

At this time, in the year of 1889, the Orange Grove sugar plantation consisted to two hundred acres or one hundred and seventy arpents of land. The Orange Grove sugar plantation was bounded on the north side by the land of Decuir, on the south side by the land of Olivier, and on the east side by the land of Frilot. This was the same property that was sold to the State of Louisiana on the date of June 12, 1885 for unpaid taxes assessed against the land for the year of 1884, and the owner of the property was Mrs. Eugene Barthlemy Olivier, who was the mother of Robert A. Olivier.

Robert A. Olivier, of the Parish of Iberia, was the last and the highest bidder for the property of the Orange Grove sugar plantation. The property of the Orange Grove plantation was sold to Robert A. Olivier for the sum of the original price of $16.67 with a 20% charge added. Also, there was a charge of $16.60 for unpaid taxes added to the purchase price. Also, Robert A. Olivier paid the sum of $28.12, which amount was for the unpaid State of Louisiana and Iberia Parish taxes assessed against the Orange Grove sugar plantation. The costs of the sale were also added and paid by Robert A. Olivier. The aggregate sum of $48.00 bid was paid by Robert A. Olivier for the property of the Orange Grove sugar plantation.

On May 26, 1885, the property of the Orange Grove sugar plantation was sold to the State of Louisiana for unpaid taxes assessed against the property for the year of 1884. These unpaid taxes were accrued in the name of H. Segura, who was serving as overseer of the Orange Grove sugar plantation.

On October 8, 1889, Joseph A. Breaux, a notary public in the town of New Iberia, acted as agent for the Mutual National Bank of New Orleans in a conveyance and mortgage with vendee, John Newton Pharr, of the Parish of St. Mary. The Mutual National Bank of New Orleans conveyed all their rights, titles, and interest in the Orange Grove plantation property to Captain John Newton Pharr.

This sale was made for the consideration and the purchase price of $9,500.00. Therefore, the amount of $4,750.00 or half of the total purchase price was payable on the date of January 2, installments. The amount of $9,500.00 was to bear interest at the rate of 8% from the date. This purchase price represented three promissory notes that were made by the former owners of the Orange Grove plantation, Eugene Abraham, Robert Alfred Olivier, and Warren R. Farmer. Also, these three promissory notes were drawn and subscribed in soldio to their own order, and endorsed in the blank. The first promissory note was for the amount of $4,700.00, and it was payable on the date of January 1, 1889. The other two promissory notes were for the amount of $2,075.00, and these two promissory notes were due in two years after the date.

It was understood by the agent for the Mutual National Bank of New Orleans, Joseph A. Breaux, that the Mutual National Bank of New Orleans transfers only title to the Orange Grove plantation as the title was acquired by the Mutual National Bank of New Orleans at a sheriff sale on the date of June 2, 1888. Also, no other warranty to the Orange Grove sugar plantation was given by the Mutual National Bank of New Orleans.

Before the signing of this act of mortgage and conveyance, it was well understood that the purchaser, Captain John Newton Pharr, of the Parish of St. Mary, assume all responsibility in the costs, and liability in the injunction suit, entitled, "John Newton Pharr versus the Mutual National Bank of New Orleans." This civil lawsuit was relative to the sale of certain buildings, such as the sugar house, and the machinery, which was situated on the Orange Grove sugar plantation.

Also, the purchaser, Captain John Newton Pharr, of Fairhope Plantation, which was situated in the Parish of St. Mary, binded and obligated himself to accept delivery of the Orange Grove plantation from the Mutual National Bank of New Orleans. Captain John Newton Pharr held the Mutual National Bank free and harmless in the premises with the bank's intention to sell the rights, titles, and interest to the Orange Grove plantation, as acquired at a sheriff sale on the date of June 2, 1888.

The delivery of the Orange Grove sugar plantation to Captain John Newton Pharr is to be before the expiration of the current year of 1888. Also, the time of the delivery of the Orange Grove plantation shall not affect the promissory note, which was made by Robert Alfred Olivier and Eugene Abraham to the Mutual National Bank of New Orleans for rent of the Orange Grove sugar plantation during the current year.

On the date of May 22, 1888, John T. Hardie, who was serving as President of the Mutual National Bank of New Orleans, took cognizance of the fact by virtue of an act of mortgage, which was passed by P. Laurent Renoudet, a notary public in the town of New Iberia, whereby Joseph A. Breaux, acted on behalf of the Mutual National Bank as their agent, sold and conveyed the plantation, known as Orange Grove plantation without warranty to the rights, title, and interest of the Orange Grove plantation, which was situated in the Parish of Iberia. These were the rights, titles, and interest which the former owners of the Orange Grove plantation, Eugene Abraham, Robert Alfred Olivier, and Warren R. Farmer had in and to the Orange Grove plantation. This recognition of cognizance of the fact of the act of mortgage was passed in the City of New Orleans, Louisiana at the office of Faher, and in the presence of these two witnesses, John C. Schoen, and Fred Schoen. John C. Schoen and Fred Schoen were the two witnesses to this cognizance by the President of the Mutual National Bank.

On the date of October 8, 1888, the Mutual National Bank of the United States of America was the vendor with Joseph A. Breaux, who was acting as agent, conveyed the property comprising the Orange Grove sugar plantation to Warren R. Farmer, Robert Alfred Olivier, both residents of Iberia Parish, Louisiana, and Eugene Abraham, of the City of New Orleans, Louisiana. Also, sold besides all the property was all the rights, titles, and interest of the Orange Grove sugar plantation.

At this time, in the year of 1888, the Orange Grove sugar plantation consisted on three separate tracts of land. Tract one measured nine arpents fronting Bayou Teche with a depth of forty arpents. The lines of tract one closed in the rear so as to make the width of the tract five or six arpents. Tract one was bounded below or on the lowerside by the lands of the heirs of Baptiste Collette, and above by the lands of tract two of the Orange Grove sugar plantation, and in the rear by lands formerly belonging to the succession of Charles Napoleon Olivier. Tract two of the Orange Grove plantation was situated between Bayou Teche and the public road, known as the Old Spanish Trail. This public road leads from the town of New Iberia to the town of Jeanerette. Tract two measured six arpents fronting Bayou Teche with an irregular depth of five arpents. Tract two was bounded above by land formerly belonging to the succession of Charles Napoleon Olivier, and below by land of tract one of the Orange Grove plantation, and in the rear by the public road, or the Old Spanish Trail. The third tract of land, which comprised the Orange Grove plantation was situated on the public road. Tract three had a frontage along Bayou Teche of six arpents with a depth of thirty six arpents. Tract three was bounded above by the lands belonging to the succession of Charles Napoleon Olivier, and below or on the lowerside by the lands of tract two of the Orange Grove plantation.

This sale was made and accepted for the price and consideration of $9,500.00. The purchase price was payable at one half or the amount of $4,750.00 on January 1, 1889. The balance of the purchase price or the amount of $4,750.00 was payable in two equal annual installments. The balance of the purchase price or the amount of $4,750.00 was conditioned to bear interest at the rate of 8% from date. The purchasers, Warren R. Farmer, Robert Alfred Olivier, and Eugene Abraham furnished their three promissory notes, which promissory notes were drawn and subscribed by them to their own order, and endorsed by them in the blank. The first promissory note was for the amount of $4,750.00, and it was payable on January 1, 1889. The other two promissory notes were each for the amount of $2,375.00, and these two promissory notes were due in one and two years from the date of October 8, 1889, and October 8, 1890.

It was well understood that Joseph A. Breaux, who was acting as agent for the Mutual National Bank, only transferred the title, which title was acquired by the bank at the sheriff sale, which was held on the date of January 3, 1888. No other warranty to the property of the Orange Grove plantation was transferred.

Also, before signing the act of sale, it was well understood and stipulated that the purchasers, Robert Alfred Olivier, Warren R. Farmer, and Eugene Abraham, assumed all responsibility, costs, and liability in the injunction suit instituted by Captain John Newton Pharr against the Mutual National Bank. This suit was relative to the use of certain buildings and machinery, which were situated on the Orange Grove plantation. The purchasers, Robert Alfred Olivier, Warren R. Farmer, and Eugene Abraham agreed to bind and obligate themselves to take the place of the Mutual National Bank in this suit, and to hold the Mutual National Bank free and harmless in the premises. Before the signing of the act of conveyance, it was agreed that the delivery of the Orange Grove plantation before the expiation of the current year or 1888 shall not in any way affect the note given by Robert Alfred Olivier and Eugene Abraham to the Mutual National Bank for rent of the Orange Grove plantation during the coming year.

It was well understood also that the vendees, Robert Alfred Olivier, Eugene Abraham, and Warren R. Farmer shall sell their rights, title, and interest in the improvements claimed by third persons on the property comprising the Orange Grove plantation. If the purchasers, Robert Alfred Olivier, Eugene Abraham, and Warren R. Farmer, successfully defend themselves against these claims they shall be the owners, but otherwise they shall not be the owners.

John T. Hardie, who was serving as President of the Mutual National Bank of New Orleans and acting in his capacity as the President of the Bank, declared that he has this day taken cognizance of this notarial act, which notarial act was passed before P. Laurent Renoudet, a notary public in the town of New Iberia, Louisiana. Joseph A. Breaux, who was acting as agent for the Mutual National Bank, sold and conveyed without warranty to Eugene Abraham, Robert Alfred Olivier, and Warren R. Farmer, all the rights, titles, and interest of the Mutual National Bank in and to the property comprising the Orange Grove plantation. Also, John Hardie, the President of the Mutual National Bank, confirmed the actions of the agent of the bank, Joseph A. Breaux, in the premises of the Orange Grove plantation,a nd grants the same validity in law of the Orange Grove plantation,a nd grants the same validity in law and in fact as if he himself (John T. Hardie) had been personally present and acting in his capacity as President of the Mutual Bank of New Orleans, Louisiana.

On the date of June 24, 1889, at the residence home of Therese Laura Dalcour Olivier, who was the legal and lawful wife of Eugene B. Olivier, and which residence home was situated on the Orange Grove plantation, the execution of an act of right of way took place. This right of way agreement was between the parties with Mrs. Eugene Barthlemy Olivier, as the vendor, and Morgan's Louisiana & Texas Railroad & Steamship Company as the vendee.

The railroad tracks of the Morgan's Louisiana & Texas Railroad passes through the property of the Orange Grove plantation, which property belongs to Mrs. Eugene Barthlemy Olivier. The right of way for the railroad tracks of the Morgan's Louisiana & Texas Railroad was taken for granted, but this railroad track greatly enhances the adjoining property of Eugene Barthlemy Olivier.

Morgan's Louisiana & Texas Railroad Company plans to erect a section house and a flag station on the property of the Orange Grove plantation. Therefore, Mrs. Therese Laura Dalcour Olivier, who was the wife of Eugene Barthlemy Olivier, proposed to Morgan's Louisiana & Texas Railroad in consideration of such a location to grant and transfer one acre of the lands of Mrs. Eugene Barthlemy Olivier to the right of way, and the public road, which public road leads from Vantress Store across the Grand Marais.

The managers of Morgan's Louisiana & Texas Railroad Company have erected on the lands of Therese Laura Dalcour Olivier, who was the wife of Eugene Barthlemy Olivier, a section house building and in front a flag station. In consideration of $1.00 one dollar, which was paid in cash at the execution of this act of right of way. Therese Laura Dalcour Olivier, the wife of Eugene Barthlemy Olivier, has acknowledged the receipt of $1.00 and has granted acquittal.

Now, Therese Laura Dalcour Olivier sells and delivers with full warranty of title, and full subrogation of any rights in full to the property to Morgan's Louisiana & Texas Railroad. This right of way is situated on the Orange Grove plantation.

The piece of property that the right of way was situated on consisted of one acre of land, which had a front of four hundred and eighteen feet on the north side of the right of way with the depth of one hundred and four feet and six inches. This one acre tract of land was bounded on the north and the east sides by the lands of Therese Laura Dalcour Olivier, on the south side by the right of way of Morgan's Louisiana & Texas Railroad Company, and on the west side by the public road, which public road leads from of the Orange Grove plantation, which was the property of Therese Laura Dalcour Olivier, who was the wife of Eugene Barthlemy Olivier, and the Orange Grove plantation was situated on the west side of Bayou Teche in the newly created Parish of Iberia.

At this time, in the year of 1888, Alphonse Tertrou was the holder of a mortgage from Therese Laura Dalcour Olivier, the wife of Eugene B. Olivier. This mortgage bore on the property, which comprised the Orange Grove plantation. This transfer of the one acre of land to Morgan's Louisiana & Texas Railroad was authorized to be cancelled from this mortgage by Alphonse Tertrou. Also, Alphonse Tertrou authorized the clerk of court of the Parish of Iberia to erase and cancel this one acre tract of land from this mortgage. This mortgage was still in force on the other tracts of land, which comprised the Orange Grove plantation.

On April 16, 1888, Deputy Sheriff Charles A. Judice, of the Parish of Iberia, conveyed to the Mutual National Bank of the City of New Orleans, the property comprising the Orange Grove plantation, which was situated in the Parish of Iberia. The Orange Grove plantation was sold at a sheriff sale.

The Orange Grove plantation comprised these three tracts of land, which were situated in the Parish of Iberia, in the State of Louisiana. Tract one of the Orange Grove plantation measured nine arpents fronting along Bayou Teche, with a depth of forty arpents. In the rear of tract one, the depth of the lines closing in made the width of the tract of land five or six arpents. Tract one was bounded below by the lands belonging to the heirs of Baptiste Collette, and above by the lands of tract two of the Orange Grove plantation, and in the rear by land formerly belonging to the succession of Charles Napoleon Olivier.

Tract two of the Orange Grove plantation was situated between Bayou Teche and the public road, or the Old Spanish Trail. This public road leads from the town of New Iberia to the town of Jeanerette. Tract two measured six arpents fronting Bayou Teche with an irregular depth of five arpents. Tract two was bounded above by land formerly belonging to the succession of Charles Napoleon Olivier, and below by lands of tract one of the Orange Grove plantation, and in the rear by the public road, or the Old Spanish Trail. Also, situated on tract two was the plantation home of the Orange Grove plantation. The Orange Grove plantation home was the residence home of Mr. and Mrs. Eugene Barthlemy Olivier and their family.

Tract three of the Orange Grove plantation was situated on the public road. Tract three had a frontage of six arpents on the public road, or the Old Spanish Trail, with a depth of thirty six arpents. Tract three was bounded above by the lands belonging to the succession of Charles Napoleon Olivier, and below by tract two of the Orange Grove plantation.

of the Orange Grove plantation.

On the date of April 11, 1885, the advertisement of the upcoming sheriff sale of the Orange Grove plantation appeared in the New Iberia Enterprise, which was a newspaper that was published in the town of New Iberia, Louisiana. This advertisement stated that the Orange Grove plantation, which consisted of 1,100 acres of land and improvements. In the year of 1885, the Orange Grove plantation was bounded on the north side by the lands of Decuir, on the south side by the lands of Olivier, and on the east side by the lands of Frilot. The sheriff sale of the Orange Grove plantation was to take place on the date of June 12, 1885 at the courthouse doors of the Iberia Parish Courthouse, which was situated on Main Street in the town of New Iberia, Louisiana.

This advertisement of the public auction of the Orange Grove plantation could only occur after all the legal formalities that were required by the laws of the State of Louisiana were met. These legal formalities were specified in Article 96 of the General Assembly of the State of Louisiana, and these legal formalities were approved on July 5, 1882. This was an act, which was to provide an annual revenue for the State of Louisiana by levying an annual tax upon all property situated in the State of Louisiana. The property that was taxed was property that was not exempted by the Constitution of the State of Louisiana from taxation.

The reason that the Orange Grove plantation was seized by the sheriff of the Parish of Iberia was for the unpaid taxes levied against the Orange Grove plantation in the year of 1884. Mrs. Eugene B. Olivier was the owner of the Orange Grove plantation, but she was unable to pay the taxes for the year of 1884. According to the assessment rolls of the Parish of Iberia, the taxes levied against the Orange Grove plantation were $12.10 for Louisiana State taxes, $4.90 for Iberia Parish taxes, interest was thirty five cents, and the costs was $9.30, which makes a total amount of taxes and delinquent payments being the sum of $16.65.

On the date of June 2, 1885, George Robertson, who was serving as Deputy Sheriff and Ex-Officio Tax Recorder of the Parish of Iberia, sells and conveys the property and the improvements of the Orange Grove plantation to the State of Louisiana. The reason that the State of Louisiana was the purchaser was because the State of Louisiana was the last and the highest bidder for the Orange Grove plantation. Also, the State of Louisiana acquired through this conveyance all the rights, titles, and interests that Mrs. Eugene Barthlemy Olivier had to the Orange Grove plantation.

The reason that Mrs. Eugene Barthlemy Olivier lost her sugar plantation, Orange Grove, was because she was unable to pay the $16.65 taxes, which were assessed on the property of the Orange Grove plantation for the year of 1884. But, it was understood and stipulated that in conformity with the laws of the State of time within one year after the date of this conveyance, or the date of June 2, 1886. To redeem the property of the Orange Grove plantation, Mrs. Eugene Barthlemy Olivier would have to pay the $16.65 in unpaid taxes, which taxes were assessed against the Orange Grove plantation in cash, and a 20% charge with costs added.

On the fifteenth day of the month of March of the year of 1882, Eugene Barthlemy Olivier, a planter of the Parish of Iberia, acted by virtue of a general power of attorney for his wife, Therese Laura Dalcour Olivier, in executing a lease agreement with John Newton Pharr. This act of general power of attorney is recorded in the miscellaneous records of the Parish of Iberia. Also, Prosper D. Olivier, a farmer and a resident of the Parish of Iberia, was also a partner in the leasing of this tract of land to John Newton Pharr with Eugene Barthlemy Olivier.

Eugene B. Olivier agreed to lease to John Newton Pharr, a tract of land, which tract of land comprised a portion of the farm situated on the south side in the back of Grand Marais. Also, this tract of land fronted the Grand Marais or the Great Swamp. This tract of land was bounded on both the east and the west sides by lands belonging to Eugene Barthlemy Olivier, the lessor, on the south side by the public road, and on the north side by the Grand Marais or the Great Swamp. This tract of land measured six arpents.

This lease agreement was made on the conditions that John Newton Pharr, the lessee, will build a sugar mill on this leased tract of land. The sugar mill will be built to accommodate a large capacity, which will be required to grind and process the sugar cane from the Orange Grove plantation, and the sugar cane form the neighbors of the Orange Grove sugar plantation. Also, this lease is made free of any costs to the lessors, Eugene B. Olivier and Prosper D. Olivier. This lease is to remain in full force for as long as the sugar mill shall remain on the tract of land. It was understood that all works, buildings, and materials erected on the tract of land are to remain the property of the lessors, Eugene B. Olivier and Prosper D. Olivier, and subject at any time to the exclusive control of Eugene B. Olivier and Prosper D. Olivier.

On the date of February 10, 1882, Mary Radar Bussy, who was the widow of the late Doctor Harvey Bussy, of the Parish of Iberia, conveyed the undivided half share and interest of the partnership of Bussy, Pharr, and Olivier to Captain John Newton Pharr, a resident of the Parish of St. Mary. In regards to this act of conveyance, Mary Radar Bussy acted in her own name and in the name of her minor children, Ellen Bussy, Frank Bussy, Mary Bussy, Jesse Bussy, and Maude Bussy.

A family meeting of the Bussy family was also held on the date of February 10, 1882, to discuss the selling of the share and

A family meeting of the Bussy family was also held on the date of February 10, 1882, to discuss the selling of the share and interest of the partnership. This partnership was formed between Captain John Newton Pharr, of the Parish of St. Mary, Mistress Therese Laura Dalcour Olivier, who was the wife of Eugene B. Olivier, and Doctor Harvey Bussy. At this family meeting, the major children of the late Doctor Harvey Bussy, Florence Bussy, who was the wife of Doctor C. A. McGowan, of the town of Jeanerette, Louisiana, John C. Bussy, also of the town of Jeanerette, Louisiana, and Edward Bussy were in favor of conveying their portion of the partnership interest and share to Captain John Newton Pharr.

The property that was sold at this act of conveyance was an undivided half share and interest in the Orange Grove sugar house and improvements, which are contained in the sugar house. The Orange Grove sugar house was situated on a tract of land that comprised the Orange Grove plantation, and this tract of land was bounded on the north side by the lands of Olivier, on the south side by land belonging to the heirs of Baptiste Collette, on the east side by Bayou Teche, an on the west side by the public road, which public road is known as the Old Spanish Trail.

The improvements that were contained in the Orange Grove sugar house, comprised the following articles, three boilers, four clarifiers, four piers tanks, one small pier tank, one chimney, one small pier, two sugar tanks, one doator, one rawshed, and other minor articles. Also, included in this act of conveyance was real property that was included in the Bussy share of the partnership.

This sale was made for the purchase price of $5,665.00. Captain John Newton Pharr, of the Parish of St. Mary, paid to Mary Radar Bussy the cash sum of the purchase price at the execution of this act of conveyance. Mary Radar Bussy granted acquittal for the debt and acknowledge the receipt of the purchase price being paid in full.

In regards to the purchase price, the interest of Mary Radar Bussy and her minor children were all divided evenly and distributed equally among all the heirs of the late Doctor Harvey Bussy. The share of Mary Radar Bussy to the purchase price of the sale of the one undivided half share and interest of the partnership of Olivier and Company was the amount of $2,832.50, and the share of the minor children of the late Doctor Harvey Bussy was the amount of $1,770.31. The major children of the late Doctor Harvey Bussy, Edward and John C. Bussy, and their sister, Florence Bussy McGowan, each received the amount of $354.06 as their share of the partnership of Olivier and Company. After all the distributions among the heirs of the late Doctor Harvey Bussy had been completed, the total amount came to the sum of $5,665.00, which was the purchase price.

Also, it was agreed between the parties, Mary Radar Bussy and Captain John Newton Pharr, that Mary Radar Bussy and her minor children, Ellen Bussy, Frank Bussy, Mary Bussy, Jesse Bussy, and Maude Bussy, together with the major children, John c. Bussy, Edward Bussy, and Florence Bussy, who was the wife of Doctor C. A. McGowan, withdraw from the partnership, known under the name of Olivier and Company. Also, from the date of February 10, 1882, the Bussy connection with the partnership of Olivier and Company is forever severed and the responsibility and liability ceases to exist. Captain John Newton Pharr assumes all the debts existing against the Orange Grove sugar house and the improvements, and the other property contained in the one undivided half share and interest of the late Doctor Harvey Bussy.

On the date of March 30, 1880, John Newton Pharr, a resident of the Parish of St. Mary, Doctor Harvey Bussy, a resident of the Parish of Iberia, and Mistress Therese Laura Dalcour Olivier, who was the wife of Eugene B. Olivier, all met at the office of Joseph Breaux, who was a lawyer and a notary public in the town of New Iberia, to declare that they intended to execute an act of partnership. Also, there were two witnesses, U. S. Haas and N. P. Dillion, who were present and would serve as witnesses to the act of partnership. The reason that this partnership was formed was for the purchasing of sugar cane, molasses, and syrup.

The partnership was known as Olivier & Company, and it was domiciled at the Orange Grove sugar plantation, which was situated in the Parish of Iberia. One of the essential conditions of the partnership, Olivier & Company, was that Alphonse Tertrou, who was a sugar factor in the City of New Orleans, Louisiana, was to extend the payment of all mortgage notes bearing on the Orange Grove sugar plantation. These mortgage notes were extended until the expiration of the partnership of Olivier & Company. Previously to the forming of the partnership, Olivier & Company, Therese Laura Dalcour Olivier mortgaged the Orange Grove sugar plantation in favor of her sugar factor, Alphonse Tertrou, of the City of New Orleans, Louisiana. Actually, the partnership of Olivier & Company had a tenure of six years from the date of March 30, 1880, or the expiration date of March 30, 1886.

There were several mortgage notes that were bearing on the Orange Grove plantation, and these notes were executed between Therese Laura Dalcour Olivier, who was the wife of Eugene B. Olivier, and her sugar factor, Alphonse Tertrou, of the City of New Orleans, Louisiana. The act of mortgage bore the date of April 17, 175, and there existed a series of mortgage notes that were signed by Therese Laura Dalcour Olivier, who was the wife of Eugene B. Olivier. The first mortgage note was dated February 14, 1875, and it was for the amount of $1,889.27, which note was payable on November 12, 1875. The second mortgage note was for the amount of $1,896.66, and this second note was payable on the date of November 18, 1875. The third mortgage note was for the amount of $1,894.67, on December 1, 1875. Also, the fifth mortgage note was for the amount of $1,900.31, and it was payable on the date of December 3, 1875. The sixth mortgage note was for the amount of $1,905.58, and it was payable on December 12, 1875. The seventh mortgage note was for the amount of $1,909.03, and it was due on December 18, 1875. The eighth mortgage note was for the amount of $1,913.09, and it was due on December 25, 1875 or Christmas Day. The ninth mortgage note was for the amount of $1,921.32, and it was due on January 8, 1876. The tenth mortgage note was for the amount of $1,925.50, and it was due on January 15, 1876. Also, the eleventh mortgage note was for the amount of $1,931.47, and it was due on January 25, 1876. And, the twelfth and the final mortgage note was for the amount of $1,935.63, and it was due on the date of February 1, 1876. All twelve mortgage notes were conditioned to bear interest at the rate of 8%.

Also, the partnership of Pharr & Bussy was to advance the monies necessary to put in repair the sugar house, and the machinery in the sugar house, which were situated on the Orange Grove plantation. The repairs that were to be made to the Orange Grove sugar house, and the machinery, which was situated inside the sugar house, was not to exceed the cost of $10,000.00, and no interest was to be charged for the monies advanced.

The partnership of Pharr & Bussy was composed of Doctor Harvey Bussy of Loisel Plantation, which is situated in Iberia Parish, and Captain John Newton Pharr of Fairhope Plantation, which was situated in St. Mary Parish. Also, the partnership of Pharr & Bussy existed at the Loisel plantation. Captain John Newton Pharr and Doctor Harvey Bussy became interested in pursing and advancing their business interests, which enabled them to become interested in purchasing a portion of the Orange Grove sugar plantation. The reason that Doctor Harvey Bussy and Captain John Newton Pharr were able to purchase a portion of the Orange Grove sugar plantation was because the owner of the Orange Grove sugar plantation, Therese Laura Dalcour Olivier, who was the wife of Eugene B. Olivier, had fell on hard times, and was heavily indebted to her sugar factor, Alphonse Tertrou, of the City of New Orleans. Laura Therese Olivier feared that she would be unable to pay the mortgage notes that were bearing on the Orange Grove plantation, so she was interested in pursuing business partners, who would enable the Orange Grove plantation to grow more sugar cane and produce more sugar, syrup, and molasses. Hopefully, these larger profits would help pay off the mortgage.

It was agreed between Doctor Harvey Bussy, and Captain John Newton Pharr, and Therese Laura Dalcour Olivier, that she (Mrs. Olivier) would refund to the partnership of Pharr & Bussy in equal annual payments all the monies advanced by the partnership of Pharr & Bussy for the repairs to be made to the Orange Grove sugar house, and the machinery, which was situated inside the sugar house. These payments were to come from the share of profits of house, and the machinery, which was situated inside the sugar house. These payments were to come from the share of profits of the Orange Grove sugar plantation that were to be received by Therese Laura Dalcour Olivier as her share of the partnership of Olivier & Company. The repairs and the improvements that were to be made in the Orange Grove sugar house, and to the machinery inside the sugar house were to be the property of Pharr & Bussy until the repairs and improvements were paid in full by Therese Laura Dalcour Olivier.

Also, it was agreed that Eugene Barthlemy Olivier, the husband of Therese Laura Dalcour Olivier, was to be in charge of the business of Olivier & Company. Eugene Olivier was to provide the services as manager, and he was to receive a salary at the end of each grinding season or sugar making season of $250.00.

All the syrup and the molasses that was produced by the partnership of Olivier & Company was delivered by Olivier & Company to the Loisel sugar house, which was situated on the Loisel plantation. The partnership of Pharr & Bussy furnished all the boats and tanks that were necessary to transport the syrup and molasses to the Loisel sugar house. Also, the partnership of Pharr & Bussy was to make raw brown sugar from the syrup that was transported from the Orange Grove sugar house to the Loisel sugar house. This syrup was to be made into sugar in the vacuum pair at the Loisel sugar house. Also, the partnership of Pharr & Bussy was to receive one and one fourth cents per pound of sugar that was produced in the Loisel sugar house. All the sugar, syrup, and molasses were to be sold by the partnership of Pharr & Bussy.

At the end of each year or the end of each grinding season, the profits shall be divided between the three partners in this manner, Therese Laura Dalcour Olivier shall receive one half of the profits, and Doctor Harvey Bussy and Captain John Newton Pharr received the other half of the profits. This was the net profits that were paid to the three partners, and this was determined after all the expenses, taxes, and costs had been deducted. Also, all the repairs and improvements made to the Orange Grove sugar house, and the machinery situated in the sugar house was not subject to any claim by the partnership of Pharr & Bussy until the partnership of Pharr & Bussy are paid all advances of monies for the repairs and improvements.

On the date of June 24, 1880, Therese Laura Dalcour Olivier, the wife of Eugene B. Olivier, granted a right of way agreement to the Morgan's Texas & Louisiana Railroad & Steamship Company. Eugene B. Olivier had authorized his wife, Therese Laura Dalcour Olivier, to sign the right of way agreement.

The railroad track of the Morgan's Louisiana & Texas Railroad & Steamship Company passes this angle of the property of the Orange Grove plantation, and this right of way had been taken a railroad house and a flag station. Therese Laura Dalcour Olivier, the wife of Eugene B. Olivier, had proposed to the railroad company of the Morgan's Louisiana & Texas Railroad & Steamship Company that in consideration of such a location to grant and transfer one acre of land to the right of way of the railroad and the public road leading from Vantrues store across the Grand Marais or the Great Swamp. The managers of the railroad company of Morgan's Louisiana & Texas Railroad & Steamship Company under the proposal having vested in the lands of Mrs. Eugene Barthlemy Olivier or the Orange Grove plantation a section house building and put in a flag station. The considerable for the right of way agreement was the sum of one dollar or $1.00, which was paid in cash at the execution of this right of way agreement.

Actually, the purchase of Captain John Newton Pharr into the partnership of Olivier & Company, was the start of the interest of the Pharr family in the Orange Grove plantation. It was the sons of Captain John Newton Pharr, that formed and operated the sugar enterprise, J. N. Pharr & Sons, Ltd., that was situated at the Orange Grove sugar plantation until the date of July 11, 1928.

This right of way agreement in favor of Morgan's Louisiana & Texas Railroad & Steamship Company was situated on the Orange Grove plantation, which was situated on the west side of Bayou Teche, in the newly created Parish of Iberia. At this time, the Orange Grove plantation was owned by Mr. and Mrs. Eugene Barthlemy Olivier, and the Olivier family also resided on the plantation. The Orange Grove plantation was bounded above by the public road, which was known as the Old Spanish Trail, and below by the lands of Sidoninis Collette or formerly the land belonging to the heirs of Baptiste Collette. Also, Eugene Olivier owned a sugar plantation or a sugar estate on the Island of Cuba, which was situated in the Dominica of Spain.

Before this right of way agreement in favor of Morgan's Louisiana & Texas Railroad & Steamship Company could be formally legalized and finalized, F. Alphonse Tertrou, who was the holder of a mortgage, had to be notified. On the date of July 7, 1880, F. Alphonse Tertrou was notified of the right of way agreement, which consisted of one acre of land of the Orange Grove plantation. The reason that F. Alphonse Tertrou had to be notified was because he was the holder of a mortgage, which mortgage was bearing on the Orange Grove plantation. Alphonse Tertrou agreed that it was all right to transfer one acre of land of the Orange Grove plantation to Morgan's Louisiana & Texas Railroad & Steamship Company, and also he authorized and empowered the Recorder of Mortgages in the office of the clerk of court of the Parish of Iberia to cancel and erase the mortgage, which mortgage was in favor of Alphonse Tertrou and against Therese Laura Dalcourt Olivier, the wife of Eugene B. Olivier, from the mortgage records of the Parish of Iberia. The mortgage that was bearing on the one acre tract of land that was transferred to Morgan's Louisiana & Texas Railroad & Steamship

Company was cancelled and annulled.

Mrs. Therese Laura Dalcour Olivier, the wife of Eugene B. Olivier, conveyed and sold with full warranty of title and full subrogation of rights in the property to Morgan's Louisiana & Texas Railroad & Steamship Company.

The tract of land that was conveyed to the Morgan's Louisiana & Texas Railroad & Steamship Company was a one acre tract of land, which had a front of four hundred and eighteen feet on the north side of the right of way of the railroad by the depth of one hundred and four feet and six inches. This tract of land was bounded on the north and the east sides by the lands of Therese Laura Dalcour Olivier, on the south side by the railroad right of way, and on the west side by the public road, which road leads from the Kraus store to the Grand Marais or the Great Swamp.

On the date of April 23,1875, Therese Laura Dalcour Olivier, who was the legal and lawful wife of Eugene Barthlemy Olivier, mortgaged her sugar plantation, known as the Orange Grove plantation to her sugar factor, F. Alphonse Tertrou, of the City of New Orleans, Louisiana. This act of mortgage was executed before Joseph A. Breaux, a notary public in the Parish of Iberia. The property of the Orange Grove sugar plantation was mortgaged to secure the payment of promissory notes identified with an act of mortgage, which act of mortgage was annulled and cancelled. This mortgage was executed in accordance with directions issued from F. Alphonse Tertrou, who was the sugar factor of Mrs. Olivier, and was responsible for the selling and marketing of the raw brown sugar, the molasses, and the syrup, which was manufactured on the Orange Grove sugar plantation. Also, P. Laurent Renoudet, who was a notary public and a lawyer in the town of New Iberia, acted as the agent for Alphonse Tertrou in this matter.

At this time, Therese Laura Dalcour Olivier, the wife of Eugene Barthlemy Olivier, mortgaged the property, which comprised the Orange Grove plantation, and the improvements on the plantation. Also, mortgaged was the sugar house, and the machinery that was situated in the Orange Grove sugar house. The machinery consisted of boilers, sugar kettles, and the sugar mill engine. Also, the growing crop of sugar cane that was planted and being cultivated on the Orange Grove plantation was mortgaged. Besides the property, the sugar house and the machinery, and the crop of sugar cane, all the farming implements, and the cattle and livestock that were on the Orange Grove plantation were also mortgaged. The land of the Orange Grove plantation was to remain mortgaged until the final payment of the promissory notes and the eventual release of the mortgage.

The mortgage that was executed in the year of 1874 between Therese Laura Dalcour Olivier, the wife of Eugene B. Olivier, and Alphonse Tertrou, of the City of New Orleans, was payable in the months of November and December of the year of 1874, and in the month of January of the year of 1875. The amount of $1,891.60 was payable on the date of November 12, 1874, and the amount of $118.20 was payable on the date of November 23, 1874. The amount of $1,800.00 was payable on the date of November 28, 1874. The amount of $1,900.58 was payable on the dates of December 12, 18, and 25 of the year of 1874. The amount of $1,921.21 was payable on the date of January 15, 1875. Also, the amount of $1,931.00 was payable on the date of January 25, 1875, and the amount of $1,936.63 was due on the date of February 1, 1876. All mortgage payments were to be made at the office of Alphonse Tertrou in the City of New Orleans, Louisiana. All of these promissory notes were conditioned to bear 8% interest until the full and the final payment of the notes. This was the mortgage that was executed on the date of March 21, 1874 between Therese Laura Dalcourt Olivier, the wife of Eugene B. Olivier, and her sugar factor, Alphonse Tertrou, of the City of New Orleans, Louisiana, to secure the payment of $18,606.25.

On the date of May 21, 1875, Therese Laura Dalcour Olivier, the wife of Eugene B. Olivier, mortgaged her plantation, known as Orange Grove, which was situated in the newly created Parish of Iberia, to secure the payment of $22,914.00. The sum of $22,914.00 was borrowed solely for the benefit of the separate advantage and benefit of the separate property of Therese Laura Dalcourt Olivier.

Therese Laura Dalcour Olivier, the wife of Eugene B. Olivier, appeared in the presence of Judge Henry Train, who was serving as Judge of the Third Judicial District court, which was holding court in the Parish of Iberia. After Judge Henry Train had examined Therese Laura Dalcourt Olivier, in the judge's chambers at the Iberia Parish Courthouse, he (Judge Train) made the execution that this mortgage being executed was separate and apart from the property of her husband, Eugene Barthlemy Olivier. Mrs. Olivier desired to borrow money for her separate use. Judge Henry Train authorized Therese Laura Dalcourt Olivier to borrow the sum of $22,914.00 for her separate use, and incur all liability of this mortgage.

On the date of March 11, 1873, Therese Laura Dalcour Olivier, who was the wife of Eugene Barthlemy Olivier, executed a mortgage in favor of her sugar factor, Alphonse Tertrou, of the City of New Orleans, Louisiana. The amount of this mortgage was $16,000.0, and the borrowed money was subject to bear interest at the rate of 8%. Therese Laura Dalcour Olivier had mortgaged the property, which comprised the Orange Grove plantation as collateral.

The payments of the promissory notes for this mortgage were, the first promissory notes was for the amount of $1,500.00, and it was payable on the date of November 5, 1873, the second promissory note was for the amount of $1,500.00, and it was due on the date of November 18, 1873, the third promissory note was for the amount of $1,500.00, and it was due on the date of December 5, 1873, the fourth promissory note was for the amount of $1,500.00, and it was payable on the date of December 15, 1873, the fifth promissory note was for the amount of $1,500.00, and it was due on the date of December 20, 1873, the sixth promissory note was for the amount of $1,500.00, and it was payable on the date of January 5, 1874, and the seventh promissory note was for the amount of $1,500.00, and it was due on the date of February 20, 1874, and the eighth promissory note was for the amount of $1,500.00, and it was due on the date of February 25, 1874.

On the date of April 20, 1872, Eugene Barthlemy Olivier acted as the agent for his wife, Therese Laura Dalcour Olivier, at the execution of an act of mortgage. This act of mortgage was executed in the presence of Joseph A. Breaux, who was a notary public in the town of New Iberia, Louisiana. The reason that Eugene Barthlemy Olivier was able to act as the agent for his wife, Therese Laura Dalcour Olivier, was by virtue of an act of power of attorney.

The act of power of attorney was executed before Patrick Burke, who was a notary public in the town of New Iberia, Louisiana. In granting this power of attorney to her husband, Eugene Barthlemy Olivier, Therese Laura Dalcour Olivier declared that she had constituted and appointed in her place, and stead her husband, Eugene Barthlemy Olivier, of the Parish of Iberia, as her true and lawful attorney for her and in her name, and for her use, to ask, demand, due for, recover, and renew all such debts, goods, wares, deeds, accounts, and other demands. Also, Eugene Barthlemy Olivier, the husband of Therese Laura Dalcour Olivier, was responsible for all the debts owed by his wife, Therese Laura Dalcour Olivier, as well as pay all claims and discharge all sums of money owed to Therese Laura Dalcour Olivier. By this act of power of attorney, Eugene Barthlemy Olivier could sell, mortgage, and purchase property and other goods in the name of his wife, Therese Laura Dalcour Olivier.

An act of general power of attorney gives another person the right to act as attorney in fact for another person in matters, such as buying, selling, transfer, comprises, and mortgages of all real property and movable property, and also be sued for the other person and in the name of the other person.

Therese Laura Dalcour Olivier, the wife of Eugene Barthlemy Olivier, was indebted to Alphonse Tertrou, her sugar factor, of the City of New Orleans, Louisiana, for the sum of $11,000.00. To pay this loan back, Mistress Olivier contracted seven promissory notes, which promissory notes were made to her own order and endorsed by Mrs. Therese Laura Dalcour Olivier.

The first promissory note was for the amount of $1,500.00, and it was payable on the date of November 15, 1872, the second promissory note was for the amount of $1,500.00, and it was due on the date of December 5, 1872, the third promissory note was for the amount of $2,000.00 and it was due on the date of December 25, 1872, the fourth promissory note was for the amount of $1,500.00 and it was due on the date of January 15, 1873, the fifth promissory note was for the amount of $1,500.00, and it was due on the date of January 25, 1873, the sixth promissory note was for the amount of $1,500.00 and it was payable on the date of February 12, 1873, and the seventh promissory note was for the amount of $1,500.00 and it was payable on the date of February 25, 1873. All seven promissory notes were conditioned to bear interest until the maturity of the promissory notes.

All seven promissory notes were in the favor of Alphonse Tertrou, the sugar factor of the Orange Grove plantation. Alphonse Tertrou owned the sugar factor firm of A. Tertrou & Company, which was situated in the City of New Orleans. P. Laurent Renoudet, a notary public in the town of New Iberia, Louisiana, acted as the agent for Alphonse Tertrou and the firm of A. Tertrou & Company, of the City of New Orleans, Louisiana. Actually, the holder of the seven promissory notes was the firm of Alphonse Tertrou & Company of the City of New Orleans. This mortgage bore on tract one, tract two, and tract three of the Orange Grove sugar plantation, which was situated in the newly created Parish of Iberia.

The amount of $7,000.00 of the amount mortgaged is actually due at this time, but $400.00 are for advances to be made during the present year for Therese Laura Dalcour Olivier to cultivate and harvest her present crop of sugar cane on her plantation, which is known as Orange Grove. The proceeds from the harvesting of the crop of sugar cane is to be made to pay off the mortgage payments at the firm of A. Tertrou & Company in the City of New Orleans, Louisiana. There is a two and one half per cent charge due the firm of Alphonse Tertrou & Company for the handling of the selling and marketing of the sugar cane crop, and its products from the Orange Grove plantation. This two and one half percent charge will be deducted from the total proceeds of the harvesting of the sugar cane crop.

On the date of April 8, 1879, an act of mortgage was executed between Mistress Therese Laura Dalcour Olivier, who was the wife of Eugene Barthlemy Olivier, and Alphonse Tertrou, the sugar factor, which mortgage was bearing on the Orange Grove sugar plantation. For the execution of this act of mortgage, Eugene Barthlemy Olivier acted as the agent and attorney in fact for his wife, Therese Laura Dalcourt Olivier. The reason that Eugene Barthlemy Olivier was able to act as the agent for his wife was because he was granted by virtue of an act of power of attorney, the right to represent Mrs. Olivier as her agent and attorney in fact.

In the act of power of attorney, it was stated that Eugene Barthlemy Olivier had the right to buy, sell, transfer, compromise, and mortgage, be sued for, and be sued by his lawful wife, Therese Laura Dalcourt Olivier. This power of attorney was given to Eugene Barthlemy Olivier by his wife, Therese Laura Dalcour Olivier.

Eugene B. Olivier declared and acknowledged that his principal and constituent, Therese Laura Dalcourt Olivier, is a resident of the newly created Parish of Iberia, and is truly indebted unto the commercial firm of A. Tertrou & Company, which commercial establishment operates as a sugar factor. The true sum of the indebtness is the amount of $2,000.00. For the reimbursement of this sum of $2,000.00, Therese Laura Dalcour Olivier has furnished her four promissory notes, which promissory notes are made to her own order and endorsed by Therese Laura Dalcour Olivier.

The first promissory note of the four promissory notes was for the amount of $581.87, and the first promissory note was payable on the date of January 1, 1871. The second promissory note is for the amount of $653.95, and this second promissory note was payable on the date of January 1, 1871. The third promissory note is for the amount of $400.00, and this promissory note was payable on the date of February 1, 1871. Lastly, the fourth and the final promissory note was for the amount of $360.88 and the fourth promissory note was payable on the date of February 15, 1871. The property that this mortgage was bearing on was the property comprising the Orange Grove plantation.

On the date of August 11, 1869, Therese Laura Dalcour Olivier conveyed unto Thomas M. Floyd a tract of land, which was situated on the east side of Bayou Teche. This act of conveyance was executed at the residence home of Mrs. Therese Laura Dalcour Olivier, and the Olivier residence home was situated on the west side of Bayou Teche, on the Orange Grove plantation.

The tract of land that was sold to Thomas M. Floyd was a part of the Orange Grove plantation, which Mrs. Olivier had received from her husband, Eugene Barthlemy Olivier. This tract of land measured one and one fourth arpents fronting on Bayou Teche by a depth of forty arpents. Also, this tract of land was bounded on the north side by the lands of Eugene B. Walet, on the south side by Bayou Teche, on the east side by land belonging to the heirs of Baptiste Collette, a free man of color, and on the west side by the land belonging to the estate of the late Charles Napoleon Olivier.

At the execution of this act of conveyance, Thomas M. Floyd, the vendee, paid unto Therese Laura Dalcour Olivier, the vendor, or the seller, the sum of $700.00, which was paid in cash money.

Also, Thomas M. Floyd promises to deliver unto Therese Laura Dalcour Olivier five hundred fence posts and five hundred pieux, which movable property is situated on the tract of land that was sold. For the purchase price of $700.00, which was paid in cash, Mrs. Olivier grants full discharge and acquittal of the debt unto Thomas M. Floyd.

On the date of March 10, 1869, Therese Laura Dalcour Olivier, who was the legal and lawful wife of Eugene Barthlemy Olivier, executed a conventional mortgage with the A. Tertrou & Company of the City of New Orleans, Louisiana. The A. Tertrou & Company was the sugar factor for the Orange Grove plantation, and the sugar factor was employed by Therese Olivier. A sugar factor acted as a broker for the marketing and the selling of the raw sugar and its products, such as molasses and syrup from the plantation. It was a common practice for the plantation owners to execute mortgages in favor of their sugar factors. Usually mortgages, which were in favor of the sugar factors, used as collateral the proceeds, which were to be received from the harvesting of the current year's crops, which were grown on the plantation, and the products of the crop, such as brown sugar, molasses, and cane syrup which were processed on the plantation. When borrowing money from their sugar factor, the plantation owner received borrowed money usually for the expenses for the making of the current year's crop. Then, this borrowed money or principal was paid back using the proceeds from the harvesting of the crop, plus a charge for the services of the sugar factor for the selling and marketing of the product.

Therese Laura Dalcour Olivier was indebted to the A. Tertrou & Company of the City of New Orleans, for the use of $1,123.90. This sum of $1,123.90 was loaned to Therese Laura Dalcour Olivier for the purpose of making her crop of sugar cane for the year of 1868 on the Orange Grove plantation. For the reimbursement of the loaned amount or $1,123.90, Therese Olivier made and subscribed two promissory notes. These two promissory notes were drawn by Therese Olivier, and these two notes were also endorsed by her. Both promissory notes were dated March 10, 1869. The first promissory note was for the amount of $568.18, and this note was due on the date of January 1, 1879. Also, the second promissory note was for the amount of $555.72, and this final promissory note was due on the date of March 10, 1870. Each one of the two promissory notes were payable at the office of A. Tertrou & Company, which company was situated in the City of New Orleans. Also, each one of the two promissory notes were conditioned to bear interest at the rate of 8% per annum from maturity until paid in full. Both promissory notes were delivered to Alfred Duperier, who was serving as the agent for the A. Tertrou & Company of the City of New Orleans.

In executing this conventional mortgage with her sugar

factor, A. Tertrou & Company, Therese Olivier used as collateral the real property, which comprised the Orange Grove plantation. At this time, in the year of 1869, the Orange Grove plantation consisted of tract one, which was situated in the Parish of St. Mary, on the west side of Bayou Teche. Tract one of the Orange Grove plantation had a frontage of nine arpents along Bayou Teche with a depth of forty six arpents. The lines of tract one were closing in and running to the rear of the tract, so as to give this tract a width of five arpents. Tract one was bounded below or on the lowerside by lands belonging to the estate of Baptiste Collette, a free man of color, and above or one the upperside by lands of Therese Laura Dalcour Olivier in part, and in part by land belonging to the estate of the late Charles Napoleon Olivier.

Tract two of the Orange Grove plantation was situated between Bayou Teche and the public road, which road is known as the Old Spanish Trail. Tract two of the Orange Grove plantation measured six arpents fronting Bayou Teche by an irregular depth of four arpents. Tract two was bounded below or on the lowerside by land belonging to the estate of the late Charles Napoleon Olivier, in the rear by the public road, which road is known as the Old Spanish Trail.

Therese Laura Dalcour Olivier received ownership of the Orange Grove plantation from her husband, Eugene Barthlemy Olivier, through an act of Dation en paienment. This Dation en paienment was recognized by a decree of the Third Judicial District Court, which was holding court in the Parish of St. Mary.

The property of the Orange Grove plantation remains mortgaged in the favor of A. Tertrou & Company of the City of New Orleans, until Therese Olivier pays the full and final payment of this conventional mortgage. Also, Therese Laura Olivier is not to alienate or encumber the mortgaged premises of the Orange Grove plantation. The A. Tertrou & Company is the holder of the two promissory notes. Also, the A. Tertrou & Company served as the sugar factor for the Orange Grove plantation, for which purposes the A. Tertrou & Company sold and marketed the raw sugar and the products of raw sugar, which were produced on the Orange Grove plantation.

On the date of May 13, 1867, Eugene Barthlemy Olivier conveyed his plantation, known as Orange Grove, to his legal and lawful wife, Therese Laura Dalcour Olivier through an act of Dation en paienment. This act of Dation en paienment was executed before James G. Parkerson, a notary public in the Parish of St. Mary. The reason that Eugene Olivier conveyed his plantation to his wife, Therese Olivier, was because he (Eugene Olivier) had to be out of the State of Louisiana, and the country of the United States of America for quite sometime, while he (Eugene Olivier) managed his sugar plantation or sugar estate on the Island of Cuba in the Dominica of Spain.

In the year of 1867, the real property, which comprised the Orange Grove plantation consisted of tract one, which was lying and situated on the west side of Bayou Teche in the then Parish of St. Mary. Tract one was bounded above or one the upperside by lands belonging to the estate of the late Colonel Charles Napoleon Olivier, and the deceased wife of Colonel Charles Napoleon Olivier, Felicite Chretin Olivier, and below or on the lowerside by land belonging to Jean Baptiste Collette, a free man of color. Also, tract one had a frontage on the west side of Bayou Teche of fifteen arpents, and this tract had a width of land running back from Bayou Teche to the public road, known as the Old Spanish Trail. This tract of land had an irregular depth from the public road to Bayou Teche ranging from three to four arpents. Tract one had a frontage along the public road or the Old Spanish Trail of nine arpents, and this ran back from the public road to the ordinary depth of forty arpents or about thirty seven acres from the public road to the side lines next to the land belonging to the estate of the late Colonel Charles Napoleon Olivier. This tract closed in to about five arpents, which gave a superficial area on the west side of the public road or the Old Spanish Trail as originally acquired by Eugene Barthlemy Olivier from his father, Colonel Charles Napoleon Olivier. At this time, in the year of 1867, tract one of the Orange Grove plantation was valued at the sum of $6,000.00.

The public road was known as the Old Spanish Trail. At this time, the Old Spanish Trail was traveled frequently by residents, travelers, and peddlar going from the town of Franklin to the town of New Iberia, or from the town of New Iberia to the town of Franklin. The Old Spanish Trail was the main road, which led through the bayou country of the State of Louisiana to the Acadian country of the State of Louisiana. Actually, the Old Spanish Trail began in the town of St. Augustine, Florida and it ended in the town of San Diego, California.

Tract one of the Orange Grove plantation had been purchased by Eugene Barthlemy Olivier from his father, Colonel Charles Napoleon Olivier, in the year of 1856. The lines of tract one were established by a survey made of the property at that time, in the year of 1856, and this survey was not kept in the public records of the Parish of St. Mary.

Tract two of the Orange Grove plantation was lying and situated on the east side of Bayou Teche, in the then Parish of St. Mary. Tract two had a frontage of one and one fourth arpents along Bayou Teche by the ordinary depth of forty arpents. Tract two was bounded above or on the upperside by land belonging to the estate of the late Colonel Charles Napoleon Olivier, and below or on the lowerside by land belonging to Jean Baptiste Collette, a free man of color. At this time, in the year of 1867, tract two of the Orange Grove plantation was valued at the price of $600.00.

Also, conveyed through this act of Dation en paienment to

his wife, Therese Laura Dalcour Olivier, Eugene Barthlemy Olivier relinquishes his rights and ownership to ten shares of the New Orleans, Opelousas, & GW Railroad. These ten shares of the New Orleans, Opelousas, & GW Railroad Company were valued at $6.00.

Eugene Olivier also relinquished the rights of inheritance in the estate of the late Colonel Charles Napoleon Olivier and his second wife, F. Chretin Olivier, his deceased parents.
The value of the inheritance rights of Eugene Barthlemy Olivier was the sum of $5,000.00.

Also, the livestock on the Orange Grove plantation were transferred and conveyed through this Dation en paienment to Therese Laura Dalcour Olivier, who was the wife of Eugene Barthlemy Olivier. The livestock consisted of six mules, which mules were valued at $15.00 each mule, and the total value was the sum of $900.00, ten American horses, and these horses were valued at $65.00 through $150.00 for each American horse, and the total value of the ten American horses was the sum of $215.00, one creole horse, which was valued at $35.00, one American mare, which was valued at $50.00, fourteen milk cows and calves, and each cow and calf were valued at $30.00 each, and the total valuation of the fourteen milk cows and calves were $420.00, and three /two years old, and ten yearling calves, which were valued at $10.00 each, and the total valuation was $130.00.

The farm equipment was also conveyed through the Dation en paienment to Therese Laura Dalcour Olivier, and this farm equipment consisted of seven plows, and a lot of gear, which was valued at $9.00, ten harness, which were valued at $5.00, six carpenter tools and agricultural implements, which were valued at $25.00, one double gear, which was valued at $25.00, one rifle, which was valued at $15.00, and one saddle and bridle, which was valued at $25.00.

This personal property of Eugene Barthlemy Olivier was transferred to his wife, Therese Laura Dalcour Olivier, through this Dation en paienment, which personal property consisted of one gold watch and chain, which was valued at the price of $150.00.

Also, the furniture and fixtures that were situated in the residence home of Mr. and Mrs. Eugene Barthlemy Olivier, which residence home was situated on tract one of the Orange Grove plantation. Later in the years, the part of tract one where the Olivier residence home was situated because tract two of the Orange Grove plantation. The furniture consisted of twenty one pieces of household furniture which included seven armories, which were valued at $20.00 each armoire, and the total value of the seven armoires was the sum of $140.00, one bookcase and miscellaneous library, which was valued at the price and sum of $50.00, one desk, which was valued at the price of $20.00, two bureaus with marble tops covered with glass, and each bureau was valued at the price of $20.00 each. The total valuation was the sum of $40.00, four washstands, and the washstands had marble tops. These washstands were valued at the purchase price of $15.00 each washstand, and the total valuation of the four washstands was the sum of $60.00. One side board, which was situated in the dining room of the Orange Grove plantation home, and this side board was valued at the price of $40.00. Also, this side board had a marble top. One ice box, which was valued at the price of $12.00, three toilets, which were valued at the price of $50.00, three sofas, which were valued at the price of $60.00, one center table, which was valued at the price of $10.00, one large glass, which was valued at the price of $60.00, and two clocks, which were valued at the price of $50.00. Also, the furniture that was situated in the plantation home of the Orange Grove plantation consisted of eight bedsteads, bedding, and linen, which were valued at the sum of $800.00, one couch and one child's crib, which were valued at the price of $357.00, four hair cushion rockers, which were valued at the price of $10.00 each rocker, and the total valuation of the four rockers was the sum of $40.00, one invalid chair, which was valued at the price of $10.00, one sewing machine, which was valued at the price of $75.00, two small tables and four large table, which were valued at the price of $42.00, one safe, which was valued at the price of $5.00, one lot of candlesticks, lamps, hand irons, shovels, tongs, and fire fenders, which were valued at the price of $30.00, four cane rockers, which were valued at the price of $15.00, one office chair, which was valued at the price of $5.00, twelve parlor chairs, which were valued at the price of $24.00, three dozen or thirty six chairs, which were valued at the price of $1.00 each chair, or the total valuation of $36.00, a lot of crockery and glassware, and wash basins, which was valued at the price of $50.00, one stove and cooking utensils, which were valued at the price of $20.00, one lot of silverware, tea pot, sugar dish, two salt pans and spoons, two large soup ladles, two large sauce spoons, eight large spoons, fifteen large forks, thirteen desert spoons, nineteen desert forks, eighteen tea spoons, one butter knife, sugar tongs, and a sugar stand, which were all valued at the price of $240.00, and one wagon with wooden axles on the wheels, which was valued at the price of $240.00. The total valuation of the movable property that was situated on the Orange Grove plantation at this time of the Dation en paienment was the total amount of $16,087.00.

The reason that Eugene Barthlemy Olivier agreed to execute a Dation en paienment to his wife, Therese Laura Dalcour Olivier, was because he (Eugene Olivier) was indebted to his wife for the sum of $74,991.35. To reduce the state of indebtness of Eugene Barthlemy Olivier to his wife, Therese Laura Dalcourt Olivier, so far as in the power of Eugene Barthlemy Olivier and availing himself of the provisions of Article Number 2421, Section Second of the Civil Code of the State of Louisiana, it was the wish and the intention of Eugene Barthlemy Olivier to make a Dation en paienment

The reason that Eugene Barthlemy Olivier agreed to execute a Dation en paienment to his wife, Therese Laura Dalcour Olivier, was because he (Eugene Olivier) was indebted to his wife for the sum of $74,991.35. To reduce the state of indebtness of Eugene Barthlemy Olivier to his wife, Therese Laura Dalcourt Olivier, so far as in the power of Eugene Barthlemy Olivier and availing himself of the provisions of Article Number 2421, Section Second of the Civil Code of the State of Louisiana, it was the wish and the intention of Eugene Barthlemy Olivier to make a Dation en paienment to his wife.

The Dation en paienment made by Eugene Barthlemy Olivier to his wife, Therese Laura Dalcour Olivier, was made for the purpose of replacing to her, (Therese Laura Dalcour Olivier) the funds or the money, property, and effects that he (Eugene Barthlemy Olivier) had alienated for his own use and purpose since their marriage. Actually, the Dation en paienment was a partial repayment of money, property, and effects to Therese Laura Dalcour Olivier. After the Dation en paienment, Eugene B. Olivier was still indebted to his wife, Therese Laura Dalcour Olivier. At this time of the Dation en paienment in the year of 1867, the property, which comprised the Orange Grove plantation was free from all mortgages. The execution of this Dation en paienment was held at the Orange Grove plantation home. The two witnesses to the execution of this Dation en paienment were Joseph Olivier, the half brother of Eugene B. Olivier, and Joshua Smith.

On the date of July 19, 1860, Joseph Marcel Dencause made a conveyance of a slave to Eugene Barthlemy Olivier. This slave was named Peter, and he was a mulatto man. Peter, the slave, resided and worked on the Orange Grove plantation.

On the date of November 30, 1856, Alexander Olivier, the brother of Eugene Barthlemy Olivier, made a conveyance of slaves to Eugene Olivier. This sale of the slaves was held at the residence of Alexander Olivier and his wife, Amanda Lastrapes Olivier, which residence was located at Bayou Chone in the Parish of St. Martin.

The slaves that were conveyed to Eugene Olivier were Hector, alias Daniel, a mulatto man slave, and he was 20 years old. Hector was sold to Eugene B. Olivier for the purchase price of $1,300.00. Another slave, Joe, a black slave man, and he was age 25 years old. Joe was sold for the purchase price of $1,000.00. Lastly, the slave woman, Martha, who was age 22 years old, and was sold to Eugene B. Olivier for the purchase price of $900.00. All three of these slaves were brought to the Orange Grove plantation to reside and work.

At the time of the conveyance of the slaves, Alexander Olivier was married to Amanda Lastrapes, but Amanda Lastrapes Olivier was not of the age of majority, meaning she was a minor. Under the laws of the State of Louisiana, Amanda Lastrapes Olivier was unable to renounce her legal mortgage, which she held on the property of her husband, Alexander Olivier. Amanda Lastrapes Olivier had a legal mortgage on these three slaves. The renunciation of the dotal, matrimonial, and paraphernal rights to these three slaves would have to be made by Amanda Lastrapes Olivier when she reached the age of majority,and this renunciation was an obligation.

A view of the former Joseph Devezin Olivier plantation home, now the home of Mr. and Mrs. Numa LaBiche

Orange Grove

A view of the former overseer home on the Orange Grove plantation. This home was formerly occupied by Mr. and Mrs. Jacob Segura Landry of New Iberia, Louisiana.

OLIVE BRANCH PLANTATION

On April 14, 1909, John Newton Pharr sold to John C. Bussy at a cash sale, a sugar plantation, known as Olive Branch Plantation, which was situated in Iberia Parish, Louisiana. Olive Branch Plantation consisted of two hundred and fifty six arpents of land and three fourths hundredths of an arpent. The Olive Branch Plantation was situated on the east side of Bayou Teche. It was bounded above or on the upperside by the lands of Henry Train, in front by Bayou Teche, and in the rear by the land Mrs. Joseph Devezin Olivier, and on the lowerside or below by the lands of Pharr and Bussy.

Besides the immovable property of the Olive Branch Plantation, the movable property that was situated on the plantation was also sold. The movable property consisted of mules, carts, and implements of husbandry. The implements of husbandry consisted of hoes, shovels, rakes, scythes, and plows, and these implements were used in the cultivation of the crop of sugar cane. Also, the seed cane, which is sometimes called plant cane, was also conveyed. The seed cane is the cane that is planted in the spring or the fall for the next year's crop. At this time, in the year of 1909, the farmers planted sugar cane in either the spring or the fall for their nest year's crop.

On January 27, 1898, John Newton Pharr purchased at a cash sale the plantation, known as Olive Branch, from the commercial firm of Erman & Cahn, which firm was represented by Leon Cahn. The commercial firm of Erman & Cahn was owned by Messrs Abraham Erman and Leon Cahn, and it was located in the City of New Orleans, Louisiana.

On December 19, 1895, Erman and Cahn, a commercial firm from the City of New Orleans, Louisiana, purchased the Olive Branch plantation from the heirs and legal representatives of John Cowley and Edward Kissack. Mr. P. Laurent Renoudet, a notary public, prewsided at the execution of the act of conveyance, on the behalf of the legal heirs and representatives of John Cowley and Edward Kissack. The legal heirs and representatives of Edward Kissack and John Cowley were John Cowley, William Cowley, Marion Cowley, and Edward H. Cowley. Erman and Cahn was a commercial firm, which was owned by Messrs. Abraham Erman and Leon Cahn. Lazard Kling represented the firm of Erman and Cahn at the execution of this deed of conveyance of the Olive Branch Plantation.

On December 24, 1895, or on Christmas Eve, Messrs. Abraham Erman and Leon Cahn purchased at a cash sale the moveable property that was located on the Olive Branch Plantation from J. B. Lobdell. Felix J. Dreyfrus served in his capacity as the notary public for the execution of this act of conveyance. Leon Cahn represented the commercial firm of Erman & Cahn. Also, Leon Cahn accepted delivery of the movable property, which was situated on the Olive Branch Plantation, on the behalf of the commercial firm of Erman & Cahn.

The movable property consisted of 11 mules, 1,800 barrels of corn, 60 tons of hay, 4 - three wheel mules carts, two cultivators, two rotary hoes, one lot of gears, one lot pack saddle and bridles, eighty two mule plows, 4-one mule plows, and 5-four mule plows.

This sale of the movable property that was situated on the Olive Branch Plantation was sold for the price of $2,410.00, which was paid in cash money. J. B. Lobdell acknowledged to have received the $2,410.00 in cash money from Leon Cahn.

Also, on April 11, 1895, J. B. Lobdell conveyed movable property at a cash sale to Messers J. & L. Dreyfrus. P. Laurent Renoudet served in his capacity as the notary public for the execution of the act of conveyance for this movable property. This sale was made and accepted for the price of $5,500.00, which was paid in cash at the execution of the act of conveyance.

The movable property that was sold by J. B. Lobdell to J. & L. Dreyfrus consisted of one American Bay mare mule that was seven years old, and named Emma. The mare mule, Emma, measured fifteen hands high. Another American Bay horse mule, that was four years old, and the name of this mule was Beaumont. The American Bay horse mule, Beaumont, measured sixteen hands high. Another American black mare mule, that was eight years old, and this mule was named Delta. The American black mare mule, Delta, measured sixteen hands high. Another American black horse mule, that was five years old, and this mule was named Scott. The mule, Scott, was a new mule on the Olive Branch Plantation.

On August 14, 1893, John B. Lobdell and Deliah Cowley, the widow of John Cowley, entered into an agreement for the eventual purchase of Olive Branch Plantation by John B. Lobdell. Mrs. Deliah Cowley was the natural tutrix and guardian of her minor children, Mary, Edward, John, and William. Mrs. Deliah Cowley and her children were all residents of the City of Galveston, which is situated in the State of Texas. John B. Lobdell was a resident of Iberia Parish, but he was originally from Lobdell, Louisiana, which is a small town, situated near the City of Baton Rouge, Louisiana.

In this agreement, Deliah Cowley, who was the widow of John Cowley and her children, John, William, Mary, and Edward Cowely, were known as parties of the first part, and John B. Lobdell was known as the party of the second part. The parties of the first part, Deliah Cowley and her children, William, Edward, John, and Mary bind and obligate themselves to sell with full warranty for the price and sum of $7,000.00 to John B. Lobdell, the plantation known as, Olive Branch Plantation.

The contract of the sale of the Olive Branch Plantation was signed by the part of the first part, Deliah Cowley, the widow and the children, John, William, Mary, and Edward in February of the year of 1895, or as soon as the children, who are now minors become of age. When the children become of age, they will be able to dispose of their shares of the Olive Branch Plantation. Therefore, John B. Lobdell, the party of the second part, obligates himself to pay $2,260.53 in cash money at the signing of this contract by Deliah Cowley, the widow, and the children, Edward, William, John, and Mary Cowley. Then $334.57 is to be paid in cash on the day that the act of conveyance is executed and signed for the transfer of the interests, rights, and privileges to the Olive Branch Plantation. Then, John B. Lobdell will pay an additional sum of $2,000.00 in cash money at the last mentioned time. Then for the balance of $3,000.00, John B. Lobdell will furnish his four certain promissory notes. Each one of the four promissory notes will be for the sum of $750.00, which will bear interest at 8% interest per annum from the date until paid. These four promissory notes will be payable in one, two, three, and four years after the date of the sale. The four promissory notes will be secured by a vendor's lien and special mortgage on the property of the Olive Branch Plantation.

Also, it was understood that the party of the first part, Deliah Cowley and the children, John, William, Edward, and Mary Cowley, will waive and relinquish any and all rights and rents, particularly those rents in the years of 1893 and 1894, and the party of the first part will give John B. Lobdell, the party of the second part, the free use of the Olive Branch Plantation until the proposed sale of the plantation is signed and executed by all the parties. Also, the party of the first part, the Cowley's, will not be able to collect any rent from John B. Lobdell for the use of the plantation from this 14th day of August in the year of 1895 or until the sale is executed.

Also, it was agreed that Deliah Cowley, now acting for herself and for the children, and the holder of a mortgage bearing upon the Olive Branch Plantation, hereby waives any and all rights that she may have or has to the Olive Branch Plantation. On the date that the proposed sale is complete, this mortgage will be cancelled and erased from the mortgage books in the recorder's office, now the office of the clerk of court. This agreement was executed and signed in Galveston, Texas.

The Olive Branch Plantation in the year of 1893, was described as a certain tract of land, measuring four arpents fronting Bayou Teche, with a depth of forty arpents. The total land area was one hundred and sixty arpents of land. The boundaries of the Olive Branch Plantation were on the upperside or above the lands of tract two of the Olive Branch Plantation, and below or on the lowerside the lands of Joseph Devezin Olivier, the front of the plantation faces Bayou Teche, and in the rear of the plantation are the lands belonging to Charles Olivier. This is tract one of the Olive Branch Plantation.

Tract two of the Olive Branch Plantation measured two and one fourth arpents fronting Bayou Teche, with a forty three arpent depth. The total land area of tract two was ninety six arpents and seventy five hundredths of an arpent of land. The boundaries of tract two of the Olive Branch Plantation were on the upperside the plantation of Henry Train, and below by tract one of the Olive Branch Plantation, the front of tract two fronts Bayou Teche, and in the rear of tract two are situated lands belonging to Charles Olivier.

The Olive Branch Plantation consisted of two separate tracts of land, which formed the plantation. Also, sold with these two tracts of immovable real estate belonging to the Olive Branch Plantation were the movable property, such as implements of husbandry, which are used in the farming of sugar cane.

On August 6, 1892, John Cowley sold or conveyed to Deliah Cowley his one fifth undivided interest in the Olive Branch Plantation, which was cultivated as a sugar plantation. Also, conveyed were all the improvements, such as ditches and bridges, as well as all the machinery, which was situated on the plantation. Also, on July 27, 1892, William J. Cowley conveyed his one fourth interest in the Olive Branch Plantation to Deliah J. Cowley.

On August 31, 1892, by a virtue of the Honorable Court, it was ordered that the heirs of John Cowley be put in possession and full ownership of the Olive Branch Plantation. This property was bequeathed to William, Mary, Edward, and John Cowley by the succession of John Cowley. On February 9, 1891, the succession of Edward Kissack bequeathed this property to the heirs of John Cowley.

On February 19, 1885, Edward Kissack repurchased the Olive Branch Plantation at a sheriff sale, and Deputy Sheriff Charles A. Judice presided over the sheriff sale. The Olive Branch plantation came to be seized by Deputy Sheriff Charles A. Judice, of the Parish of Iberia, in the matter of the civil law suit of Edward Kissack, the plaintiff, and W. H. Webb and E. Richardson, the defendants. At this time, the Olive Branch Plantation consisted of three separate tracts of land.

The first tract of land measured four arpents fronting Bayou Teche with a depth of forty arpents. The boundaries of tract one were on the upperside by tract two of the Olive Branch plantation, and below by the lands of Joseph Devezin Olivier. Tract two of the Olive Branch Plantation measured four arpents fronting Bayou Teche, with a depth of forty arpents. Tract two was bounded above by the lands of Henry Train, and below by tract one of the Olive

Branch plantation, and the land in the rear of tract two belonged to the estate of Charles Olivier. Both tracts one and two consisted of one hundred and sixty arpents of land.

Tract three measured two and a fourth arpents fronting Bayou Teche, and the depth was forty three arpents. The total land area of tract three was ninety six arpents of land and seventy five hundredths of an arpent of land. Tract three was bounded above by the lands of Henry Train, and below by tract two of the Olive Branch plantation. The front of tract three fronted Bayou Teche, and the land in the rear of tract three belonged to the estate of Charles Olivier.

Also, conveyed at this sheriff sale was all the buildings that were situated on the Olive Branch plantation, and the improvements, such as ditches, drainage, fences, and bridges. The movable property sold was eight American mules, five carts, seven plows, one loft harness, and one ferry with chain attached. This one ferry with the chain attached was used to cross Bayou Teche from the east to the west side and vice versa. Also, one lot of seed cane, which was to be used to plant next season's sugarcane crop, and implements of husbandry, which implements were used in the cultivation of the crop of sugar cane and other crops grown on the Olive Branch plantation.

On December 12, 1884, in a civil lawsuit being tried at the courthouse in Iberia Parish, Louisiana, Edward Kissack stated in open court that W. H. Webb and Edmund S. Richardson were indebted to him (Edward Kissack) for the sum of $5,500.00 with interest at 7% per annum from November 30, 1881. On November 30, 1881, W. H. Webb and Edmund S. Richardson gave Edward Kissack five promissory notes that were made payable to the order of Edward Kissack. Each of the five promissory notes were for the sum of $1,100.00, and the notes were made due one, two, three, and four years from the date of November 30, 1881. These notes were given in a sale and mortgage of the conveyance of the Olive Branch plantation to W. H. Webb and Edmund S. Richardson from Edward Kissack. A vendor's privilege and lien was placed on the property of the Olive Branch plantation to secure the payment of these notes. Evidently, W. H. Webb and Edmund S. Richardson did not pay any of these notes to Edward Kissack, which were the grounds of the civil law suit.

On November 12, 1884, Edward Kissack had an order signed by Judge Fred L. Gates and that order was sent to Messrs W. H. Webb and Edmund S. Richardson. This was an order of seizure and sale, which was issued out of the Honorable Court of Iberia Parish. Also, this order of seizure and sale directed the Sheriff of Iberia Parish to seize and sell the property belonging to the Olive Branch sugar plantation.

The terms and conditions of the order of seizure and sale

The terms and conditions of the order of seizure and sale stated that $3,500.00 with interest at 7% from November 30, 1881, and all costs payable in cash. W. H. Webb and Edmund S. Richardson could keep their ownership of the Olive Branch sugar plantation, if they met these terms and conditions of paying $3,500.00 with the 7% interest from November 30, 1881, and also all costs, with cash to pay the sum, and on these credit terms $1,100.00 due on November 30, 1885, and $1,100.00 due on November 30, 1886. That is paying half the balance of the purchase price on November 30, 1885, and the other half of the purchase price on November 30, 1886.

Also, conveyed on this date were all the livestock that were on the Olive Branch plantation, the carts to haul cane, and the agricultural implements, such as hoes, scythes, rakes, shovels, and plows.

On December 19, 1881, at a cash sale, John Cowley conveyed to Edward Kissack, a portion of land that would become a part of the Olive Branch plantation. This tract of land measured eighty arpents with a depth of forty arpents. It was bounded on the south by Bayou Teche, on the west by the lands of Henry Train, and on the east by the land of Edward Kissack.

In the early 1880's, Edward Kissack, a native of England, and a wealthy man acquired the tracts of land that would comprise the future Olive Branch plantation. Edward Kissack invested in a sugar plantation, because sugar cane was a money making crop, and the sugar planters were making money. At this time there was no federal government interferences with the sugar industry. The sugar that was grown and processed, here in Louisiana, was sold throughout the United States.

On January 19, 1882, Victor Vautrin conveyed a tract of land to Edward Kissack. This tract of land was situated on the east side of Bayou Teche, and it measured three arpents fronting Bayou Teche with a depth of forty arpents. The total area of this tract of land was one hundred and twenty arpents of land. The boundaries of this tract of land were above or on the upperside by the lands of the widow of Pierre Darby, and below or on the lowerside by the lands of the heirs of Rosette Boutte, who was deceased at this time.

On February 5, 1881, Valcourt Olivier acquired at a cash sale from Victor Vautrin a tract of land. This tract of land was situated on the east side of Bayou Teche, and it fronted Bayou Teche with an arpentage of three arpents and eleven feet. Also, sold to Valcourt Olivier were all the buildings that were situated on this tract of land at this time.

The movable property that was sold included five American mules, three creole horses, one American mare, three carts, and all the implements of husbandry, which included hoes, rakes, shovels, and scythes.

On January 16, 1873, Reginald Shawcross purchased this tract of land from Emory E. Norton, who was serving in his capacity as assignee in the bankruptcy case of the estate of S. O. Nelson & T. A. Nelson, bankruptcy case number 45 of the United States District Court for the District of Louisiana. Emory E. Norton declared that in an act of sale at the auction, which he caused to be made by Nicholas J. Henry, Esquire, a licensed auctioneer, in the City of New Orleans, Louisiana, on the 10th and the 11th days of the month of January of the year of 1873. The auction of the property was held at the Merchant and Auctioneer Exchange, which was situated in the City of New Orleans, between Canal Street and the Custom House Street. The reason for this auction to be held was because of a virtue, dated April 25, 1871, and that was issued out of the United States District Court for the Eastern District of Louisiana. Judge E. K. Donnell was the judge, who issued the virtue, and was also the judge sitting on the bench for this bankruptcy case of the bankrupt estate of S. O. Nelson and T. A. Nelson, both bankrupt individually and as members of the commercial firm of S. O. Nelson & Company. This virtue caused a public sale to be made of the property that formerly belonged to S. O. Nelson.

This tract was sold to Reginald Shawcross for the sum and price of $5,200.00. Also, Reginald Shawcross was the last and the highest bidder on the whole for this tract of land. Reginald Shawcross was a native Englishman and a merchant, who was doing business in the City of New Orleans, Louisiana. It was on one of his business trips that Reginald Shawcross purchased this tract of land that formerly belonged to S. O. Nelson of the City of New Orleans, Louisiana. Reginald Shawcross was present at the Merchant and Auctioneer Exchange, in the City of New Orleans, to make this conveyance and accept delivery of the conveyed property.

The property purchased by Reginald Shawcross was a tract of land that was situated on the east side or east bank of Bayou Teche, which tract had a frontage on Bayou Teche of three arpents and eleven feet, with a depth of forty arpents in parallel lines. The total land area of this tract of land was one hundred twenty arpents of land. The boundaries of this tract of land were above by the lands of the widow of Pierre Darby, and below by the lands belonging to the estate of Rosette Boutte, a free woman of color, now deceased. Also, twenty arpents from Bayou Teche were excluded from this purchase, but these twenty arpents originally went with this tract of land. This sale was made and accepted for the price of $5,200.00, which Reginald Shawcross has paid in American cash money to the vendor, Emory E. Norton. Emory E. Norton has acknowledged full receipt and granted full acquittal of the purchase price. It is understood that the vendor, Emory E. Norton, sells and transfers the property without any warranty or guarantee of title and encumbrances (incumbrance), if there are any, which exist on the property that have not been discharged. Reginald

Shawcross has discharged Emory E. Norton, the vendor, from all liabilities and responsibilities in the premises of this property of the warranties and incumbrance (encumbrances).

On July 7, 1871, Mistress Celeste Olivier Pellerin, the wife of Louis Pellerin and the half sister of Joseph Devezin Olivier, conveyed to Edward Kissack a tract of land that would comprise a part of the future Olive Branch plantation. The tract of land that was purchased from Celeste Olivier Pellerin was known as tract two of the Olive Branch plantation. This tract had a frontage of two and one fourth arpents fronting Bayou Teche, by a depth of forty arpents. The total land area of this tract of land was ninety arpents of land. This tract of land was purchased by Joseph Devezin Olivier and Joseph Alcide Darby, jointly, and this tract of land was situated on the east bank of Bayou Teche. Joseph Devezin Olivier was the half brother of Celeste Olivier Pellerin, and Joseph Alcide Darby was the first cousin to Joseph Devezin Olivier, as their mothers were sisters. On January 1, 1872, Edward Kissack assumed the use and delivery of the land and buildings that were situated on the tract of land.

Joseph Alcide Darby was the son and youngest child of Constance DeBlanc Darby and her husband, Francois St. Marc Darby. Joseph Devezin Olivier was the youngest son of Celeste DeBlanc Olivier, and her husband, Charles Napoleon Olivier. Celeste DeBlanc and Constance DeBlanc were sisters. Also, Celeste DeBlanc Olivier was the first wife of Charles Napoleon Olivier, and she died or passed away prior to December of the year of 1826. Celeste and Constance DeBlanc were the daughters of Louis Charles DeBlanc. Also, a sister of Charles Napoleon Olivier's was married to Chevalier DeBlanc, who was the brother of Celeste DeBlanc or Mrs. Charles Napoleon Olivier, and Constance DeBlanc, or Mrs. Francois St. Marc Darby. Also, the term, "Devezin," is a French title of the Olivier family, and the full name of Charles Napoleon Olivier was Charles Napoleon Olivier, Sieur Devezin. The Olivier family was descended from French nobility.

On June 27, 1870, R. Hulton Ker conveyed a tract of land to Edward Kissack, and this tract of land became known as tract one of the Olive Branch plantation. This act of conveyance was executed by Joseph Breaux, a notary public in the City of New Iberia. This tract of land measured four arpents fronting Bayou Teche, with a depth of forty arpents. The total land area of this tract of land was one hundred and sixty arpents of land. This tract of land was bounded on the upperside or above by the lands of Alexander Pellerin and below or on the lowerside by the lands of Joseph Devezin Olivier, and the land in the rear of this tract of land or tract one of the Olive Branch plantation belonged to the estate and heirs of Charles Napoleon Olivier.

On April 22, 1871, Joseph Alcide Darby conveyed to R. Hulton Ker a tract of land, which was situated on the eastern side of Bayou Teche, in the newly created Parish of Iberia. This tract of land measured four arpents fronting Bayou Teche, with a depth of forty arpents, which gave a total land area of one hundred and sixty arpents. Also, all the buildings that were situated on the property was sold. This tract was bounded on the upperside or above by the lands of Alexander Pellerin, and below or on the lowerside by the lands of Joseph Devezin Olivier, and the land in the rear of this tract belonged to the estate and heirs of Charles Napoleon Olivier.

Tracts one and tract two of the Olive Branch plantation were at one time a part of the lands that were purchased by Joseph Alcide Darby from Joseph Devezin Olivier. Tract three of the Olive Branch plantation was at one time a part of the Nelson Place or St. Mary's Plantation. At one time, tract three of the Olive Branch plantation belonged to the lands that belonged to Stephen Owens Nelson, both individually and as a part of the commercial firm of S. O. Nelson & Company of the City of New Orleans, Louisiana. The name of the plantation Olive Branch was given to the three tracts of land that were purchased by Edward Kissack, after he had purchased these tracts of land. Edward Kissack named the plantation, "Olive Branch," because the plantation, he hoped would be a very good investment and very fruitful.

In the month of September of the year 1867, Joseph Devezin Olivier, a half brother to Celeste Olivier Pellerin, conveyed a tract of land to Celeste Olivier Pellerin, a half sister of Joseph Devezin Olivier's, and the wife of Louis Pellerin. The witnesses to this act of conveyance were Joseph Alcide Darby, a first cousin of Joseph Devezin Olivier's and Francois A. Dumartrait. This act of conveyance was executed in the presence of Barthlemy d'Augueseau de la Houssaye, a notary public. This parcel of land was situated on both the east and the west sides of Bayou Teche, in the Parish of St. Mary, now Iberia Parish. This tract of land was bounded above by the lands of Joseph Devezin Olivier, the present vendor, and on the west side by the lands of Zenon Olivier, a free man of color, and Edouard Simon, a free man of color, these were the boundaries of the land belonging to this tract, that were situated on the east bank of Bayou Teche. The boundaries of the land belonging to this tract, that were situated on the west bank of Bayou Teche were above or on the upperside by the lands of Joseph Devezin Olivier, and on the east side by the lands belonging to the succession of Charles Napoleon Olivier, the father of Joseph Devezin Olivier. Both the tract of land on the east side and the tract of land on the west side of Bayou Teche were forty arpents or thirty four American acres in depth. The price of this tract of land was $2,000.00, which was paid in cash, and full acquittal of the purchase price was given by Joseph Devezin Olivier. This was the tract of land that would be the future tract two of the Olive Branch plantation. Celeste Olivier Pellerin, the wife of Louis Pellerin, purchased this tract of land as an investment opportunity.

On September 10, 1867, Joseph Devezin Olivier conveyed to Joseph Alcide Darby, his first cousin, a tract of land, which was situated in the Parish of St. Mary, now Iberia Parish. Present a this act of conveyance were Celeste Bellechase Olivier, the wife of Joseph Devezin Olivier, the vendor or the seller of the property, as well as Francois M. Dumartrait and Alexander Pellerin, who would serve in their capacity as witnesses to this act of conveyance. This piece of property or this tract of land would become the future tract one of the Olive Branch plantation.
Joseph Devezin Olivier conveyed five hundred arpents of land to Joseph Alcide Darby, which land was situated on both the east and the west sides or banks of Bayou Teche. Also, these five hundred arpents of land fronted Bayou Teche on the east and west sides.

At this at of conveyance on September 10, 1867, Celeste Bellechase, the lawful wife of Joseph Devezin Olivier, was taken out of the hearing presence of her husband, and carefully read and explained the terms of this act of conveyance by the notary public, who was executing this act of conveyance. Celeste Bellechase Olivier was authorized and assisted by her husband, Joseph Devezin Olivier. Celeste Bellechase Olivier declared in the presence of the notary public that she was familiar and aware of the terms and conditions of this act of conveyance. The reason this was done was because Celeste Bellechase Olivier by the laws of the State of Louisiana, at this time in 1867, had a legal mortgage on all the property owned by her husband, Joseph Devezin Olivier. This mortgage was to protect Celeste Bellechase Olivier's dotal, paraphernal, and matrimonial rights and privileges that she may have in the title, interest, and rights to the property owned by her husband. Celeste Bellechase Olivier also renounced all her dotal, paraphernal, and matrimonial rights that she may have in these five hundred arpents of land being conveyed by her husband, Joseph Devezin Olivier, to J. Alcide Darby.

Before Olive Branch Plantation was sold to Joseph Alcide Darby by Joseph Devezin Olivier, this plantation formed a part of the Joseph Olivier plantaton, the Darby Brothers plantation, and the Gravier plantation.

The former home of Joseph Wade Gajan, and his wife, Maxie Gajan, which stands on land, which once formed the homesite tract of St. Mary's Plantation.

The railroad track as it runs through the former plantations of John Ermant Leitmeyer, Stephen Owens Nelson, Louis and Charles DeBlanc, and Jean Baptiste Despanets DeBlanc

THE FRILOT PLANTATION

On the date of September 5, 1888, Anastasie Frilot Brettmeyer, who was the wife of Eugene Brettmeyer, made a conveyance to the Mount Zion Methodist Church, which is situated on the Old Spanish Trail. The full name of the Mount Zion Methodist Church was the Mount Zion Methodist Episcopal Church. Actually, the Mount Zion Methodist Episcopal Church, is situated on property that once comprised the plantation owned by both Rosetta Boutte', a free woman of color, and the wife of Claude Frilot, and Leon Frilot. Rosetta Boutte' and Claude Frilot were the maternal grandparents of Anastasie Frilot Brettmeyer, who was the person conveying the property to the Mount Zion Methodist Episcopal Church. Also, Anastasie Frilot was the daughter of the late Marie Louisa Aimee Frilot, who a daughter of Rosetta Boutte' and Claude Frilot. Anastasie Frilot married Leon Frilot, a free man of color.

The Mount Zion Methodist Episcopal Church still exits today, and it is located on the Old Spanish Trail, between the town of New Iberia and the town of Jeanerette, Louisiana. It is a white wooden building, and a very quaint looking church.

At the execution of the act of conveyance for the property of the Mount Zion Methodist Episcopal Church, the people representing the Mount Zion Church were Edward Price, Eloi Boutte', Amas Green, Harusogene Daniels, Issac and Albert Marshall. Albert Marshall and Harusogene Daniels did not know how to write their names or sign their names, so they made their x marks on the paper of the act of conveyance.

The property conveyed to the Mount Zion Methodist Episcopal Church was a lot of ground, that contained one half arpent fronting the Old Spanish Trail, by the depth of two arpents. The boundaries of the property, on which the Mount Zion Church stands, were on the north side by the public road or the Old Spanish Trail, on the south side by the property of Anastasie Frilot Brettmeyer, who was the seller, on the east side by the property of Louis Frilot, who was a cousin of Anastasie Frilot Brettmeyer, and on the west side by the property of Anastasie Frilot Brettmeyer.

The lot of ground where the Mount Zion Methodist Episcopal Church was to be constructed was to be used and maintained as a place of divine worship. Also, the lot was for the use of the minister of the Mount Zion Church, and the members or the congregation of the Mount Zion Methodist Episcopal Church. This lot of ground was subject to the ministerial appointments of at the church. From time to time, authority of the conference of the Mount Zion Church, and the conferences are within the bounds of where the possession of the church is situated. This lot would become church property, under the authority of the church.

The sale price of this lot of ground was $100.00. It was paid in a sale and mortgage. Part of the payment of $50.00 was paid in cash at the execution of the act of conveyance. The balance of the purchase price was to be paid by a promissory note, which was furnished by the purchasers. The promissory note was for the sum of $50.00. Also, this promissory note was drawn to the order of the purchasers and endorsed by the purchasers. The payment of this promissory note was due on January 10, 1889. Also, this promissory note was conditioned to bear interest at the rate of 8% per annum from the date of September 5, 1883 or until paid in full. All state, parish, and other taxes were paid by Anastasie Frilot Brettmeyer, the seller, of the lot of ground.

On the date of December 18, 1883, Eugene Brettmeyer, the husband of Anastasie Frilot, executed an act of Dation en paienment to his legal wife, the former, Anastasie Frilot. A Dation en paienment is a legal act, which a person can convey or sell to another person real estate property, movable property, brands, proceeds of crops, livestock, and rights, titles, and interest. Usually, an act of Dation en paienment was executed to repay or reimburse another person for money owed. If a husband had the use of his wife's land for the making of crops and the raising of livestock, and the proceeds from these businesses were used by the husband for his own uses, then the husband was indebted to his wife for the use of her land for the making of his crops and the raising of his livestock. Dation en paienment means paid in full. Usually, the act of Dation en paienment was executed to repay someone in full for the use of land or movable property. It was a common practice, in the late 1800's, for a husband to convey all of his land to his wife through an instrument of Dation en paienment.

So on the date of December 18, 1883, Eugene Brettmeyer admitted that he was truly indebted to his legal wife, Anastasie Frilot Brettmeyer, for the total sum of $800.00. This was the sum made from the sale of twenty five heads of cattle. In the year of 1869, Leon Frilot, a free man of color, and the father of Anastasie Frilot Brettmeyer, and the father-in-law of Eugene Brettmeyer, made a gift of some of the twenty five cattle to his daughter, Anastasie. Also, in the year of 1873 or 1874, Alcide Frilot, the brother of Anastasie Frilot Brettmeyer, made another gift of some more cattle to his sister, Anastasie Frilot Brettmeyer. In the year of 1873 or 1874, the cattle that were given to Anastasie Frilot Brettmeyer, were from the estate of the late Leon Frilot, who was the father of both Alcide Frilot and Anastasie Frilot Brettmeyer. Also, after both gifts were given, Eugene Brettmeyer, the husband of Anastasie Frilot, sold the cattle for the amount of $800.00. Eugene Brettmeyer admitted he was indebted to his wife, Anastasie Frilot Brettmeyer, for the sum of $800.00, which actually was a legal mortgage on the property of Eugene Brettmeyer. This

$800.00 was the amount realized from the sale of the herd of twenty five cattle, which were sold by Eugene Brettmeyer. Also, the sum of $800.00, which was the purchase price of the cattle was used for the benefit and desires of Eugene Brettmeyer. Eugene Brettmeyer used the $800.00 for his own use. In order to reduce the amount of Eugene Brettmeyer's indebtness to his wife, Anastasie Frilot Brettmeyer Eugene Brettmeyer makes a Dation en paienment to his wife of the following property: one case of beer, one bar wheels, one case of milk, one half barrel of whiskey, a bill of B. H. Flashpaltor, which bill was dated September 13, 1883, one half barrel of Honey Peach wine as per bill of Flashpaltor, which bill was dated the date of September 13, 1883, one box # SS Flint, one box graham crackers, one box vermicelli, as per bill of J. M. & Taylor, which bill was dated the date of October 16, 1883, one sack of salt, one can of lard, one case of beer, as per bill, which bill was dated the date of October 24, 1883, one barrel of rice, one box of cheese, one box of yeast powder, one box of peaches, and one can of syrup.

The act of Dation en paienment was under the provisions of Section 2 of Article 244 of the Revised Civil Code of the State of Louisiana. It was the wish of Eugene Brettmeyer to make this act of Dation en paienment to his wife, Anastasie Frilot Brettmeyer. At the execution of this act of Dation en paienment, Anastasie Frilot Brettmeyer was present and accepting the rights, titles, and interest to the movable goods.

On December 3, 1882, Anastasie Frilot, who was the wife of Eugene Brettmeyer, sold to Albert Theodule and Jeanne Green, a tract of land, which was situated on the east side of Bayou Teche, in the newly created Parish of Iberia, which was formerly St. Mary Parish. This sale was made for the purchase price of $90.00, which was paid in current money at the execution of the act of conveyance.

This tract of land measured one half of an acre in front along Bayou Teche, with a depth of six acres. The tract of land was bounded on the north side and the east side by the lands of Alcide Frilot, on the south side by Bayou Teche, and on the west side by the lands of Valcourt Olivier. This tract of land contained nine arpents of land.

On the date of August 27, 1869, Joseph Devezin Olivier conveyed to Leon Frilot, a free man of color, a portion of the tract of land that he (Joseph Devezin Olivier) purchased at a public auction of the sale of the estate of the late Rosette Boutte', who was a free woman of color. This tract of land that was conveyed to Leon Frilot, a free man of color, and the son-in-law of the late Rosetta Boutte', was situated on both the west and the east banks of Bayou Teche, in the then Parish of St. Mary. This tract of land had a frontage along Bayou Teche of four arpents on the west side, and a frontage of six arpents along Bayou Teche on the east side. The tract of land on both the east and the west sides of Bayou Teche had a depth of forty arpents of land. The total land area of the tract of land, which was situated on the east bank of Bayou Teche, was two hundred and forty arpents of land. Also, the tract of land, which was situated on the west bank of Bayou Teche had a total land area of one hundred and sixty arpents of land. This tract of land was situated at about four and one half miles below the town of New Iberia, Louisiana. This tract of land was bounded on the north side by land owned by Louis Frilot, a free man of color, on the south side by the lands of Zenon Olivier and others, and on the east side by the lands belonging to the estate of the late Colonel Charles Napoleon Olivier.

The purchase price of this tract of land was the sum of $7,400.00. This purchase price was represented by two promissory notes, which were drawn by Joseph Devezin Olivier, and made to the order of Eugene Barthlemy Olivier, who was acting as administrator of the estate of the late Rosetta Boutte', a free woman of color. Both promissory notes were dated the date of January 28, 1860. Each one of the promissory notes was for the amount of $3,700.00. These two promissory notes bore on the property conveyed by Joseph Devezin Olivier to Leon Frilot, a free man of color. Both, Eugene Barthlemy Olivier and Henry Train served in their capacity as witnesses to the execution of the act of conveyance.

Madame Celeste Bellechase Olivier was taken out of the hearing and physical presence of her husband, Joseph Devezin Olivier, to renounce her dotal, paraphernal, and matrimonial rights to this property that was being conveyed.

Marie Louise Aimee Frilot, a free woman of color, was a daughter of the late Rosetta Boutte' and Claude Frilot. Also, she was married to Leon Frilot, and of this marriage three children were born, Alcide, Adelaide, and Anastasie Frilot. Marie Louise Aimee Frilot departed life in the month of November of the year of 1854. Marie Louise Aimee Frilot left property that belonged to the community of acquets and gains that existed between her and her husband, Leon Frilot, a free man of color. Also, the three children of the late Marie Louise Aimee Frilot, Adelaide, Anastasie, and Alcide were the legal representatives and heirs of their deceased mother.

In the spring of the year of 1856, Hyacinth Olivier, a free man of color, filed a petition in the probate court in St. Mary Parish, Louisiana to open the estate of the late Marie Louise Aimee Frilot, a free woman of color. Hyacinth Olivier stated that Marie Louise Aimee Frilot died intestate or leaving no written will and last testament. When a person dies intestate, the property that belongs to the estate of the deceased person, will be disposed of to the heirs of the deceased person according to the laws of descent and distribution. Also, on May 7, 1856, Hyacinth Olivier was named administrator of the estate of the late Marie Louise Aimee Frilot. Also, Hyacinth Olivier stated that Marie Louise Aimee Frilot was the wife of Leon Frilot, a free man of color, and the mother of Alcide, Adelaide and Anastasie Frilot.

On May 14, 1856, Eugene Barthlemy Olivier and Dubriel Olivier, a free man of color, and a brother-in-law of the late Marie Louise Aimee Frilot, swore in front of Judge Albert Voorhies, of St. Mary Parish, that they will conduct a true appraisal of the property belonging to the estate of the late Marie Louise Aimee Frilot, late of the Parish of St. Mary.

Also, on the date of July 23, 1856, Leon Frilot, the husband of the late Marie Louise Aimee Frilot, was appointed administrator of the estate of the late Marie Louise Aimee Frilot. Leon Frilot stated that since the death of his wife, Marie Louise Aimee Frilot, he had not had the time to conduct an inventory of the property belonging to the estate.

According to the laws of the State of Louisiana, an inventory and appraisal of the property belonging to an estate had to be conducted. Usually, a judge of the probate court would issue a writ to conduct an inventory. Also, the judge of the probate court would appoint someone to conduct an inventory, and that person would go to the last place of residence of the deceased person on a designated date to conduct an inventory of the property. Usually, there were two or more other persons, who would aid the person conducting the inventory. Also, there were witnesses to the persons conducting the inventory. Usually, a public sale or auction would follow the conducting of an inventory of the property.

On September 18, 1856, a family meeting was held to decide what to do with the property that was inherited by the three minor chidden of the late Marie Louise Aimee Frilot, and to appoint tutors to the minor children. This meeting was composed of Dubriel Olivier, a free man of color, Zenon Olivier, a free man of color, Jean Delile, a free man of color, and they were the uncles of the minor children, Pierre Boutte', a free man of color, was another uncle by marriage of the three minor children, Zenon Decuir, also attended this meeting. Zenon Olivier, fils or junior, and Dominque Olivier, both free men of color, were second cousins of the three minor Frilot children. This family meeting was composed of the nearest relatives of the minor children, and these relatives were all residents of St. Mary Parish, Louisiana. The results of this family meeting was to sell all the property belonging to the estate of the late Marie Louise Aimee Frilot. The tutors of the three minor children were also named, and they were Dubriel Olivier, a free man of color, Zenon Olivier, a free man of color, Pierre Boutte', a free man of color, Zenon Olivier, fils or junior, Dominque Olivier, a free man of color, and Jean Delile, a free man of color.

On October 23, 1856, Barthlemy de la Houssaye, a notary public in St. Mary Parish, went to the last residence of the late Marie Louise Aimee Frilot, a free woman of color, to conduct an inventory and appraisal of the property belonging to the estate. Also, Barthlemy de la Houssaye had in his possession a virtue of a writ, which was issued out of the probate court in St. Mary Parish, on the date of September 23, 1856, which writ authorized Barthlemy de la Houssaye to sell at public auction the property belonging to the estate of the late Marie Louise Aimee Frilot. The terms of the sale of the public auction were that the dwelling house was to be sold on a six and twelve month credit from the date of the sale, the slaves were to be sold on a credit of one, two, and three years from March 1, 1858, and the balance of the movable property was to be sold. Any property sold under the sum of $10.00 was to be paid in cash, and all property sold over the sum of $10.00 was to be paid on a credit basis, which was due in the month of March of the year of 1857. The purchasers making credit purchases were to give their promissory notes.

The inventory of the property that belonged to the estate of the late Marie Louise Aimee Frilot, the deceased wife of Leon Frilot, in the year of 1856, consisted of these slaves: William, a slave griffee man, age 34 years of old, and valued at the sum of $1,250.00, and Onezime, a colored slave man, age 32 years old, and valued at the sum of $1,380.00.

This was the livestock that were being raised on the Frilot plantation, in the year of 1856, three gentle work oxen, which were valued at $15.00 each, and the total valuation was the sum of $45.00, two gentle milk cows and calves, which were valued at $20.00, each cow and calf, and the total valuation was the sum of $40.00, two gentle milk cows, valued at $15.00 each cow, and the total valuation was the sum of $30.00, one bull that was four years old, and this bull was valued at the sum of $10.00, one four year old beef cow, that was valued at $15.00, a lot of three young calves, each calf was valued at $8.00 each, and the total valuation of the lot of three calves was the sum of $24.00, one young creole horse that was four years old, and the total value of the creole horse was the sum of $10.00.

The dwelling house, where the late Marie Louise Aimee Frilot, and her husband, Leon Frilot, and their three children, Alcide, Adelaide, and Anastasie, lived was valued at $50.00.

The movable property that was on the Frilot plantation,in the year of 1856, consisted of one corn crib that was valued at $16.00, and one old ox cart that was valued at $15.00. This was all of the property that was accounted for and valued at the inventory that was taken in the year of 1856.

There was another inventory that was taken on the date of June 2, 1860, and this inventory was taken by Barthlemy de la Houssaye, a notary public in St. Mary Parish. Also, Zenon Olivier, fils of junior, and Baptiste Collette, both free men of color, aided Barthlemy de la Houssaye in the conducting of this inventory of the estate of the late Marie Louise Aimee Frilot, a free woman of color. The inventory that was taken on the date of June 2, 1860 consisted of this real estate property, one tract of land, that was lying and situated in the Parish of St. Mary. This tract of land was situated on the west side of Grand Marais or the Great Swamp. Also, this tract of land contained one hundred and forty eight arpents of land, and the total valuation of the tract of land was the sum of $2,800.00.

There was another tract of land, which was situated in St. Mary Parish, and this tract of land was also on the west side of Grand Marais or the Great Swamp. This tract of land contained eighty arpents of land. Also, this tract of land was bounded above by the lands of Zenon Olivier, a free man of color, and on below by the lands of Mulot Boutte', a free man of color. The total valuation of this tract of land was the sum of $700.00.

There was also a town lot, which was situated in the town of New Iberia. This town lot measured one hundred feet across the front. The boundaries of this town lot were on the lowerside by the town lot of Hycinthe Olivier, a free man of color, and above by the street separating the property of the town lot from the town lot of Leopold Smith. The total valuation of the town lot was the sum of $225.00.

These two slaves were also on the plantation of Leon Frilot, in the year of 1860, a slave woman, named Jeanne, with her child, Melaide, who was a girl slave child of fifteen months old. The total valuation of both the slave woman, Jeanne, and her child, Melaide, was the sum of $1,575.00.

There were three notes that belonged to the estate of the late Marie Louise Aimee Frilot, a deceased free woman of color, and the late wife of Leon Frilot, a free man of color. One promissory note was in favor of Leon Frilot and hand signed by Camille Hebert and E. Montagne, fils (junior) for the amount of $300.00. This promissory note was due in three months from the date of April 7, 1860. Also, this promissory note was conditioned to bear 8% interest from the date of April 7, 1860.

There was another promissory note dated the date of March 8, 1860, and this mortgage note was signed by Falville Kerlegan and Marcelin Carriere. Also, this promissory note was in favor of Leon Frilot. The amount of the notes was the sum of $220.00, with 8% interest per annum from maturity. This promissory note was payable or due on May 20, 1860.

The third promissory note was hand signed by Jacob Feund, and this mortgage note was dated the date of March 30, 1860. The amount of this mortgage note was the sum of $225.00. Also, this promissory note was payable on demand. The third promissory note was conditioned to bear 8% interest per annum from the date of March 30, 1860.

This was the livestock that was situated on the plantation of Leon Frilot, and his deceased wife, Marie Louise Aimee Frilot, at this time when the second inventory was conducted. There were eleven gentle cows, and three of the cows had calves. The value of each cow was $15.00, and the total valuation of the eleven cows as the sum of $165.00. Also, there were three pair of work oxen, and the work oxen were valued at $40.00 a pair of oxen. The total valuation of the work oxen was the sum of $120.00. There were six calves and these six calves were one years old. The value of each one year old calf was the sum of $6.00 a head, and the total valuation was the sum of $42.00. Also, on the plantation were eight creole mares and the creole mares were valued at $14.00 a mare. The total valuation of the eight creole mares was the sum of $112.00. These eight creole mares were unbroken. There were seven creole horses, and these seven creole horses were gentle. Each one of the seven creole horses were valued at $40.00 a head, and the total valuation of the seven creole horses was the sum of $280.00. There was one stud horse, and the stud horse was unbroken. This stud horse was valued at the sum of $25.00. Also, there was one gentle stud horse, and this gentle stud horse was valued at the sum of $30.00. On the Frilot plantation, were two wild creole horses, and these two wild creole horses were valued at $30.00 each horse. The total valuation of these two wild creole horses was the sum of $60.00. A small vachery of cattle also belonged to this estate, and this small vachery of cattle was situated in the Parish of Calcasieu. This small vachery of cattle contained seventy heads of cattle. The total valuation of this small vachery of cattle was the sum of $700.00. Another small vachery of cattle belonged to this estate, and this vachery of cattle contained fifty heads of cattle. This small vachery of fifty heads of cattle was situated in the Parish of St. Mary. The total valuation of the small vachery of fifty heads of cattle was the sum of $500.00. This was all the livestock that was situated on the Frilot plantation in the year of 1860.

This was the movable property that belonged to the estate of the late Marie Louise Aimee Frilot, and this movable property was situated on the Frilot plantation. One ox cart, which was valued at the price of $6.00, one horse cart, which was valued at the price of $4.00, one lot of plantation tools, which was valued at the price of $22.00, one lot of pieuxs, and these pieuxs were valued at the price of $162.00, one lot of kitchen utensils, which was valued at the price of $4.00, one bed and bedding, which was valued at the price of $14.00, one bed and bedding, which was valued at the price of $8.00, one bed and bedding, which was valued at the price of $8.00, one armoire, which was valued at the price of $8.00, one bureau, which was valued at the price of $5.00, two small cypress tables, which was valued at the price of $1.50, one cypress dining table, which was valued at the price of $2.00, one safe, which was valued at the price of $4.00, one safe, which was valued at the price of $1.50, one dozen or twelve chairs, which were valued at the price of $15.00, one rocker, which was valued at seventy five cents, and one clock, which was valued at the price of $10.00.

The total valuation of the property that belonged to the estate of the late Marie Louise Aimee Boutte' Frilot, a free woman of color, and the late wife of Leon Frilot, a free man of color, was the sum of $8,393.75. This inventory was conducted in the presence of Jules Porison and Francois Lanet, who served in their capacity as competent witnesses. Also, this inventory was conducted at the residence of Leon Frilot.

On the date of July 25, 1860, the total amount in the account for the estate of the late Rosetta Boutte', who was the deceased wife of Claude Frilot, contained the amount of $35.00, which was retained for expenses of homologation as per account. Also, in the month of October of the year of 1860, the notice of the probation of the will and last testament of Rosetta Boutte' was printed in the Attakapas Register, a weekly newspaper. On the date of August 15, 1866, a draft was made to the account by Jules Germain Olivier for the two promissory notes of Auguste Landry, which two promissory notes were owed to the estate of Rosetta Boutte'. Each one of the two promissory notes was for the amount of $21.66, and each promissory note was conditioned to bear interest at the rate of 8% from the date of July 25, 1861, and July 25, 1862. Also, on the date of August 15, 1866, Joseph Devezin Olivier made a purchase from Falville Kerlegan, a grandson of the late Rosetta Boutte', in the amount of $7,400.00. In the month of November of the year of 1866, Ursin Olivier paid his two promissory notes to the account of the estate of the late Rosetta Boutte'. Each one of the promissory notes was for the amount of $26.66, which made a total amount deposited to the account of the estate of the late Rosetta Boutte' a total of $53.33. Also, on the date of the 1st of November, in the year of 1866, Leon Frilot paid to the account of the estate of the late Rosetta Boutte' his two promissory notes, which was the total amount of $54.00.

Now, the account of the estate of the late Rosetta Boutte' had a balance of the amount of $7,129.47. This amount of $7,129.47 was to be partition among the heirs of Rosetta Boutte'. Also, this amount of $7,129.47 was represented by ten promissory notes. The share or portion of this amount, which was rightfully given to Falville Kerlegan, who represented one tenth share, or the sum of

$712.95. This total amount of the sum of $7,129.47 resulted from the purchase by Joseph Devezin Olivier, and having now subsequently been assigned by Falville Kerlegan to the heirs of the estate of the late Rosetta Boutte'.

Each one of the heirs of the estate of the late Rosetta Boutte', received a one tenth amount of the total amount of the sum of $7,129.47 or the sum of $712.95. From this total amount, Arthemis Frilot received an amount of $717.94. Modeste Frilot, a daughter of the late Rosetta Boutte, and Modeste Frilot Olivier, the deceased wife of Zenon Olivier, children each received one seventh of the one tenth share of their deceased mother's, share in the estate of their late grandmother, Rosetta Boutte.
The children of Modeste Frilot Olivier were Louis Olivier and he received the sum of $102.56, Zenon Olivier and he received the sum of $102.56, Clemarie Olivier and he received the sum of $102.56, Dominque Olivier and he received the sum of $102.56, Avelaide Rosa Olivier, who was the wife of Zenon Decuir,
received the sum of $102.56, Joseph Alphonse Olivier received the sum of $102.56 as his share in the estate of his grandmother, Rosetta Boutte', and Justin Cylore Olivier received the sum of $102.56 as his share in the estate of his maternal grandmother, the late Rosetta Boutte'.

Antoine Frilot, a son of the late Rosetta Boutte' and Claude Frilot, received the sum of $717.94 as his share in the estate of his deceased mother.

Adeline Frilot, a daughter of the late Rosetta Boutte' and Claude Frilot, and the widow of Pierre Olivier, received the sum of $717.94 as her share in the estate of her deceased mother.

Azelie Frilot, a daughter of Rosetta Boutte' and Claude Frilot, and the widow of Honore' Olivier, received the sum of $717.94 has her share in the estate of her deceased mother.

Marie Frilot, another daughter of the late Rosetta Boutte' and Claude Frilot, and the widow of Celestin Rochon, received the sum of $717.94 as her share in the estate of her deceased mother.

The children of Francois Frilot, a deceased son of the late Rosetta Boutte' and Claude Frilot, inherited their father's, Francois Frilot's, share in the estate of their deceased paternal grandmother, Rosetta Boutte'. The children of Francois Frilot were Francois Frilot, Junior or fils, and he received the sum of $120.56 as his share in the estate of his late grandmother. Franuilet Frilot received the sum of $102.56 as his share of his late paternal grandmother's estate, Hilarie Frilot received the sum of $102.56 as her share in the estate of her paternal grandmother, Prosper Frilot received the sum of $102.56 as his share of the estate of his late paternal grandmother, Rosetta Boutte', Rose Aimee Frilot received the sum of $102.56 as her share in the estate of her paternal grandmother, Eloi Frilot received the sum of $102.56 as his share in the estate of his paternal grandmother, and Carmelite Frilot received the sum of $102.56 as her share in the estate of her paternal grandmother.

The two daughters of the late Louis Frilot, a deceased son of the late Rosetta Boutte' and Claude Frilot, Leonnie and Lennie Frilot, both inherited each one half of their late father's share in the estate of the late Rosetta Boutte'. Leonnie Frilot, who was the wife of Frances Leonelle, received the sum of $102.56 as her share of the estate of her paternal grandmother. Also, Lennie Frilot, who was the wife of Joseph Philip Frilot, received the sum of $102.56 as her share of the estate of her late paternal grandmother. Both Leonnie and Lennie Frilot were both over the age of majority. At this time, in the 1860's, the age of majority in the State of Louisiana, was twenty one years of age.

There were other children, who were heirs of the late Louis Frilot, a deceased son of Rosetta Boutte' and Claude Frilot, but these children were all considered minors under the laws of the State of Louisiana. They also inherited a portion of their late father, Louis Frilot's, share in the estate of the late Rosetta Boutte'. The minor children of the late Louis Frilot were Louis Frilot, Junior or fils, and he received the sum of $102.56 as his share, Crabie Frilot received the sum of $102.56, Cora Frilot received the sum of $102.56 as her share in the estate of her paternal grandmother, Hilaire Frilot received the sum of $102.56 as their share in the estate of their paternal grandmother, and Cathiche Frilot received the sum of $102.56 from the estate of the late Rosetta Boutte.'

These were the heirs of the late Phillpe Aimee Frilot, Phillpe Aimee Frilot, Junior or fils, who received the sum of $102.56 as his share in the estate of his paternal grandmother, Rosetta Boutte', Rosetta Frilot, who received the sum of $102.56 as her share in the estate of her deceased paternal grandmother, Pierre Frilot who received the sum of $102.56 as his share in the estate of his paternal grandmother, Louis Frilot received the sum of $102.56 as his share in the estate of his paternal grandmother, Charles Frilot received the sum of $102.56 as his share in the estate of his paternal grandmother, and Aspasie Frilot, a deceased child of Phillpe Aimee Frilot, and a grandson of the late Rosetta Boutte', was represented by his children in the estate of his paternal grandmother. Phillpe Aimee Frilot, pere or Senior, was a deceased son of the late Rosetta Boutte' and Claude Frilot. Under Louisiana law, when a parent dies, in this case the father, Phillpe Aimee Frilot, the children inherit equally a portion of their deceased parent's share in the estate. When an estate is probated, the proceeds of the estate are divided equally among all the heirs.

As a result of the public auction of the real estate and the movable property belonging to the estate of the late Rosetta

Boutte', a free woman of color, which was held on the date of January 25, 1860, at the doors of the St. Mary Parish Courthouse, which was situated in the town of Franklin, Louisiana, there existed indebtness to the estate. This indebtness was in the form of promissory notes, which the purchasers had given for the credit payment of their purchases.

There were two promissory notes made to the order of Leon Frilot, a free man of color, and the son-in-law of the late Rosetta Boutte', and Hilarie Decuir. These promissory notes were in soldio. Also, these promissory notes were dated the date of January 25, 1860. Each one of the promissory notes was for the amount of $1,286.66. One promissory note was payable on the date of July 25,1861, and the other promissory note was payable on the date of July 25, 1862.

There were two more promissory notes made to the order of Zenon Olivier, fils or junior, and Charles Frilot, both were free men of color. These two promissory notes were in soldio. Also, both promissory notes were dated the date of January 25, 1860. Each one of the two promissory notes was for the amount of $833.33. The first promissory note was payable on the date of July 25, 1861, and the second promissory note was payable on the date of July 25, 1862.

Also, there were two more promissory notes made to the order of Pellerin Olivier and Jules Germanin Olivier, and these two promissory notes were in solido. These two promissory notes were dated the date of January 25, 1860. Each one of the promissory notes was for the amount of $833.33. The first promissory note was payable on the date of July 25, 1861, and the second promissory note was payable on the date of July 25, 1862.

Two more promissory notes were given by purchasers at the public auction of the estate of the late Rosetta Boutte,' and these two promissory notes were made to the order of A. H. Brown, and Edouard Perrot. Both of these promissory notes were in soldio. These two promissory notes were dated the date of January 25, 1860. Each one of these two promissory notes were for the amount of $228.33. The first promissory note was due on the date of July 25, 1861, and the second promissory note was due on the date of July 25, 1862.

Two more promissory notes were made in favor of the estate of the late Rosetta Boutte', a free woman of color, and these two promissory notes were made to the order of Leon Frilot and Zenon Olivier, both free men of color. Both promissory notes were in soldio. These two promissory notes were dated the date of January 25 1860. Each promissory note was for the amount of $683.33. The first promissory note was due on the date of July 25, 1861 and the second promissory note was due on the date of July 25, 1862.

There were two more promissory note that were made in favor of the estate of the late Rosetta Boutte', a free woman of color, for purchases made at the public auction sale. These two promissory notes made to the order of Charles Frilot and Zenon Olivier, fils or junior, both were free men of color. Also, both promissory notes were made in solido. These two promissory notes were dated the date of January 25, 1860. Each one of the two promissory notes were made for the amount of $166.66. The first promissory note was due on the date of July 25, 1861, and the second promissory note was due on the date of July 25, 1862.

Two more promissory notes were made in favor of the estate of the late Rosetta Boutte', and these two promissory notes or mortgage notes were made in favor of Marie Rochon, who was a free woman of color and Clemarie P. Olivier, a free man of color. These two promissory notes were in solido. Both promissory notes were dated the date of January 25, 1860. Each one of the two promissory notes were for the amount of $666.66. The first promissory note was due on the date of July 25, 1861, and the second promissory note was due on the date of July 25, 1862.

There were two more promissory notes in favor of the estate of the late Rosetta Boutte', who was a free woman of color, and these two mortgage notes were made to the order of Jacob Freund. These two promissory notes were both dated the date of January 25, 1860. All of the above promissory notes or mortgage notes were all conditioned to bear 8% interest.

Fils was a French term that was used frequently in the 1800's, and the term, "fils," means junior in the French language. Therefore, the French term, "pere," means senior, and this term was also used frequently in the 1800's. An example of the usage of pere and fils would be this Jean Baptiste LeMoyne de la Houssaye, pere, designates the father, and Jean Baptiste LeMoyne, fils, designates the son. Today, the English terms junior and senior are used in place of the French terms.

On the date of January 25, 1860, a public succession sale was held to dispose of some of the real estate property and other movable belonging to the estate of the late Rosetta Boutte', late of the Parish of St. Mary. Prior to the public succession sale, the Sheriff of St. Mary Parish, William Haifleigh, had to advertise this forthcoming public succession sale of real estate property and movable property belonging to the estate of the late Rosetta Boutte' in both the French and the English languages. These notices of advertisement advertising the public succession sale were posted on the doors of the St. Mary Parish Courthouse, which was situated in the town of Franklin, Louisiana. Also, Sheriff William Haifleigh had to advertise this public succession sale in the weekly newspaper, the Planters Banner. The advertisement stated the date of the public succession sale, in this case the date was January 25, 1860, the place of the public succession sale, in this case it was held at the doors of the St. Mary Parish Courthouse, which was situated in the town of Franklin, Louisiana, what property was being sold and the time of the public sale, which was usually 11:00 A. M. through 4:00 P. M., and the terms of the sale, whether the purchasers had to pay cash or furnish credit with good collateral. Usually the terms of most public sales was anything purchased under $500.00 was sold for cash, and anything over $500.00 was sold on terms of credit of one year credit. The Planters Banner was the weekly newspaper, which was printed in the town of Franklin, Louisiana.

On the date of January 25, 1860, the public succession sale was held, and lot one of the Rosetta Boutte plantation was auctioned off to be sold to the last and highest bidder. All public sales were conducted on an auction like basis, with the purchasers bidding the sale price. Lot one of the Rosetta Boutte' plantation was situated and lying in the Parish of St. Mary, and it was situated on the west side of Bayou Teche. Lot one had a frontage of four and one half arpents fronting Bayou Teche with a depth of forty arpents. Lot one was bounded on the upperside by the lands of Leon Frilot, a free man of color, and below by the lands of Joseph Devezin Olivier. Also, all the buildings situated on the property were included in the sale. The sale price of lot one was $4,900.00. The last and highest bidder was Joseph Devezin Olivier, who was neighbor of the late Rosetta Boutte's, and he owned a sugarcane plantation below the plantation of Rosetta Boutte.

Lot two of the Rosetta Boutte plantation was situated opposite lot one. Lot two was situated on the east side of Bayou Teche in the Parish of St. Mary. Also, lot two had a frontage along Bayou Teche of five arpents with a depth of forty arpents. Lot two was bounded above by the lands of Leon Frilot, a free man of color, and below by the lands of Joseph Devezin Olivier. Also, all the buildings that were situated on lot two were included in this sale. The purchase price of lot two was the sum of $1,200.00. Joseph Devezin Olivier was the last and the highest bidder for lot two, and lot two was sold to him. Also, the neighbor on the upperside or above of the Rosetta Boutte' plantation was Leon Frilot, who was the son-in-law of the late Rosetta Boutte', being the husband of her daughter, Marie Louise Aimee Frilot, and the father of three grandchildren of the late Rosetta Boutte', who were Alcide Frilot, Adeline Frilot, and Anastasie Frilot.

Another son of the late Rosetta Boutte and Claude Frilot, Phillpe Aimee Boutte' had died before the estate of the his late mother, Rosetta Boutte', had been probated and settled. In the State of Louisiana, whenever a son or a daughter predeceases one of their parents or both of their parents, then the children of that son or daughter or the grandchildren of the parents will inherit

their deceased parent's share in the estate of their grandparents. If the grandchild of the parents or the child of the son or daughter is an only child, then that child will inherit the whole share of their deceased parent. Also, the deceased parent's share in the estate will be divded equally among all children, for example, if a son has five children, then each child of the son will inherit a one fifth share of their grandparents estate. The way a share of an estate is divided will depend on how many children the deceased parent left, and the share is usually divided equally among all the children. If an estate is worth the sum of $100,000.00 and there are four children, then each child will receive the sum of $25,000.00 as their share of the estate. If one of the four children are deceased, then the children of the deceased child will inherit
their parent's interest into the estate of a deceased grandparent.

The son of Rosetta Boutte' and Claude Frilot, Philip Aimee Boutte' had died prior to the death of his mother. Also, Rosetta Boutte' had named her son, Philip Aimee Frilot, for her brother, Philpeau Boutte'. Philpeau Boutte' was the brother of Rosetta Boutte, and the maternal uncle of Phillpe Aimee Frilot. Philpeau Boutte' owned the Boutte' plantation, which was situated above the plantation of his sister, Rosetta Boutte'.

Another son of the late Rosetta Boutte' and Claude Frilot, Aspasie Frilot had passed away prior to the death of his mother. Aspasie Frilot was represented in the settlement of the estate of the late Rosetta Boutte', a free woman of color and the wife of Claude Frilot, by his children, or the grandchildren of the late Rosetta Boutte' and Claude Frilot. The legal heirs and children of Aspasie Frilot were Edmund Frilot and he received the sum of $51.28 as his share in the estate of the late Rosetta Boutte', and Prosper Francis Frilot, who was the other son of Aspasie Frilot and he received the sum of $51.28 as his share in the estate of the late Rosetta Boutte.' In this case, the share of Aspasie Frilot was divided among his two children. Each child of Aspasie Frilot received one half of their deceased father's interest or share in the estate of the late Rosetta Boutte.'

Aimee Frilot, the deceased wife of Leon Frilot, and another daughter of the late Rosetta Boutte' and Claude Frilot was also represented by her three children, who were Alcide Frilot, Adelaide Frilot, and Anastasie Frilot. The full name of Aimee Frilot was Marie Louise Aimee Frilot. Also, Alcide, Anastasie, and Adelaide were the grandchildren of the late Rosetta Boutte' and Claude Frilot. Alcide Frilot received the sum of $34.18 as his share in the estate of his late grandmother, Adeliade Frilot received the sum of $34.18 as their share in the estate of their late grandmother, Rosetta Boutte', and Anastasie Frilot received the sum of $34.18 as her share in the estate of her late grandmother, Rosetta Boutte.' In this case, the share of the late Marie Louise Aimee Frilot, the deceased wife of Leon Frilot, and a daughter of the late Rosetta Boutte', was equally divided among her three children, with each child receiving a one third interest of the share of their late mother, Aimee Frilot, in the estate of their late grandmother, Rosetta Boutte.'

The following grandchildren of the late Rosetta Boutte' a free woman of color, and the deceased wife of the late Claude Frilot, also inherited from the estate left by Rosetta Boutte', through their parents, who were the children of the late Rosetta Boutte.' The children of Francoise Frilot a free man of color, and a son of the late Rosetta Boutte', were Irmaillete Frilot, Hilaire Frilot, and Prosper Frilot. All of the children of the late Francoise Frilot, were residents of the Parish of St. Mary, and the State of Louisiana. Also, a daughter of the late Francoise Frilot, Rose Aimee Frilot, who was a free woman of color, and the wife of Jean LaCour of the City of New Orleans inherited a portion of her deceased father's share in the estate of the late Rosetta Boutte.'

These were other grandchildren, who inherited from the estate of the late Rosetta Boutte', Louise Frilot, who was the wife of Iramuis Lemelle, Leonnie Frilot, who was the wife of Joseph Phillpe Frilot, and Louis Frilot. All these grandchildren were residents of St. Landry Parish, which is situated in the State of Louisiana.

The children of the late Louis Frilot, a grandson of the Rosetta Boutte' were Coralie, Cora, Hilaire, and Catachie Frilot. All these grandchildren were minors and were under the natural tutorship of their mother, Aimee P. Frilot, who was the widow of the late Louis Frilot. All these grandchildren and their mother were residents of the Parish of St. Landry.

The children of Phillpe Aimee Frilot, a free man of color, were Phillpe Aime Frilot, who was a resident of the Parish of St. Mary, Pierre Frilot, who was a resident of the Republic of Mexico, Louis Frilot and Charles Frilot, who were both residents of the Parish of St. Mary, which is situated in the State of Louisiana.

The children of Aspasie Frilot, who was the deceased wife of S. B. Raymond, were Edmond Francois Raymond, and Prosper Francoise Raymond, and both these children were residents of the Parish of St. Martin, which is situated in the State of Louisiana.

The children of the late Aimee Frilot, who was the deceased wife of Leon Frilot, were Alcide, Adelaide, and Anastasie Frilot. These three children were all minors, who were under the natural tutorship of their father, Leon Frilot. Anastasie Frilot would become the wife of Eugene Brettmeyer.

Honville Kerlegan was another grandson of the late Rosetta Boutte', and he was the son of Marie Frilot, who was a daughter of

Rosetta Boutte and Claude Frilot. Marie Frilot was married to Jean Baptiste Kerlegan of the Parish of St. Martin, which is situated in the State of Louisiana.

The total assets belonging to the estate of the late Rosetta Boutte' that was to be partitioned was the total amount of $9,546.97. This was how the assets belonging to the estate of the late Rosetta Boutte'' was divided. The children of the late Rosetta Boutte' received these sums of money, Arthemise Frilot, who was a child of the late Rosetta Boutte' and Claude Frilot, received the sum of $867.90, Modeste Frilot, who was a daughter of the late Rosetta Boutte' and Claude Frilot, received the sum of $867.90, Antoine Frilot, who was a son of the late Rosetta Boutte' and Claude Frilot, received the sum of $867.00, Adeline Frilot, who was a child of the late Rosetta Boutte' and Claude Frilot, received the sum of $867.00, Azehi Frilot, who was a child of the late Rosetta Boutte' and Claude Frilot, received the sum of $867.00, Hortense Frilot, who was a daughter of the late Rosetta Boutte' and Claude Frilot, received the sum of $867.00, and Leon Frilot, who was a son of the late Rosetta Boutte' and Claude Frilot, received the sum of $867.00. These grandchildren, who were the heirs and legal representatives of the late Rosetta Boutte' and Claude Frilot inherited portions of their deceased parent's share in the estate of the late Rosetta Boutte'.

The children of Francoise Frilot, who was a deceased son of the late Rosetta Boutte' and Claude Frilot. were Prosper Frilot and he received the sum of $123.98 from the estate of his late paternal grandmother, Rosetta Boutte', Eloi and he received the sum of $123.98 as his share in the estate of his late grandmother, Rosetta Boutte', and Rose Aimee Frilot and she received the sum of $123.98 as her share in the estate of her late grandmother, Rosetta Boutte'.

The children of the late Louis Frilot, who was another son of the late Rosetta Boutte' and Claude Frilot were Coralie Frilot and she received the sum of $123.98 as her share in the estate of the her late paternal grandmother, Rosetta Boutte, Cora Frilot and she received the sum of $123.98 as her share in the estate, Cathehe Frilot and he received the sum of $123.98 as his share in the estate, and Hilaire Frilot and he received the sum of $123.98 as his share in the estate. These four children were all minors, who were under the legal age or under the age of majority, and were under the tutorship of their mother, Anne Paillet Frilot, who was the widow of the late Louis Frilot.

These were the children of the late Phillpe Aimee Frilot, and they were Rosetta Frilot and she received as her share in the estate of her late grandmother, Rosetta Boutte', the sum of $123.98, Pierre Frilot and he received the sum of $123.98 as his share in the estate, and Louis Frilot and he received the sum of $123.98 as his share in the estate of his late grandmother.

These were the children and the legal heirs and representatives of the late Aspasie Frilot, who was a deceased daughter of the late Rosetta Boutte' and Claude Frilot, and these children were Alcide Frilot and he received the sum of $41.32 as his share of the estate of his maternal grandmother, Anastasie Frilot and she received the sum of $41.32 as her share of the estate of her maternal grandmother, and Adelaide Frilot and she received the sum of $41.32 as her share of the estate of her late maternal grandmother. At the time of the probating of the estate of the late Rosetta Boutte', all three of these children were minors, who were under the age of majority, and these three minor children were all under the natural tutorship of their father, Leon Frilot.

The only child of the late Marie Frilot, who was a deceased daughter of the late Rosetta Boutte' and Claude Frilot. was Honville Kerlegan. Marie Frilot was married to Jean Baptiste Kerlegan. Honville Kerlegan received the sum of $67.00 as his share of the estate of his late maternal grandmother, Rosetta Boutte'.

The division of the assets of the estate belonging to the late Rosetta Boutte', late of the Parish of St. Mary, nowdays Iberia Parish, was done in the month of July of the year of 1860.

There were slaves that were residing and working on the plantation that was formerly owned by the late Rosetta Boutte', a free woman of color. These slaves were sold at the public auction of the belonging of the estate of Rosetta Boutte', which was held on the date of January 25, 1860. The slaves were sold as follows: Charles, a slave man, who was age 65 years old. The slave man, Charles, had a defective complaint of laryngitis. Also, the slave man, Charles was addicted to alcohol being an alcoholic. The slave man, Charles was sold to Charles Frilot, a free man of color, for the purchase price of $500.00. Baptiste, a slave man, and he was 58 years of age. Baptiste was subject to rheumatism or modern day arthritis. The slave man, Baptiste, was sold to Aaron W. Brown for the purchase price of $1,850.00. Numa, a slave man, and he was 50 years of age. The slave man, Numa, was subject to rheumatism. The slave man, Numa, was sold to Zenon Olivier, a free man of color, for the purchase price of $1,180.00. Jim, a colored slave boy, and Jim was about eleven years of age. The slave boy, Jim, was sold to Pellerin & Olivier for the purchase price of $2,500.00. Moda, a colored slave boy, and Moda was fourteen years of age. The slave boy, Moda, was sold to Zenon Olivier, a free man of color, for the purchase price of $1,850.00. Ellen, a slave woman, and her son, Albert, were also sold. The slave woman, Ellen, was about twenty nine years of age, and her son, Albert, was eleven years of age. The slave woman, Ellen, and her son, Albert, were sold to Jacob Frenund for the purchase price of $3,500.00. Then the slave woman, Adelaide, who was fifty six years of age. The slave woman, Adelaide, was subject to rheumatism, and she had a swelling on one leg. The slave woman, Adelaide, was sold to Hortense Rochon, a free woman of color, for the purchase price of $400.00. Christine, a colored slave girl, and she was eleven years of age. The slave girl, Christine, was sold to Hortense Rochon for the purchase price of $1,600.00. Jane, a colored slave woman, and her child, Melaide, were also sold. The slave woman, Jane, was twenty seven years of age, and her child, Melaide, was one years old. Also, the slave woman, Jane, had been lame or crippled for some time. The slave woman, Jane, and her child, Melaide, were sold to Leon Frilot, the son-in-law of the late Rosetta Boutte', for the purchase price of $2,050.00. Justine, a colored slave woman, and her two children were also sold at the public auction of the belonging to the estate of the late Rosetta Boutte', which was held on the date of January 25, 1860. The two children of the slave woman, Justine, were Francoise and Euzebee. The slave woman, Justine, was twenty two years of age, and her two children, Francoise was age fifteen months old, and Euzebee was an infant slave, who was age one month old. Justine, the slave woman, and her two children, Francoise and Euzebee, were sold to Leon Frilot. The purchase price of the slave woman, Justine, and her two children, Francois and Euzebee, was the sum of $3,860.00.

Also, all the livestock that was on the plantation of the late Rosetta Boutte' was sold at the public auction, which was held on the date of January 25, 1860. The livestock that was sold consisted of one creole horse, which was sold for the purchase price of $51.00, one creole horse, which was sold for the purchase price of $72.00, one creole horse, which was sold for the purchase price of $25.00, one creole horse, which was sold for the purchase price of $85.00, one creole horse, which was sold for the purchase price of $50.00, one creole horse, which was sold for the purchase price of $85.00, and one creole horse, which was sold for the purchase price of $50.00. Another creole horse, which was sold for the purchase price of $25.00, to Zenon Olivier, a free man of color. Another creole horse, which was sold for the purchase price of $21.00, to Hypolite Bayard. Another creole horse, which was sold for the purchase price of $25.00, to Philemon Boutte'. Another creole horse, which was sold for the purchase price of $19.00, to N. Miller. Another creole horse, which was sold for the purchase price of $18.00, to N. Miller. Another creole horse, which was sold for the purchase price of $20.00, to N. Miller. Another creole horse, which was sold for the purchase price of $15.00, to L. Decuir. Another creole horse, which was sold for the purchase price of $8.00, to L. Decuir. Three mares and colts that were sold to E. Ratier for the purchase price of $36.00. Four colts, that were two years old, was sold to Charles Frilot, a free man of color, for the purchase price of $3.00. Eleven mares were sold to Prosper Francois for the purchase price of $137.50, and each mare sold for the purchase price of $12.50. One pair of work oxen for the first choice was sold to Leon Frilot for the purchase price of $55.00. One pair of work oxen, of the second choice, was sold to Edmund Francois for the purchase price of $50.00. Another pair of work oxen, of the third choice, was sold to Edmund Francois for the purchase price of $50.00. Another pair of work oxen, of the fourth choice, was sold to Edmund Francois for the purchase price of $56.00. One beef cow was sold to Edmund Francois for the purchase price of $17.00. One pair of bulls was sold for the purchase price of $25.00 to Edmund Francois. One cow and calf was sold to Edmund Francois for the purchase price of $18.50. One cow was sold to Edmund Francois for the purchase price of $20.00. One cow was sold to Edmund Francois for the purchase price of $13.00. Another pair of work oxen, of the fifth choice, was sold to Marcelin Carriere for the purchase price of $41.0.00. One beef cow was sold to E. Theriot for the purchase price of $18.00. Another pair of bulls was sold to Alfred "Alphe" Bayard for the purchase price of $26.50. Another pair of bulls was sold for the purchase price of $18.00 to Mr. Broussard. Another pair of bulls was sold for the purchase price of $16.00 to Louis Frilot, a free man of color, and a son of the late Rosetta Boutte' and Claude Frilot. Three calves were sold for the purchase price of $12.00 to Louis Frilot. One Yearling was sold to Louis Frilot for the purchase price of $7.00. One cow and calf was sold to E. N. Ratier for the purchase price of $23.00. Another cow and calf was sold to Charles Frilot, a free man of color and a son of the late Rosetta Boutte' and Claude Frilot, for the purchase price of $16.00. Another cow and calf was sold to Charles Frilot for the purchase price of $11.50. Two cows and calves were sold to Hypolite Bayard for the purchase price of $13.00. One cow was sold to Charles Frilot, a free man of color, for the purchase price of $23.00. Two heifers were sold to Jacob Freund for the purchase price of $23.00. One lot of wild cattle was sold to Jules Vincent for the purchase price of $235.00. One herd of sheep was sold for the purchase price of $79.05, and each sheep was sold for the purchase price of $1.55. One lot of hogs was sold to E. N. Ratier for the purchase price of $27.00. Also, the brand that was used to brand the cattle on the plantation of Rosetta Boutte' was sold to Jules Vincent for the purchase price of $115.00.

The movable property that was situated on the plantation of the late Rosetta Boutte' at the time of this public auction on the date of January 25, 1860 consisted of one lot of corn that was sold for the purchase price of $23.00 to Leon Frilot, one lot of hay, which was sold for the purchase price of $11.00, to Leon Frilot, a free man of color, one lot of fodder, which was sold for the purchase price of $51.00, to Leon Frilot, one ox cart, which was sold for the purchase price of $11.00, to Leon Frilot, one lot of yokes, which was sold for the purchase price of $1.00, to Leon Frilot, one lot of chains, which was sold for the purchase price of $2.25, to Leon Frilot, one lot of agricultural sundries or miscellaneous agricultural equipment was sold for the purchase price of $1.50, one lot of cow peas or black eyed peas was sold to Joseph Devezin Olivier, who was the owner of the Olivier plantation and a neighbor of the late Rosetta Boutte', for the purchase price of $5.00, one lot of ploughs was sold to Dubriel Soliet for the purchase price of $6.00, one lot of harrows was sold to Desire Borel for the purchase price of $5.25, one sofa was sold to Leon Frilot for the purchase price of fifty cents, one lot of scythes was sold for the purchase price of $1.00, which was paid in cash, two X saws was sold to Charles Darby, a free man of color, for the purchase price of $6.50, one lot of hoes and spades was sold for the price of $1.00, which was paid in cash money, the tools were sold to Leon Frilot for the purchase price of $2.25, one lot of potatoes was sold to Leon Frilot for the purchase price of $3.25, one corn mill was sold to Leon Frilot for the purchase price of $1.00, one lot of sheep skins was sold for sixty five cents, one lot of chains was sold for fifty cents, one armoire and the contents of the armoire were sold for the purchase price of $115.00 to Marie Therese Frilot, one bedstead and the bedding was sold to Marie Therese Frilot for the purchase price of $25.00, one pair of candlesticks was sold to Marie Therese Frilot for the purchase price of $1.50, one bedstead and the bedding was sold to Jules Vincent for the purchase price of $27.00, one old prep was sold to Ursin Olivier for the purchase price of $3.50, one pair of blankets was sold to Zenon Olivier for the purchase price of $2.00, one blanket and the counterpane was sold to Zenon Olivier for the purchase price of $2.00, one small table was sold to Charles Frilot for the purchase price of $1.25, one looking glass was sold for the purchase price of $3.00 to Charles Frilot, one bowl and pitcher was sold for the purchase price of $3.00 to Charles Frilot, two pair of armoirs were sold to Zenon Olivier a free man of color and the son-in-law of the late Rosetta Boutte, for the purchase price of $2.00, one dozen silver spoons and silver forks were sold to Mr. Fusilier for the purchase price of $181.00, and this purchase money was advanced, one table was sold to Leon Frilot for the purchase price of $2.00, one lot of tumblers was sold to Leon Frilot for the purchase price of sixty five cents, one clock was sold for the purchase price of $5.00 to Leon Frilotr, one lot of crockery and knives was sold for the purchase price of $1.50 to Leon Frilot, one safe was sold to Leon Frilot for the purchase price of $3.00, one large jar was sold to Leon Frilot for the purchase price of $3.50, one lot of buckets was sold to Leon Frilot for the purchase price of $2.25, twelve or one dozen chairs and rocking chairs were sold to Charles Darby for the purchase price of $8.50, one lot of pots was purchased by Jacob Freund for the purchase price of $3.25, one lot of jars and sundries was sold to Jacob Freund for the purchase price of $2.25, and one table was sold for cash.

On March 29, 1859, Charles M. Olivier, who served as recorder in the Parish of St. Martin, which is situated in the State of Louisiana, was authorized and empowered to take and conduct an inventory of the property, which was situated in the Parish of St. Martin, belonging to the estate of the late Rosetta Boutte', a free woman of color.

On the date of April 4, 1859, Charles M. Olivier, the

recorder of the records in the Parish of St. Martin, proceeded to take an inventory and estimation of all the property lying in the Parish of St. Martin belonging to the estate of the late Rosetta Boutte', a free woman of color. Charles M. Olivier swore and gave the oath to take an inventory of an estate to Richard Easterly and Louis S. Gary, both of St. Martin Parish. This swearing in and administering an oath to persons who were authorized and empowered to conduct and take an estimation inventory of an estate was required by the laws of the State of Louisiana.

The inventory of the estate of property lying in St. Martin Parish belonging to the estate of Rosetta Boutte', a free woman of color, consisted of a town lot lying in the town of St. Martinville, Louisiana. This town lot comprised a part of the property that was a plantation formerly owned by George Berard. This town lot consisted of lot #22 of the plat. Also, this town lot was bounded on the north side by Honore Street, on the south side by lot # 44, on the east side by lot # 41, and on the west side by Vivian Street. There was another piece of property situated in the town of St. Martinville, and this property adjoined the first lot, being lot # 22. This second lot was bounded on the north side by lot # 42, on the south side by Hyacinth Street, on the east side by lot # 43, and on the west side by Vivian Street. Also, there was a third piece of property, and this third piece of property was situated in the same place as the other two pieces of property. This third piece of property was lot # 59. Lot # 59 was bounded on the east side by Vivian Street, and on the north side by Honore Street. The appraised valuation of these three town lots lying in the town of St. Martinville, which is situated in the State of Louisiana, was the sum of $425.00. This was all the property which was situated in the Parish of St. Martin belonging to the estate of the late Rosetta Boutte'.

These were the heirs and the legal representatives to the estate of the late Rosetta Boutte', a free woman of color, and these heirs were recognized as her legal heirs under the laws of the State of Louisiana. The children of Rosetta Boutte' sired by Claude Frilot, were Arthemis Frilot, who was now residing in the Republic of Mexico and was represented in these proceedings of the estate of his late mother, Rosetta Boutte' by Abner D. Minor, Modeste Frilot, a free woman of color, who was married to Zenon Olivier, a free man of color, and a daughter of the late Rosetta Boutte'. Modeste Frilot Olivier, and her husband, Zenon Olivier, both resided in the then Parish of St. Mary, which is situated in the State of Louisiana, and the actual site of their residence was not far from her mother's, Rosetta Boutte', plantation. Antoine Frilot was a son of the late Rosetta Boutte', and a free man of color, and he was residing in the Republic of Mexico. Adeline Frilot, who was a free woman of color, and a daughter of the late Rosetta Boutte'. Adeline Frilot was the widow of Pierre Olivier, a free man of color, and she was a resident of the Parish of St.

Mary, which is situated in the State of Louisiana. Azelie Frilot, a free woman of color, and a daughter of the late Rosetta Boutte'. Azelie Frilot was the widow of Honore Olivier, and a resident of St. Landry Parish, which is situated in the State of Louisiana. Marie Hortense Frilot was another daughter of the late Rosetta Boutte', and the widow of the late Celestin Rochon. Marie Hortense Frilot Rochon, a free woman of color, was a resident of the Parish of St. Mary, which is situated in the State of Louisiana. Leon Frilot, a free man of color, and a son of the late Rosetta Boutte'. Leon Frilot was a free man of color, and he was a resident of the Parish of St. Mary, which is situated in the State of Louisiana.

By the year of 1859, Rosetta Boutte' had passed away leaving her twelve children to inherit her estate, which included immovable property and movable property. On the date of February 23, 1859, Judge John Parkerson, who was serving as judge in the Parish of St. Mary, authorized an inventory of the immovable property and the movable property belonging to the estate of the late Rosetta Boutte'' to be conducted and taken. Oscar Olivier and Leonce de la Houssaye were both appointed by the 14th District Court in St. Mary Parish to conduct and take an inventory of the immovable and the movable property belonging to the estate of the late Rosetta Boutte'.

On the date of March 11, 1859, Oscar Olivier and Leonce de la Houssaye went to the last place of residence of the late Rosetta Boutte' to conduct and take this inventory. This inventory was ordered by a writ, which was issued out of the 14th District Court of St. Mary Parish, and this writ was signed by Judge John Parkerson, who was judge of St. Mary Parish and ordered the inventory to be taken.

The immovable property that was found by Oscar Olivier and Leonce de la Houssaye at the plantation of the late Rosetta Boutte' consisted of a tract of land, which was situated on the west side of Bayou Teche, in the then Parish of St. Mary, now Iberia Parish. This tract of land had a frontage of four and one half arpents on Bayou Teche, with a depth of forty arpents. this tract of land was bounded on the upperside by the land of Leon Frilot, a free man of color, and on the lowerside by land belonging to Joseph Devezin Olivier. This included all the buildings and the improvements, which were situated on the tract of land. This tract of land was valued at the sum of $3,600.00. There was another tract of land, which was situated immediately opposite the above mentioned tract of land, and this tract of land was situated on the east side of Bayou Teche, in the then Parish of St. Mary, now Iberia Parish. This tract of land had a frontage of five arpents on Bayou Teche, and a depth of forty arpents. This tract of land was bounded on the upperside by the land of Leon Frilot or the Frilot Plantation, and on the lowerside by the land of Joseph Devezin Olivier or the Olivier Plantation. This tract of land, which was situated on the eastern side of Bayou Teche, with all the improvements on the tract of land was valued at the sum of $5,000.00.

Also, there were thirteen slaves belonging to the estate of the late Rosetta Boutte', a free woman of color. These slaves were Charles, a slave man, about 65 years of age, and valued at the price of $200.00. Baptiste, a slave man age 58 years of age, and valued at the sum of $500.00. Jim, a slave teenage boy, age sixteen years old, and valued at the sum of $1,200.00. Media, a teenage slave, age fourteen years of age, and valued at the price of $800.00. Albert, a boy slave of eleven years old, and valued at the price of $600.00. Adelaide, a slave, who was age 56 years old, and valued at the price of $300.00. Ellen, a slave woman, who was 29 years of age, and Ellen was valued at the price of $1,400.00. Jane, a slave woman, who was 27 years of age, and she had two children, Christine, a girl slave child of eleven years of age, and Milaide, a slave toddler, who was age one years old. The slave woman, Jane, and her two children, Christine and Milaide, were valued at the price of $2,000.00. Justine, a slave woman, who was age 27 years old, her child, Francois, who was a toddler of age fifteen months old. The slave woman, Justine, and her child, Francoise, were valued at the price of $1,700.00. Numa, a slave man, who was age 50 years of age, and valued at the price of $800.00. All these slaves were owned by Rosette Boutte', a free woman of color, during her lifetime. Also, all these slaves were residing and working on the plantation of Rosetta Boutte' at the time of her death.

The movable property that was also on the plantation of Rosetta Boutte' at the time that the inventory was taken and conducted. This movable property consisted of eight creole work horses, and each creole work horse was valued at the price of $30.00 each, and the total valuation was the price of $240.00, a twenty one head herd of wild creole mares and colts, and the value of each mare and colt was the sum of $7.00, which made the total valuation the sum of $147.00, a sixty head herd of sheep, which was valued at the price of $90.00, eight work oxen, and each work oxen was valued at the price of $20.00, which made the total valuation of the eight work oxen the sum of $160.00, seven milk cows and calves, which was valued at the price of $112.00, four milk cows and calves, which was valued at the price of $11.00 for each cow and calf, and the total valuation of the four milk cows and calves was the sum of $44.00, six yearlings, which was valued at the price of $7.00 each yearling, and the total valuation of the six yearlings was the sum of $42.00, two beavers, which were valued at the price of $32.00, and each beaver was valued at the sum of $16.00, four horses, that were three years old, and were valued at the sum of $10.00 each horse, or the total valuation was the sum of $40.00, and five hogs that were valued at the sum of $16.00. Also, a lot of wild cattle consisting of a herd of thirty nine wild cattle was valued at the sum of $39.00 for the whole herd of wild cattle. Each one of the cattle of the wild herd was valued at the price of $1.00 per cattle. This wild herd of thirty nine cattle was situated in the Parish of Calcasieu, which is situated in the State of Louisiana. This was all the livestock that belonged to the estate of the late Rosetta Boutte', of the Parish of St. Mary.

This other movable property was also on the plantation of Rosetta Boutte' at the time that this inventory was conducted. This property consisted of one ox cart with wood axles was valued at the price of $10.00, one lot of old plows were valued at the price of $10.00, one lot of coupling chains were valued at the price of $2.00, one lot of scythes were valued at the price of $15.00, one lot of hoes were valued at fifty cents, one lot of harrows were valued at $3.00, one lot of ox yokes were valued at $2.00, two cross cut saws were valued at $5.00, two old spades were valued at twenty cents, one lot of fodder was valued at $12.00, one lot of corn or 300 barrels of corn were valued at the price of $150.00, twelve bales of cotton and each bale of cotton was valued at forty cents, which made the total valuation of the twelve bales of cotton, the total price of $4.80, one lot of sweet potatoes or 30 barrels of sweet potatoes were valued at the price of $30.00, one lot of kitchen furniture was valued at the price of $5.00, one dozen or twelve silver spoons were valued at the price of $30.00, and the one dozen silver spoons were a large size silver spoon, one old safe and a lot of crockery was valued at the price of $4.00, one dozen or twelve chairs were valued at the price of $6.00, one rocking chair was valued at the price of $3.00, one clock was valued at the price of $3.00, one dining table was valued at the price of $1.00, one looking glass was valued at the price of $1.00, one armoire and the contents of the armoire with a bed spread were valued at the price of $40.00, three beds and bedding was valued at the price of $35.00, one old press was valued at the price of $1.50, one small table was valued at the price of $1.50, one large jar was valued at the price of $3.00, and one lot of carpenter tools was valued at the price of $5.00.

There was also a promissory note that was drawn and signed by Francois Frilot, a free man of color, which promissory note was payable to his mother, Rosetta Boutte', a free woman of color. This promissory note was dated the date of May 24, 1862, and it was for the amount of $1,276.00. Also, this promissory note was payable in two years from the date of May 24, 1862 or payable on the date of May 24, 1864.

The total amount of the inventory of the estate of the late Rosetta Boutte' amounted to the sum of $21,565.25. This inventory of the belongings of the estate of Rosetta Boutte' was conducted in the presence of Clemarie Olivier, a free man of color, and Jean Baptiste Collette, a free man of color. Before her death in the year of 1858, Rosetta Boutte' had appointed Louis St. Marc Darby, who was now a resident of the City of New Orleans, Louisiana, as the testamentary executor of her estate. Louis St. Marc Darby declined the office of testamentary executor of the estate of the late Rosetta Boutte'', a free woman of color. At one time, Louis St. March Darby and his brother, Francois Optat Darby, owned the sugar plantation, which was situated next door to the plantation of Rosetta Boutte', and at the time of the death of Rosetta Boutte', the Darby plantation was owned by Joseph Devezin Olivier. On the date of February 23, 1859, Judge John G. Parkerson appointed Eugene Barthlemy Olivier as dative testamentary executor of the estate of the late Rosetta Boutte'. Eugene Barthlemy Olivier was the owner of the Orange Grove plantation, which was situated up the road from the plantation of Rosetta Boutte.'

Rosetta Boutte' had been a slave woman, who was emancipated and granted her freedom on the date of August 30, 1838 by her former slave owner. Being granted her freedom by an act of emancipation or maumission, Rosetta Boutte' was considered a free woman of color. She owned a plantation, which consisted of property, slaves, livestock, crops, and movable effects.

On the date of September 3, 1819, Claude Frilot contracted a special mortgage in favor of his wife, Rosetta Boutte', a free woman of color. Claude Frilot mortgaged unto Rosetta Boutte' a tract of land and two slaves. This act of mortgage was passed in the Parish of St. Mary and the judge of the Parish of St. Mary presided at the execution of the act of mortgage. The amount of money that Claude Frilot borrowed from Rosetta Boutte', a free woman of color, was the amount of $3,876.00, which was subject to bear 10% interest per annum from the date of September 3, 1819.

Also, on the date of September 3, 1839, Claude Frilot, the mortgagee, decided to convey or to sell to Rosetta Boutte', his wife, and the mortgagor, the tract of land and the two slaves that were mortgaged in the favor of Rosetta Boutte'. The reason that Claude Frilot made this decision to sell the land and the two slaves was because he (Claude Frilot) was unable to pay Rosetta Boutte' the amount of $3,876.00. So a fair and equivalent settlement of the debt of $3,876.00 which Claude Frilot owed Rosetta Boutte' was to convey or to sell the tract of land and the two slaves. In the consequence of this conveyance of the land and the two slaves, Rosetta Boutte' declared that in the sale or conveyance of the land and the two slaves, she (Rosetta Boutte') will release, discharge, and acquit Claude Frilot of the money owed or the debt in her favor on the tract of land and the two slaves. So an act of conveyance was executed between Claude Frilot and his wife, Rosetta Boutte'. On the date of September 3, 1839, Judge John Wilkerson presided over the execution of the act of conveyance between Claude Frilot and Rosetta Boutte of the land and the two slaves. Also, Hilaire Carlin and Joseph Allain served in their capacity as witnesses to the execution of the act of conveyance. Because Claude Frilot was indebted to Rosetta Boutte' for the amount of $3,876.00, and was unable to pay the sum plus all the accumulated interest on the principal, by conveying this tract of land and the two slaves was a fair settlement of his (Claude

Frilot's) account with Rosetta Boutte'.

This tract of land that was mortgaged and eventually sold or conveyed to Rosetta Boutte' was situated on both sides of Bayou Teche, which was situated in the Parish of St. Mary, now Iberia Parish. Also, this tract of land was situated on both the east and the west sides of Bayou Teche, and the depth of the tract of land on both the east and the west sides of Bayou Teche was forty arpents deep. This tract of land had a six arpent frontage on Bayou Teche, on both the east and the west sides of Bayou Teche. The boundaries of this tract of land were on the upperside or above by land belonging to Rosetta Boutte', and on the lowerside or below by land belonging to Nicholas Pelagie Loisel, who was the owner of the Loisel plantation.

The two slaves that were also mortgaged and eventually conveyed to Rosetta Boutte' were L'leville, a slave woman, who was thirty four years of age, and Charles, a slave boy, who was sixteen years of age. Both, Charles, the slave boy, and L'leville, the slave woman, were to be slaves for life in the service of Rosetta Boutte'.

On the date of January 29, 1830, Rosetta Boutte' made a formal acknowledgement of her children sired by Claude Frilot. Rosetta Boutte', a free woman of color, stated and declared that she was the natural mother of Francois Frilot, Louis Frilot, Artemuis Frilot, Modeste Frilot, Antoine Frilot, Alix Frilot, Inlmot Frilot, Hortense Frilot, Leon Frilot, Casmile Frilot, and Adeline Frilot. These twelve children were all the natural children of Rosetta Boutte' sired by Claude Frilot, who were both residents of the Parish of St. Mary, which is situated in the State of Louisiana. Also, all of these children were considered free people of color by the laws of the State of Louisiana in the year of 1830. The reason that Rosetta Boutte' made a formal and legal acknowledgement of her natural children of she and Claude Frilot would be to state who the legal heirs and representatives of her estate would be after her death. The execution of the formal document acknowledging the twelve children as the natural children and the legal heirs of Rosetta Boutte' was executed by Judge Joshua Baker, who was serving as judge of the Parish of St. Mary. The witnesses to this formal act acknowledging the twelve children as the natural heirs were Philpott Boutte, the brother of Rosetta Boutte' and Seth Lewis, both were free people of color.

In the 1800's under the laws of the State of Louisiana, a free person of color was able to legally own property, both real estate and movable, and slaves. Many free persons of color were planters and slave owners. To be considered a free person of color, under the laws of the State of Louisiana, a person had to be born of a white person and a slave or a free person of color, or two free persons of color. If a child had a white mother or father and a colored mother or father, that child was considered a mulatto under the laws of the State of Louisiana. To be considered a free person of color, a child or a person had to be given their freedom through an act of maumission by their former slave owner and set free. Many times the former slave owner gave the free person of color land, slaves, and money to help them get started in their new life. At this time, in the State of Louisiana, there were several plantations owned and operated by free people of color.

Claude Frilot died in the year of 1834, and his will and last testament was executed in the month of October of the year of 1834. On the date of September 8, 1834, Aimee Frilot, a child of Claude Frilot and Rosetta Boutte's, and a free person of color, filed a petition representing that he (Aimee Frilot) had been informed and verily believed that Claude Frilot had died or expired, and that Claude Frilot had left a last will and testament, which was deposited in the office of the Parish Judge of St. Martin Parish, or in the courthouse, which is situated in the town of St. Martinville, Louisiana. Edouard Simon was the attorney representing Aimee Frilot, a free man of color, and his brothers and sisters in these probate proceedings.

On the date of February 21, 1835, Nicholas Pelagie Loisel, who was the owner of the Loisel Plantation, was appointed and confirmed to be the testamentrix executor of the estate of the late Claude Frilot. Nicholas Loisel declined this appointment as the testamentrix executor of this estate, therefore Nicholas Loisel did not accept this appointment as executor of this estate for the reason that Nicholas Loisel did not want to be directly involved if any argument regarding inheritance of any property or slaves should result among the twelve children of Claude Frilot by Rosetta Boutte', a free woman of color. So therefore, the Judge of St. Mary Parish had to appoint another person to be the testamentrix executor of the estate of the late Claude Frilot. On the date of April 11, 1835, Edouard Simon, an attorney of St. Martinville, Louisiana, was appointed by Judge John Moore to be testamentrix executor of the estate of the late Claude Frilot.

In the early to mid 1800's, in the State of Louisiana, it was not uncommon for the neighbors to be appointed testamentrix executor or executors of wills and last testaments of their deceased neighbors. A neighbor or a person who was appointed to be testamentrix executor could either accept the office of executor for an estate, or decline the appointment. If a person declined the appointment of testamentrix executor of a will and last testament, then the Judge of the Parish, wherever the will and last testament was filed, had to appoint a new person to the office of testamentrix executor of the estate. If a judge had made an appointment to the office of testamentrix executor, it was usually a lawyer or an attorney who was appointed, because most attorneys would accept such an appointment.

Also, an inventory of the estate of the late Claude Frilot had to be conducted and taken. Therefore, Judge John Moore, of the Parish of St. Mary, appointed Colonel Charles Napoleon Olivier and Francois Malus to be the appraisers of the immovable and movable property belonging to the estate of the late Claude Frilot. At this time, an estate sale was usually conducted when the owners of neighboring plantations would come to purchase goods and property, also the family and friends of the deceased person would attend to purchase goods and sometimes property from the estate sale. Colonel Charles Napoleon Olivier and Francois Malus were duly sworn by Judge John Moore under the laws of the State of Louisiana proceeded and conducted an inventory of the estate of Claude Frilot. This inventory was conducted in the presence of Derosin Boupase and Belisaire Broussard, who were both acquainted with Francois Frilot, Aimee Frilot, and Antoine Frilot, who were sons of the late Claude Frilot.

The property discovered at the inventory of the estate of the late Claude Frilot consisted of these slaves: Alexandre, a slave man, who was age 24 years old, and valued at the sum of $1,000.00, Auguste, a slave boy, who was age 15 years old, and was valued at the sum of $800.00, Adelaide, a griffee woman slave, who was 24 years old, and her four children, Fleuce, a slave child, who was 7 years of age, Jeanne, a slave child, who was age 5 years of age, Apponsien, a slave child, who was 2 years of age, and Emma, a slave infant, who was age 2 months old, and Adelaide and her four children were valued at the sum of $480.00, Chalirette, a griffee woman slave, who was age 28 years of age, and her daughter, Genevieve, a girl slave child, who was 8 years of age, the slave woman, Chalirette and her child, Genevieve, were valued at the sum of $2,175.00, Euurluite, a slave child, who was 12 years of age, and was valued at the sum of $600.00,
Euurluite was a child of the slave woman, Chalritte, Catherine a griffe slave woman, who was 60 years of age, and valued at the sum of $300.00, and Francoise, a griffee slave man, who was 62 years of age, and he was valued at the sum of $250.00. There was no real estate property belonging to the late Claude Frilot, and the estate of Claude Frilot consisted of these twelve slaves.

On the date of March 21, 1863, it was discovered that the succession of Claude Frilot had a minor child, who was to be considered an heir of Claude Frilot's. This grandchild of Claude Frilot's had a minor interest in the estate of Claude Frilot. This grandchild of Claude Frilot's was Jean Baptiste Kerlegan, who was the child of Lucille Frilot, a daughter of the late Claude Frilot and Rosetta Boutte'. Jean Baptiste Kerlegan resided in St. Martin Parish, with his natural father, Baptiste Kerlegan, a free man of color. Jean Baptiste Kerlegan was considered a free person of color under the laws of the State of Louisiana. This child, Jean Baptiste Kerlegan will inherit from the estate and the succession of the late Claude Frilot in representation of his mother, Lucille Frilot Kerlegan, a free woman of color, who died or passed away since the succession of Claude Frilot was opened. This death of Lucille Frilot Kerlegan, caused a family meeting to be called by the judge of the Parish of St. Mary Parish to discuss the interests of the minor child in the estate, and to appoint a tutrix or tutor for Jean Baptiste Kerlegan. The tutrix or tutor was a person of the legal age and majority, who would represent a minor child or minor children in all business matters pertaining to any immovable or movable property inherited from their deceased parent or parents. Usually the tutrix or the tutor was the mother or the father of the child. Whenever a mother or a father of a child was appointed by the probate court of a parish or a county where a succession or an estate was opened, that mother or father was considered to be the natural tutor of the child. If a father was appointed as a tutor, he was called a natural tutor, but if a mother was appointed a tutor, she was called the natural tutrix. If an aunt or an uncle, cousin, friend of the family was appointed by the probate court to be a representative of a minor child, then the person appointed was referred to as a tutor of the child. When an estate involved minor children, a family meeting was held, which family meetings was presided over by the judge of the Parish court or a attorney or notary public, to discuss how the inherited assets of the minor child were to be handled and taken care of, whether the tutor would handle all business matters until the child was of the legal age or the inherited property would be sold, and the money or proceeds from the sale held by the tutor until the child reached their majority or legal age. In the case, where a tutor or tutrix mishandled the property entrusted to them by the probate court, this tutor or tutrix would or could be replaced by the probate court, if the mishandlement of the property was proved in court. The terms tutor is used for a man, and the term tutrix is used for a woman.

On the date of July 28, 1835, a partition of property, which was inherited by Rosetta Boutte', a free woman of color, and her brother, Philipeau Boutte', a free man of color, was executed. Judge John Moore of the Parish of St. Mary presided over this partition of property. It was mutually agreed between Rosetta Boutte' and Philpeau Boutte' that they would divide the property belonging to the estate of their late brother, Louis Hilaire Milo Boutte', a free man of color.

The property belonging to the estate of Louis Hilarie Milo Boutte' that was partitioned to Philipeau Boutte' was a one half interest of the tract of land, on which Louis Hilarie Milo Boutte' resided. This tract of land partitioned to Philipeau Boutte' contained one arpent of 1.650 arpents fronting Bayou Teche, with a depth of forty arpents. Also, one half of all the buildings and improvements on the tract of land, and the value of the tract of land was set at $281.25.

This movable property was partitioned to Philipeau Boutte',

hich consisted of one bay horse, which was valued at the sum of $30.00, one bay horse, which was valued at the sum of $25.00, one pair of oxen, which were valued at the sum of $20.00, and one half of a herd of horned cattle, which was valued at the sum of $170.00, and one half a herd of wild horses, which was valued at the sum of $85.00.

The property that was partitioned to Rosetta Boutte',who was the sister of the late Louis Hilaire Milo Boutte' was one half ownership of the tract of land, on which Louis Hilaire Milo Boutte' resided, and this tract of land contained one arpent or 1.650 arpents of land fronting Bayou Teche with a depth of forty arpents of land. The tract of land that was partitioned to Rosetta Boutte' was to be taken from the lowerside of the tract of land, whereas the tract of land to be partitioned to Philipeau Boutte', another brother of Rosetta Boutte', was to be taken from the upperside of the tract of land. Also, Rosetta Boutte' was given in the partition all the buildings and the improvements on the tract of land, which made the value of the tract of land the sum of $281.25.

Also, Rosetta Boutte' was given at the partition of the property belonging to the estate of Louis Hilaire Milo Boutte', one bay horse, which was valued at the sum of $36.00, one roan horse, which was valued at the sum of $10.00, one pair of oxen, which were valued at the sum of $35.00, one half of the herd of horned cattle, which was valued at the sum of $170.00, and one half of the herd of wild horses, which was valued at the sum of $85.00.

Also, Rosetta Boutte' and her brother, Philipeau Boutte', agreed that the slaves belonging to the estate of the late Louis Hilaire Milo Boutte' were to be the joint property of both Rosetta Boutte' and Philipeau Boutte'. There was also property, which was situated in the Parish of St. Martin belonging to the estate of the late Louis Hilaire Milo Boutte', which was also divided or partitioned between his sister, Rosetta Boutte' and his brother, Philipeau Boutte'.

The witnesses to the act of partition, which was contracted between Rosetta Boutte', a free woman of color, and her brother, Philipeau Boutte', a free man of color, was Colonel Charles Napoleon Olivier and Francois Malus. Francois Malus acted as the agent for the Gravier sugar plantation, which was situated below the plantation of Rosetta Boutte'.

In the year of 1793, Claude Frilot settled a mortgage in favor of Rosetta Boutte', a free woman of color. This mortgage resulted from the account of beef cattle sold. The cattle that were sold to satisfy this mortgage were fifteen beef cattle, which were sold to Pierre George for the amount of $180.00, twenty beef cattle were also sold to Pierre George for the amount of $300.00, five beef cattle were sold to Pierre George for the amount of $65.00, eighteen beef cattle were sold to M. Boutte' for the amount of $216.00, ten beef cattle were sold to Alexander Simon for the amount of $110.00, twelve beef cattle were sold to Andre Verret for the amount of $144.00, twelve beef cattle were sold to Alexandre Hebert for the amount of $121.00, eight beef cattle were sold to Paul Ambroveu for the amount of $104.00, eight beef cattle were sold to Edouard Armand for the amount $80.00, ten beef cattle were sold to Andre Martin for the amount of $101.00, twelve beef cattle were sold for the amount of $144.00, nine beef cattle were sold to Pierre Broussard for the amount of $117.00, ten beef cattle were sold to Mr. Fusilier for the amount of $160.00, and eleven beef cattle were sold for the amount of $165.00. Also, at this same time, Claude Frilot sold unto Rosetta Boutte', a free woman of color, the plantation, which Claude Frilot owned on the Old Spanish Trail.

THE GRAVIER PLANTATION

On Friday, July 20, 1832, Sheriff Benjamin Hudson, of the Parish of St. Mary, received the writ of seizure and sale of the Gravier sugar plantation, which was situated in St. Mary Parish. This writ of seizure and sale was issued by the Supreme Court of the State of Louisiana.

The Gravier sugar plantation had a frontage of fifteen arpents along Bayou Teche on both the west and the east sides of Bayou Teche. The plantation had the usual depth of forty arpents or six hundred arpents on both the west and the east sides of Bayou Teche. Also, situated on the Gravier sugar plantation was a crop of sugar cane, that was growing in the cane fields of the plantation. The plantation also had barns, slave cabins, and houses on the premises of the plantation. This plantation was bounded, in the year of 1832, on the upperside or above by the lands of Alexander Baron, and below by the lands of Louis Eloi DeBlanc. Alexander Baron was the attorney from the City of New Orleans, whom Jacques Gravier had at one time, hired as the agent of the Gravier family in the State of Louisiana

On July 28, 1832, Sheriff Benjamin Hudson served notices of the seizure and sale of the plantation of Jacques Gravier on Charles Napoleon Olivier. Also, Charles Napoleon Olivier was appointed curator by the honorable court. It was Charles Napoleon Olivier under whom the sale of the Gravier sugar plantation was to be conducted on August 1, 1832. Sheriff Benjamin Hudson advertised the Gravier sugar plantation for sale for cash. The sale of the Gravier sugar plantation was to take place at the courthouse of St. Mary Parish, which was situated in the town of Franklin, Louisiana, on Monday, September 3, 1832, between the hours of ten o'clock in the morning and four o'clock in the afternoon.

On August 20, 1832, Sheriff Benjamin Hudson was informed by the 5th District Court of the Parish of St. Mary of the further proceedings of the writ of seizure and sale of the Gravier sugar plantation. These further proceedings were to be held on October 13, 1832. Also, Sheriff Benjamin Hudson was ordered by the 5th District Court to proceed anew on the writ of Edouard Simon, an attorney representing Jacques Gravier in this suit.

On October 15, 1832, Sheriff Benjamin Hudson again seized the Gravier sugar plantation. Also, Sheriff Hudson was commanded to serve notice of the seizure and sale of the Gravier sugar plantation to Charles Napoleon Olivier, personally. Charles Napoleon Olivier was acting in his capacity as curator in this matter.

On October 18, 1832, Sheriff Benjamin Hudson advertised the Gravier sugar plantation for sale. The advertisement of any plantation before a public sale was required by the laws of the State of Louisiana. The Gravier sugar plantation was to be sold for cash. The public sale of the Gravier sugar plantation was to be held on Tuesday, November 20, 1832, between the hours of ten o'clock in the morning and four o'clock in the afternoon. The property of the Gravier sugar plantation was appraised by Winfrey Lockett, who was appointed by Sheriff Benjamin Hudson as appraiser.

On the part of the plaintiff, Alexander Baron and John Welhaver, who was appointed as appraiser by Sheriff Benjamin Hudson, and on the part of the curator, Charles Napoleon Olivier, both parties neglected to name their appraiser of the land. Both parties were justly sworn to make a just and true appraisement of the plantation, but have failed to do so.

The Gravier sugar plantation was seized again under the second writ of seizure and sale. The property of the Gravier sugar plantation was appraised at the price and sum of $9,400.00.

Also, Judge Donelson Caffery, who was serving as Judge of the Parish of St. Mary, certified that there were two mortgages bearing on the Gravier sugar plantation. It was as a result of these two mortgages made by Alexander Baron, who was the agent for Jacques Gravier, Jeanne Gravier, and Jeanne Bordier, who was the wife of Henri Petit, and who had mortgaged the Gravier sugar plantation, that a civil suit was caused to be filed by the Gravier heirs. Alexander Baron had mismanaged the financial part of the Gravier plantation, and had spent money for his own gains, not the gains of the Gravier sugar plantation. As a result, Alexander Baron was ordered to pay Gravier heirs the amount of the mortgage, which was $10,000.00, and the accrued interest at 5%, which amounted to $183.00. Also, the Gravier sugar plantation, its lands and movable property had been seized by Sheriff Benjamin Hudson, of St. Mary Parish.

Now, Judge Donelson Caffery was getting ready to proclaim the Gravier sugar plantation was for sale in a loud, clear, and audible voice at the courthouse door of the St. Mary Parish courthouse, which was located in the town of Franklin, Louisiana. Judge Caffery was saying that the lands and improvements of the Gravier sugar plantation was for sale to the last and highest bidder at this public auction. Also, the crop of sugar cane was growing on the plantation, and the crop of sugar cane was for sale. Also, no person was allowed to bet two thirds of the appraisement amount of the Gravier plantation.

On November 20, 1832, Judge Donelson Caffery had advertised the Gravier sugar plantation land for sale on a twelve month credit. The Gravier sugar plantation land was to be sold at the courthouse on Saturday, December 8, 1832, between the hours of ten o'clock in the morning and four o'clock in the afternoon.

November 21, 1832 was the sixty fifth day, since the Sheriff of St. Mary Parish, Benjamin Hudson, received the third order of the writ of seizure and sale of the land of the Gravier sugar plantation. Actually, there were three writs of seizure and sale issued out of the 5th District Court of the Parish of St. Mary for the seizure of the land of the Gravier sugar plantation, and the eventual sheriff sale. In five more days, the time did expire in which the writ of seizure and sale will have legal effect, and the writ of seizure and sale will be returned to the honorable 5th District Court of the Parish of St. Mary by Sheriff Benjamin Hudson for renewal of the writ of seizure and sale of the land, which comprises the Gravier sugar plantation. Sixty five days is the time span in which the writ of seizure and sale was suspended by the injunction. Also, on March 20, 1832, Sheriff Benjamin Hudson, of the Parish of St. Mary, certifies that the seizure and sheriff's return of the copies from the original records of the case of Jacques Gravier, Jeanne Gravier, and Jeanne Bordier Petit versus Alexander Baron.

Alexander Baron, of the City of New Orleans, answered the petition of the civil suit, which was filed for Jacques Gravier, Jeanne Gravier, Jeanne Bordier Petit, all residing in the country of France, by their attorney. Alexander Baron was serving in his capacity as their agent and representative in the State of Louisiana, in making business deals and managing their sugar plantation, which was located in the Parish of St. Mary. This suit that was filed by the Gravier heirs was pending in the 5th District Court of St. Mary Parish for quite some time. Alexander Baron insisted that the land comprising the Gravier sugar plantation was affected by a judicial mortgage in favor of Jacques Gravier, Jeanne Gravier, Jeanne Bordier Petit and against Alexander Baron. Alexander Baron was ordered to pay Jacques Gravier, Jeanne Gravier, and Jeanne Bordier Petit the amount of $10,383.00, which is the amount specified to surrender the mortgaged property to be sold to satisfy the judgement.

Also, in the suit, Alexander Baron consented to waive any rights he (Alexander Baron) may have to claim compensation for the improvements made on the land of the Gravier sugar plantation and the land affected by the plaintiff's (Jacques Gravier, Jeanne Gravier, and Jeanne Bordier Petit), mortgage while the land was in the possession of Alexander Baron.

On September 17, 1832, Noel Barthlemy LeBreton, who was serving as a clerk in the office of the Supreme Court of the State of Louisiana, stated that the copy of the suit of Jacques Gravier, Jeanne Gravier, and Jeanne Bordier Petit versus Alexander Baron was a true and original copy of the suit that was filed in the office of Noel Barthlemy LeBreton on June 23, 1832. Also, Antoine and Etienne Carraby called in the warrants or mortgages. The calling in of the warrants or mortgages was also filed in the office of Noel

Barthlemy LeBreton.

The petition of Charles Napoleon Olivier, of St. Mary Parish, Louisiana, shows respectively that Charles Napoleon Olivier appeared by the honorable court of the 5th District of the Parish of St. Mary, under Article 3369 of the Civil Code of the State of Louisiana. Charles Olivier was to serve as curator for the land relinquished by Alexander Baron on the action of a judicial mortgage of Jacques Gravier, Jeanne Gravier, and Jeanne Bordier Petit. Some difficulties have arisen since the sale of the land of the Gravier sugar plantation was suspended between Charles Olivier, who is serving as curator, and Jacques Gravier, Jeanne Gravier, and Jeanne Bordier Petit. There exists on the Gravier sugar plantation a large crop of unharvested sugar cane. This crop of sugar cane has grown ripe, and may be entirely lost if the sugar cane is not harvested. This sugar cane crop came under the title of the land to be sold. Jacques Gravier, Jeanne Gravier, and Jeanne Bordier Petit, hoped to sell their sugar plantation in a ten day time span from the advertisement of the Gravier sugar plantation for sale for cash. On October 8, 1832, Judge Joshua Lewis, of the Parish of St. Mary, issued a judge's order that the sugar cane was to be harvested and sold.

Also, Jacques Gravier, Jeanne Gravier, and Jeanne Bordier, Petit, all of the Kingdom of France, issued a petition stating that they are the legal heirs and representatives of the late Jacques Gravier, who was the original owner of this land. The petition filed was issued by Noel Barthlemy LeBreton stated that Noel LeBreton is the lawful agent and representative of the Gravier heirs. That there existed a mortgage which indebted the Gravier heirs against Alexander Baron.

On July 2, 1832, Alexander Baron was ordered to pay $10,383.00 with interest from August 8, 1832 to Jacques Gravier, Jeanne Gravier, and Jeanne Bordier Petit, or to surrender the land belonging to the Gravier sugar plantation. This property comprising the Gravier sugar plantation was to be sold to satisfy the judgment against Alexander Baron and this judgment was favor of Jacques Gravier, Jeanne Gravier, and Jeanne Bordier Petit. This judgment was issued out of the Supreme Court of the State of Louisiana in favor of the heirs of Jacques Gravier. Eventually, the land comprising the Gravier sugar plantation was purchased from the heirs of Gravier on July 31, 1835 by Francois St. Marc Darby.

In the fall of the year of 1832, at the house of John W. Jeanerette, which was situated on a neighboring plantation of the Gravier's, Judge Joshua Lewis, of St. Mary Parish, in the presence of Robert McCarty, the partner and son-in-law of Nicholas Pelagie Loisel, approved the sale of the sugar cane, which was standing in the fields of the Gravier sugar plantation, or the fields of Alexander Baron and Francois Malus. The land of the Gravier sugar plantation was commonly known to the neighbors of the Gravier sugar plantation as the land of Alexander Baron and Francois Malus. It was understood that the heirs of Jacques Gravier, who were the original owners of the plantation, had filed a suit against Alexander Baron and Francois Malus. Also, Colonel Charles Napoleon Olivier, of St. Mary Parish, had been appointed curator by Judge Donelson Caffery. Colonel Charles Olivier had informed Robert McCarty that he (Charles Olivier) was the agent or the person acting for the heirs of the late Jacques Gravier, which heirs were Jacques Gravier, Jeanne Gravier, and Jeanne Bordier Petit. The sale of the land of the Gravier sugar plantation was approved by Sheriff Benjamin Hudson, of the Parish of St. Mary, and who as well as Robert McCarty read accusations that forbid the sale. The judge approved to sell the land, and let Charles Olivier act as agent in order to make the crop of sugar cane. The large crop of sugar cane was already on the plantation and was nearly ready for harvesting. By harvesting this large crop of sugar cane, a high price for the sugar cane and molasses could be realized or made. Also, by allowing Charles Olivier to take control, and act as overseer the large crop of sugar cane would be harvested. Harvesting this crop of sugar cane would result in a high price for the sugar and molasses. Robert McCarty was the last bidder, but Francois Malus bid the highest. As a result of this, the judge assigned the crop of sugar cane to Francois Malus. As a result of this incident, Charles Napoleon Olivier did not carry off any of the sugar cane that was grown on the plantation of the heirs of Gravier.

On April 26, 1834, the jury gave a verdict for the heirs of Jacques Gravier, granting the heirs of Gravier $695.00 from the amount of the sale of the sugar cane, which was harvested from their sugar plantation. This sale of at the sugar cane was allotted in the petition filed by the heirs of the late Jacques Gravier, who were Jacques Gravier, Jeanne Gravier, and Jeanne Bordier Petit, and the full amount was subject to deductions for the expenses of attending the sale of the sugar cane. The evidence was found in favor of the plaintiff who were the heirs of Gravier. It was also ordered that the heirs of Jacques Gravier, who were Jacques Gravier, Jeanne Gravier, and Jeanne Bordier Petit, recover from Charles Napoleon Olivier, who was appointed to act as curator or representative, the sum of $695.00, which sum was subject to deductions for expenses.

Also, Benjamin Hudson, who was the sheriff of St. Mary Parish, said that on the day that the sugar cane was sold by Judge Donelson Caffery, he (Benjamin Hudson) was present at the sugar house of Francois Malus. Also, Benjamin Hudson stated that the value of the sugar cane was debated. Benjamin Hudson does not know the quality of the land that comprises the Gravier sugar plantation. Noel Barthlemy LeBreton sold the sugar and molasses, which was realized from the sugar cane, that was grown on the Gravier sugar plantation. Also, Noel LeBreton cannot say whether the bystanders, who were present at the sale of the sugar cane of the Gravier plantation were aware of the situation that existed on the plantation.

Francois Malus lived on the plantation of Alexander Baron. Also, Francois Malus served as the partner and agent, or representative of Alexander Baron. Alexander Baron was a resident of the City of New Orleans, where he practiced law. Francois Malus was a resident of the Parish of St. Mary.

Another neighbor of the Gravier sugar plantation, Nicholas Pelagie Loisel, the owner of Loisel Plantation, was acquainted with the land that comprised the Gravier sugar plantation, and the sugar cane that was in question. Nicholas Loisel was not present at the sale of the sugar cane by Judge Donelson Caffery. But, Nicholas Loisel knew that the Gravier plantation had no sugar house or sugar establishment situated on the lands that comprised the plantation. In the year of 1832, there was an early hard freeze in November, and most of the sugar planters lost sugar cane. Also, in the year of 1832, Nicholas Loisel lost some sugar cane. As a result of this early freeze or frost, no planters in the neighborhood of the Gravier sugar plantation were able to receive the sugar cane that was grown on the Gravier sugar plantation to grind in their plantation sugar houses. Because of growing and harvesting such a large crop of sugar cane, Nicholas Loisel loses sugar cane frequently from the frost.

Also, if Nicholas Pelagie Loisel was serious about purchasing a sugar plantation, he would consider the value of the sugar plantation a great deal higher of a plantation with cane all ready for planting. At this time, there was not a stalk of sugar cane on the Gravier plantation that was ready for planting. The value of a sugar cane plantation would be worth more per arpent with seed cane situated on the plantation. Also, the value of an arpent of land with seed cane would be between $50.00 and $60.00 an arpent of land.

Also, Lewis Moore, the brother of Judge John Moore, had a sugar plantation in the neighborhood of the Gravier sugar plantation, or the neighborhood of the Old Spanish Trail. About ten to fifteen arpents of the sugar cane that was grown on the Gravier sugar plantation and was taken to be ground at the sugar house of Lewis Moore. The sugar house of Lewis Moore was located on the east side of Bayou Teche. Also, a few arpents down the Bayou Teche from the sugar plantation of Lewis Moore, was a sugar plantation belonging to Thomas Houssan Thompson, known as the Rosedale Plantation. Also, there was a sugar house or a sugar establishment located lower down the Bayou Teche from the home and plantation of John W. Jeanerette, originally known as Pine Grove Plantation, now known as Beau Pre.' Sometimes the planters grind their neighbors sugar cane, but in the year of 1832 no planter due to the early freeze or frost was able to grind their neighbor's sugar cane.

Also, when a sugar planter loses his sugar cane crop from an early freeze or frost, the planter is not in a situation to grind the sugar cane of his neighbors, because the sugar cane planter is too occupied in the saving of the remains of his own sugar cane crop. In the year of 1832, the frost or the freeze, which killed the sugar cane occurred between November 15, 1832 through November 19, 1832.

Lewis Moore, the brother of Judge John Moore of St. Mary Parish, lived on a sugar plantation, which adjoined the plantation of Alexander Baron on the lowerside or below. Also, on his sugar plantation, Lewis Moore had a sugar house and all the apparatus for the making of sugar and molasses. Like all sugar planters, the sugar cane that was raised and harvested on the plantation of Lewis Moore, was ground into cane juice and made into brown sugar and molasses in the sugar house, which was situated on the plantation of Lewis Moore. Therefore, if any of the neighbors of Lewis Moore had wanted thirty or forty arpents of sugar cane ground into cane juice in the plantation sugar house, it could not have been done at Lewis Moore's place, because the sugar cane belonging to the plantation of Lewis Moore was to have been ground first at the sugar house. The neighbors of Lewis Moore, as well as Francois Malus and Alexander Baron, had hardly time to ground their own sugar cane crop into cane juice for the making of brown sugar and molasses due to the early freeze. Therefore, no sugar planter in this vicinity of the Old Spanish Trail sugar plantation neighborhood would have had time to take in another planter's sugar cane crop, and ground the neighbor's sugar cane crop into sugar cane juice. Therefore, Alexander Baron and Francois Malus lost a considerable portion of their sugar cane crop that year of 1832, and the occasion of Alexander Baron and Francois Malus losing their sugar cane crop was the occasion of the early freeze, which caused many planters to lose their sugar cane crop.

Also, Lewis Moore was present at the sale of the sugar cane that was situated and growing in the fields of the Gravier sugar plantation in the State of Louisiana. This sale of the sugar cane was ordered by the Parish Judge of St. Mary Parish, Louisiana, who was at this time in the year of 1832, Donelson Caffery. The judge of the Parish of St. Mary, Donelson Caffery met with Lewis Moore, Francois Malus, Thomas Thompson, and John W. Jeanerette at the plantation home of John W. Jeanerette, which was situated where the present day Beau Pre' cemetery now exists. This plantation home burned in the month of November of the year of 1984.

The sugar cane that was situated on the Gravier plantation, which was located on the Old Spanish Trail, had to be cut by hand as soon as the sugar cane was purchased. Francois Malus was the purchaser of the sugar cane crop that existed on the Gravier sugar plantation, and the first thing he did was to have the sugar cane cut by slave labor. This sugar cane had to be hand cut, with the slaves using cane knives to cut the sugar cane. When Francois Malus purchased the sugar cane, at the sale of the sugar cane crop that was ordered by the Parish Judge of St. Mary Parish, Judge Donelson Caffery, the sugar cane was rolled, and it was fit for seed cane or plant cane. The sugar cane had stains from frost bite on it as a result of the early frost. When the Gravier sugar plantation was offered for sale, there was no seed cane or plant cane situated on the plantation. Also, Francois Malus had other sugar cane crops besides the sugar cane crop that was situated on the Gravier sugar plantation. Francois Malus farmed a large crop of sugar cane, and he was obligated to burn one hundred and fifty arpents of sugar cane, which he (Francois Malus) could not grind. Francois Malus had only three hundred arpents of sugar cane in the year of 1832. The grinding season of the year of 1832, started on November 1, 1832, and Francois Malus began his grinding or harvest season of the year of 1832 on November 5, 1832.

Thomas Houssan Thompson, the owner and proprietor of the Rosedale Plantation, which plantation bordered the Gravier sugar plantation on the Old Spanish Trail, was also at the sale of the sugar cane crop of the Gravier sugar plantation. In the month of November of the year of 1832, Thomas Thompson was not in a situation to have purchased the sugar cane that was situated on the Gravier sugar plantation. Also, there was not any other sugar planter in this neighborhood of sugar plantations that could have profitably purchased and harvested the Gravier sugar cane. The only sugar planter in this neighborhood, who could maybe have done this was Francois Malus, who was the partner or the agent of Alexander Baron. Also, Thomas Thompson lost a portion of his sugar cane crop in the year of 1832. Thomas Thompson had a small crop of sugar cane, and he could not work and grind all of his own crop of sugar cane that year. The early frost occurred before the sale of the sugar cane that was situated on the Gravier sugar plantation.

Thomas Thompson was the owner and proprietor of the Rosedale Plantation, and the Rosedale Plantation bordered the Gravier plantation on the lowerside. Thomas Thompson did not intend to purchase the Gravier sugar plantation, when the plantation was to be sold, after the crop of sugar cane had been taken off, as rumors in the neighborhood of the Old Spanish Trail suggested. In the year of 1832, a sugar cane plantation without a crop of seed cane or plant cane would not sell for as high a selling price as a sugar cane plantation with seed cane or plant cane. The Gravier sugar plantation did not have a crop of seed cane or plant cane growing on it. The value of a sugar cane plantation is greater when the fields of the plantation are prepared for seed cane and provided with seed or plant cane, then when the fields are fallow, and not prepared or provided with seed or plant cane.

After a crop of sugar cane has been killed by the frost or or an early freeze, there exists only a short span of a few weeks, when the sugar cane juice is still good to make brown sugar and molasses. Also, after this time span has passed, the sugar cane is no longer good, and the sugar cane juice is sour. Whenever, sugar cane juice is sour, the sugar cane juice is no longer good for the making of brown sugar and molasses. Sugar cane that the juice has become sour is not even good to be used for plant cane or seed cane. Plant cane or seed cane is the sugar cane stalks that are planted in the early fall, usually middle to late August to November, to make the next year's crop of sugar cane. At this time, in the year of 1832, sugar cane was planted in the fall from September to November, and in the spring from March to May for the making of the next year's crop of sugar cane.

Sugar cane is a plant which grows from the nodules, which nodules exist on the stalk of sugar cane. The nodule is actually a ring that exists every few inches on a stalk of sugar cane. New cane plants can take root from this nodule. In planting sugar cane, the nodules of the sugar cane stalks is planted into the ground to form new sugar cane plants.

On December 5, 1832, Seth Lewis, who was serving as a clerk in the recorder's office at the St. Martin Parish courthouse, wrote Sheriff Benjamin Hudson, of the Parish of St. Mary, to state that everyone was sorry that Alexander Baron's plantation was not sold to Noel B. LeBreton at the cash sale according to the letter of instructions. Seth Lewis had recorded the letter from Noel LeBreton, which letters were written on the situation, but he did not have the time to attend to the business of his client, Noel LeBreton. It was contemplated that the Gravier sugar plantation was to be sold for the amount of $10,000.00 or higher at a one year's credit, and this sale was scheduled to take place on next Saturday. Also, Seth Lewis instructed Noel LeBreton to purchase the Gravier plantation in St. Mary Parish, in the name of Noel LeBreton for the heirs of Jacques Gravier, who were Jacques Gravier, Jeanne Gravier, and Madame Henri Petit. Noel LeBreton was also instructed to purchase the plantation for the amount of $10,000.00. W. J. Wharton was to attend the sale of the Gravier Plantation in the place of Seth Lewis. Seth Lewis was hoping that the sale of the Gravier sugar plantation would be made in the name of Noel Barthlemy LeBreton, who was the agent for the heirs of Jacques Gravier. If any other person bids over $10,000.00 of the sugar plantation, the plantation was to be sold to that person provided good security was given for the credit of the purchase price that was prescribed by the laws of the State of Louisiana. The bidding of the property of the plantation must be as high as $10,000.00.

On December 10, 1832, at the request of Charles Napoleon Olivier, who was appointed to serve as curator in this matter, Robert Trimble, a land surveyor, was employed to measure the quantity of the sugar cane that was situated on the plantation of the heirs of Jacques Gravier. Also, this same plantation was sometimes known to the neighbors in the neighborhood of the Old Spanish Trail, as the plantation of Alexander Baron and Francois Malus. This plantation was eventually sold at the insistence of the auditors of the heirs of the late Jacques Gravier, who was the true owner of the plantation. The quantity of sugar cane on the plantation measured fifteen rows or forty six arpents and thirty eights hundredths of an arpent. When the sugar cane on the Gravier plantation was measured by Robert Trimble, Charles. N. Olivier and Francois Malus were present. Also, Robert Trimble served as parish surveyor of St. Mary Parish in the year of 1832.

Previous to the year of 1834, Jacques Gravier, Jeanne Gravier, and Madame Henri Petit, all of the Kingdom of France filed a civil lawsuit against Colonel Charles Napoleon Olivier, and the basis of this lawsuit was that Charles Napoleon Olivier had taken the crop of sugar cane off their plantation illegally. On April 20, 1834, this case entitled, "The Heirs of Jacques Gravier versus Charles Napoleon Olivier," was heard in open court. The heirs of Jacques Gravier were Jacques Gravier, and Miss Jeanne Gravier of Dergerac, France, and Madame Henri Petit of Paris, France. The judge ruled in this case that it was by an order of this honorable court that the sugar cane that was situated in the fields of the Gravier plantation were ordered and sold to be taken off by Charles Olivier of St. Mary Parish. Also, the Gravier sugar plantation was ordered seized and sold by Sheriff Benjamin Hudson, of St. Mary Parish. The judge also ruled in this matter that the ordering of the sugar cane sold to be harvested was not an illegal act, and was done in accordance with the laws of the State of Louisiana. Charles Olivier had to reimburse the heirs of Jacques Gravier for the amount of money that he got from the sale of the sugar cane. Also, Charles Olivier was not responsible for any damage done to the heirs of Jacques Gravier, because he (Charles Olivier) had obtained a court order.

The civil lawsuit entitled, "Jacques Gravier, Miss Jeanne Gravier, and Madame Henri Petit versus Alexander Baron," was brought to enforce the payment of a claim of Alexander Baron against Jacques Gravier on a sugar plantation, which Alexander Baron had a one third possession. A final judgment by the honorable court was rendered in this matter, and the order directed the Gravier land to be seized and sold to satisfy the claim.

Alexander Baron did not think that he was personally bound for the debt of the mortgage which was bearing on the Gravier plantation. The amount of this mortgage was $10,383.00. Alexander Baron relinquished the property of the Gravier plantation in St. Mary Parish, when the mortgage accrued according to Article 3368 of the Louisiana Civil Code. The act of relinquishment of the lands of Gravier by Alexander Baron was done before a notary public in the presence of two witnesses, and notification had been acknowledged by Alexander Baron. Alexander Baron brought the certified Act of Relinquishment and filed this act in the office of the clerk of court of the 1st District Court of St. Mary Parish. Finally, on the petition of Alexander Baron, a curator was appointed by the honorable court to the property relinquished. Alexander Baron further showed that according to Article 3369 of the Louisiana Civil Code, the sale of the mortgage land is to be conducted under the curator, in the manner of the laws of the State of Louisiana. Also, Alexander Baron knew that Benjamin Hudson, who was then the Sheriff of St. Mary Parish, knew the premises of the Gravier plantation, and he (the Sheriff) seized the plantation and advertised the Gravier plantation for sale, as the plantation of Alexander Baron. The reason that the Gravier plantation was seized by Sheriff Benjamin Hudson was to satisfy a debt owed by Alexander Baron. When Alexander Baron relinquished the property of the Gravier sugar plantation, he was completely exonerated not only from the claim of the heirs of Jacques Gravier, but also from any further participation in the suit. The name of Alexander Baron was not to be used in the civil lawsuit.

Alexander Baron alleged that Sheriff Benjamin Hudson, of St. Mary Parish, did him great injury by using his name (Alexander Baron) in the sheriff's advertisement of the sheriff's sale of the Gravier plantation. Also, Alexander Baron alleged that Sheriff Benjamin Hudson held him (Alexander Baron) out to the public as failing to pay his debts. Alexander Baron sustained damages to the amount of $5,000.00. Alexander Baron thought that Sheriff Benjamin Hudson, of St. Mary Parish, had no authority to sell the Gravier plantation. This civil lawsuit was filed in the office of the clerk of court of St. Mary Parish on the date August 17, 1832.

On December 27, 1834, Alexander Baron was given notice that the final judgment in the civil lawsuit entitled, "Alexander Baron versus Sheriff Benjamin Hudson, of St. Mary Parish," was rendered by the District Court of St. Mary Parish. The judgment was rendered against Alexander Baron and in the favor of the heirs of Jacques Gravier, who were Jacques Gravier and Jeanne Gravier of Dergerac, France, and Madame Henri Petit of Paris, France. On the date of December 13, 1834, Alexander Baron was given ten days to appeal the lawsuit or to pay the amount he was ordered to pay. Alexander Baron was ordered to pay $3,343.42, and another $800.00 to the heirs of Jacques Gravier. This payment was to be made from the personal estate of Alexander Baron. Also, Alexander Baron was an owner of the Gravier plantation on the date of April 24, 1834.

When Sheriff Benjamin Hudson received the execution of the civil lawsuit entitled, "Alexander Baron versus Benjamin Hudson, Sheriff of St. Mary Parish," he (the Sheriff) called on Francois Malus, who was the partner of Alexander Baron, on the Gravier plantation. At this visit, Sheriff Hudson informed Francois Malus that he had received an order of seizure and sale against the plantation of the heirs of Jacques Gravier. Sheriff Benjamin Hudson thought that Francois Malus knew about the civil lawsuit before his (the Sheriff's) visit, because Francois Malus said he was expecting something of this kind. Also, Francois Malus requested witnesses and wanted a waiting period of two to three days. Alexander Baron was residing in the City of New Orleans, Louisiana, and was not living or working on the plantation. Whereas, Francois Malus was his agent and acted in the name of Alexander Baron in St. Mary Parish in matters pertaining to the plantation. At this visit, Francois Malus made some spirited remarks in regard to the proceedings of the civil lawsuit of Alexander Baron. The basis of the civil lawsuit of Alexander Baron was the order of seizure and sale. Also, Francois Malus and Alexander Baron were planning to direct the witnesses in this matter to say certain things in their favor. Alexander Baron and Francois Malus were doing what they pleased to do on the Gravier plantation, not what they were supposed to do. Also, Francois Malus would not give any information to Sheriff Benjamin Hudson at this visit.

Francois Malus lived in a dwelling house that was not on the plantation or the land that was seized. There were no improvements on the plantation, except for the dwelling house, which was situated on the west side of Bayou Teche, and the fencing that was on the land.

Francois Malus was cultivating the plantation, and he had a crop of sugar cane growing on the plantation. Also, Francois Malus appeared to exercise authority over the cultivation of the sugar cane crop. Also, wood was being cut on the Gravier plantation, but it was possible that Francois Malus had suspended the cutting of the wood. The other planters in the neighborhood thought that Francois Malus had been the manager of the plantation since the plantation was established.

Francois Malus lived on the property, which bordered the Gravier plantation on the upperside. There was a division fence between the land on which Francois Malus resided, and the land which was seized by Sheriff Benjamin Hudson of St. Mary Parish.

There was a house that was situated on the Gravier plantation, on the west side of Bayou Teche. This dwelling house was occupied by Mr. Lagarde, who was the overseer of the Gravier sugar plantation.

On December 16, 1834, the property of Alexander Baron was seized and fifty seven hogsheds of sugar was also seized. Notice of this seizure was served on Alexander Baron in the City of new Orleans, Louisiana.

On August 1, 1832, Louis Eloi DeBlanc sold to Francois Malus a certain tract of land. This tract of land had a frontage along Bayou Teche of nine arpents and twenty feet on the west side of Bayou Teche, and the ordinary depth of forty arpents. On the east side of Bayou Teche, this tract of land had a frontage of three and one half arpents with a forty arpentage depth. The tract on the east side of Bayou Teche was considered the upperside, and the tract on the west side of Bayou Teche was considered the lowerside. The boundaries of this tract of land were on the upperside by the lands of the heirs of Jacques Gravier, and on the lowerside by land owned by Lewis Moore, Junior.

The purchase price of this tract of land was $4,000.00. Francois Malus gave two promissory notes. One promissory note was for the amount of $2,000.00, and was due in the month of April of the year of 1833. The second promissory note was in the amount of $2,000.00, and was payable in the month of April of the year of 1834. The property was to remain specially mortgaged until the full and the final payment of the mortgage.

Also, it was understood between Louis Eloi DeBlanc and Francois Malus that the first promissory note in the amount of $2,000.00 should be deposited in the Bank of Louisiana, which was located at Opleousas, Louisiana. The first note should be applied to the payment of a mortgage given by Louis Eloi DeBlanc in favor of the Union Bank of Louisiana on the tract of land. The slaves belonging to Louis Eloi DeBlanc were also mortgaged. The land and the slaves were to be released when the first promissory note was due.

On December 8, 1832, Sheriff Benjamin Hudson, of the Parish of St. Mary, sold at a sheriff sale the Gravier sugar plantation, which was situated on the Old Spanish Trail, in the Parish of St. Mary. Sheriff Hudson was directed to seize and offer for sale this plantation, after a writ of seizure and sale was issued out of the 1st Judicial District Court of the State of Louisiana. This originated from a civil lawsuit, which was entitled, "Jacques Gravier, Jeanne Gravier, and Jeanne Bordier, wife of Henri Petit, as plaintiffs, versus Alexander Baron, an attorney of the City of New Orleans, Louisiana. Alexander Baron was acting as agent in the management and operations of the sugar plantation for Jacques Gravier, Jeanne Gravier, and Madame Henri Petit, of the Kingdom of France.

Prior to December 8, 1832, Sheriff Benjamin Hudson did seize the Gravier sugar plantation due to the $10,000.00 mortgage that was bearing on the plantation. This civil lawsuit was instituted on the basis of the nonpayment of the $10,000.00 mortgage, which mortgage bore on the Gravier sugar plantation, and this mortgage was instituted by Alexander Baron of the City of New Orleans. Also on this date of December 8, 1832, the Gravier sugar plantation was to be sold at the Courthouse of St. Mary Parish, which is situated in the town of Franklin, Louisiana. The terms of the sheriff sale were the land and property was to remain specially mortgage until the final payment by the purchaser. Also, the purchaser is required to give bond with good security. The interest on the purchase price for all credit sales will be 5% per annum from the date of the sale or December 8, 1832.

After carefully examining the mortgage books in the recorder's office or the modern day office of the clerk of court, Judge Donelson Caffery, a judge of St. Mary parish, certified that there existed a mortgage in the name of Alexander Baron of the City of New Orleans, Louisiana, and this mortgage was dated prior to March 21, 1829 on the Gravier sugar plantation. There also exists a conventional mortgage on the Gravier plantation, in favor of Antoine Carraby. The conventional mortgage to Antoine Carraby was dated March 21, 1829, and is for the sum of $11,700.00. The conventional mortgage is payable in annual installments.

The Gravier sugar plantation consisted of fifteenth arpents fronting Bayou Teche on both the west side and the east side of Bayou Teche. The depth of the plantation on both the east and the west sides of Bayou Teche was forty arpents, which measured six hundred arpents depth of land. Also, there was a crop of sugar cane growing on the plantation, along with livestock and slaves living on the plantation. In the year of 1832, the Gravier sugar plantation was bounded on the upperside by land owned by Alexander Baron and Francois Malus, and on the lowerside by land owned by Louis Eloi DeBlanc. The boundaries of the plantation prior to the year of 1832 were on the upperside by land that was owned by Nicholas Pelagie Loisel, the owner of the Loisel Plantation, and on the lowerside by land that was owned by Alexander Hebert.

The purchaser at the sheriff sale of the Gravier sugar plantation was Noel Barthlemy LeBreton, an attorney of the City of New Orleans, Louisiana. Noel B. LeBreton was an agent of Jacques Gravier, Jeanne Gravier, and Madame Jeanne Bordier Petit, all were residing in the country of France. The purchase price of the Gravier sugar plantation was $5,002.00. The amount of $5,002.00 or the purchase price of the Gravier sugar plantation was acknowledged and received by Sheriff Benjamin Hudson, Sheriff of St. Mary Parish. Also, Sheriff Benjamin Hudson granted acquittal of the debt of $5,002.00. The amount of $5,002.00 was used to pay and satisfy a judgment, which had been rendered by the Supreme Court of the State of Louisiana. Alexander Baron had been ordered by the Supreme Court of the State of Louisiana to pay the sum of $10,383.00, with all legal interest from August 8, 1830, and all court cost, which costs amounted to the sum of $183.86. Also, Alexander Baron was ordered to return to the court all the proceeds of the plantation to the honorable court before August 4, 1832.

On December 8, 1832, Sheriff Benjamin Hudson, sold at a sheriff sale the sugar plantation that was owned by Jacques Gravier of Dergerac, France. This sugar plantation was mortgaged by Alexander Baron, who was an attorney in the City of New Orleans, and acted as agent for Jacques Gravier. Noel Barthlemy LeBreton, an attorney in the City of New Orleans, Louisiana, purchased the sugar plantation for the price of $5,002.00. Also, Noel B. LeBreton paid the purchase price of $5,002.00 in cash, and Alexander Baron acknowledged the receipt and granted acquittal for the $5.002.00 or the purchase price of the sugar plantation.

This sugar plantation had a fifteen arpent frontage along Bayou Teche, with a depth of forty arpents, on both the east and the west sides of Bayou Teche. This was a total of six hundred arpents on the east and the west sides of Bayou Teche. At this time, in the year of 1832, this sugar plantation was bounded on the upperside or above by land owned by Francois Malus and Alexander Baron, and on the lowerside by land owned by Louis Eloi DeBlanc. Before the year of 1832, the boundaries of this sugar plantation was on the upperside or above by land owned by Nicholas Pelagie Loisel of Loisel Plantation, and on the lowerside or below by land owned by Nicholas Hebert. Also, in the year of 1832, there was a crop of sugar cane on the plantation.

Alexander Baron, an attorney of the City of New Orleans, and Francois Malus, of the Parish of St. Mary, were in partnership and owned land, which was situated in the same vicinity as the Gravier sugar plantation. Francois Malus resided on the east side of Bayou Teche, opposite the Gravier sugar plantation. Also, Francois Malus had business connections in the running of this plantation. Also, Alexander Baron, an attorney in the City of New Orleans, Louisiana, acted as the agent for Jacques Gravier in the running of the Gravier sugar plantation. But, Francois Malus and Alexander Baron schemed together to beat Jacques Gravier out of money, and they did this by mismanaging the sugar plantation. Both, Francois Malus and Alexander Baron ran the sugar plantation to benefit themselves, not Jacques Gravier. In fact, Alexander Baron had placed a $10,000.00 mortgage on the plantation, and that was the reason that Jacques Gravier could not sell the sugar plantation. With Jacques and Jeanne Gravier residing in Dergerac, France, and Madame Henri Petit residing in Paris, France, both Francois Malus and Alexander Baron thought that Jacques Gravier would not find out about how they were mishandling and mismanaging the sugar plantation. The correct arpentage that was harvested in sugar cane was never reported to Jacques Gravier.

On February 26, 1836, Sieur Jacques Gravier of Dergerac, France, conveyed or sold the property that he held a mortgage on in the State of Louisiana, which was then a part of the United States of America. The purchaser of the Gravier sugar plantation was Francois Barthlemy St. Marc Darby of the Parish of St. Martin of the State of Louisiana. Francois Barthlemy St. Marc Darby was planning on going into the business of the cultivation of sugar cane. At this time, sugarcane was a very big cash crop in the State of Louisiana.

Sieur Jacques Gravier shared ownership of the land with Miss Jeanne Gravier of Dergerac, France, and Madame Henri Petit of Paris, France. Jacques Gravier, Jeanne Gravier, and Madame Henri Petit were not present at the execution of the act of conveyance of their sugar plantation. Instead, the Graviers and Madame Petit were represented by the legal consul of Mr. Noel Barthlemy LeBreton, an attorney, who resided in the City of New Orleans, Louisiana. Also, Edouard Simon, an attorney of St. Martin Parish, Louisiana, represented Francois Barthlemy St. Marc Darby. The execution of the act of conveyance of the former Gravier sugar plantation took place in the town of St. Martinville, Louisiana. Felix de Amas, a notary public in the City of New Orleans, presided over the execution of the act of conveyance, and he wrote the act of conveyance in the French language. Also, Judge Paul Briant, who was serving as judge of the Parish of St. Martin and ex-officio notary public of the Parish of St. Martin, was the person who recorded and filed the original act of conveyance of the Gravier - Darby conveyance in the conveyance records of the office of the clerk of court of the Parish of St. Martin. The office of the clerk of court of the Parish of St. Martin is situated in the St. Martin Parish courthouse, which is situated in the town of St. Martinville, Louisiana.

The sugar plantation that Jacques Gravier, Jeanne Gravier, and Madame Henri Petit were selling or conveying was situated in the Parish of St. Mary, in the State of Louisiana. This sugar plantation had a frontage of fifteen arpents along Bayou Teche, with the ordinary depth of forty arpents on both the east and the west sides of Bayou Teche. This sugar plantation consisted on six hundred arpents of land on the east side, as well as six hundred arpents of land on the west side of Bayou Teche. Also, included with the land was all the buildings, fences, and ponds, which had been constructed on the plantation under the ownership of the Gravier family of France. Also, this sugar plantation had been surveyed by Robert Trimble, who was a land surveyor, and lived in the neighborhood where the Gravier sugar plantation was situated. Also, Robert Trimble had drawn a map of the former Gravier sugar plantation, and this map was annexed to the original act of conveyance.

Francois Barthlemy St. Marc Darby promises and obligates himself to pay unto Jacques Gravier, Jeanne Gravier, and Madame Petit, the sum of two thousand and one hundred and seventy dollars and sixty cents or $2,170.60, which was the purchase price. Also, the purchase price of the sugar plantation was to be paid in the month of April of the year of 1837.

The reason that Jacques Gravier, Jeanne Gravier, and Madame Petit were selling their sugar plantation, which was situated in the State of Louisiana, was because the $10,000.00 mortgage had been paid off. The Gravier family had been unable to sell their sugar plantation due to the $10,000.00 mortgage. Also, the sugar plantation as an investment had not paid off financially as well as the Gravier family expected it to. And the running and overseership was another problem for the Gravier's, because they had been taken advantage of by overseer and agents running the plantation, which resulted in a financial loss to the Gravier family of France.

Francois Barthlemy St. Marc Darby was a son of Jean St. Darby and his wife, Marie Francois Pellerin Darby. The father of Francois Barthlemy St. Marc Darby, Jean St. Marc Darby was an official in the militia and commander of New Iberia, Louisiana. Francois Barthlemy St. Marc Darby was born on July 21, 1708, and he was baptized at St. Martin of Tours Catholic Church in St. Martinville, Louisiana. Francois Barthlemy St. Marc Darby was married at the St. Martin of Tours Catholic Church in St. Martinville to Marie Constance DeBlanc. Marie Constance DeBlanc was the daughter of Louis Charles DeBlanc and his wife, Elizabeth Poupone D'ernville. Her Father, Louis Charles DeBlanc was an officer in the military, and had moved to St. Martinville from Natchitoches, Louisiana.

After purchasing the former Gravier sugar plantation, which was situated on the Old Spanish Trail, between the plantations of Thomas Houssan Thompson, or the Rosedale Plantation, and the plantation of Claude Frilot, Francois Barthlemy St. Marc Darby had plans to improve his newly purchased sugar plantation. Francois Barthlemy St. Marc Darby was interested in the growing of sugar cane and the processing of sugar cane into sugar products. Also, he planned to keep the Gravier sugar plantation as a sugar plantation, but he planned to build a sugar establishment or sugar house for the processing of the cane juice into raw brown sugar and molasses. Francois Barthlemy St. Marc Darby did not want to depend upon his neighbors, Louis Eloi DeBlanc, and Thomas Houssan Thompson, for the processing of his sugar cane into the juices that would make molasses and raw brown sugar. At this time, most plantations had their own sugar mill, which was situated on the plantation, and the sugar cane that was grown on the plantation was ground and processed in the plantation sugar mill. After processing the sugar cane that was grown and harvested on the plantation, the plantation sugar mills took in to grind into cane juice the sugar cane of the neighboring plantations. If an early freeze or a hurricane occurred, sometimes the planters did not get all their large crop of sugar cane processed into brown sugar and molasses in the plantation sugar mill.

On January 30, 1847, the succession sale of the real estate immovable property and the movable property of the sugar plantation belonging to the late Francois St. Marc Darby of the Parish of St. Mary was held. Francois St. Marc Darby was the husband of Mistress Constance DeBlanc Darby, and the Father of four children. The children of Francois St. Marc Darby by his marriage to Mistress Constance DeBlanc were three sons, Louis St. Marc Darby, Francois Optat Darby, and Joseph Alcide Darby, and one daughter, Mary Lodoskia Darby, who was the wife of Doctor Leonard Smith of the State of Maryland. Three of the children, Louis St. Marc, Francois Optat, and Lodoskia Darby Smith were all of the age of majority at the time of the death of their Father, Francois Barthlemy St. Marc Darby. Only, the youngest child, Joseph Alcide Darby was a minor in age.

On November 20, 1846, a family meeting of the widow, Marie Constance DeBlanc Darby and her three oldest children, Louis St. Marc Darby, Francois Optat Darby, and Mary Lodoskia Darby, who was the wife of Doctor Leonard Smith, was held. The Honorable John A. Dumartrait, who was serving as judge of the Parish of St. Mary and recorder of the Parish of St. Mary, presided over this Darby family meeting. Present at this meeting was the widow, Marie Constance DeBlanc Darby, and her oldest children, two sons, Louis St. Marc Darby and Francois Optat Darby, and the only daughter, Mary Lodoskia Darby, who was the wife of Doctor Leonard Smith of the State of Maryland. The purpose of this Darby family meeting was to discuss and deliberate the sale of the property belonging to the succession of Francois Barthlemy St. Marc Darby, but now the property belongs to the community or partnership between the widow, Constance DeBlanc Darby, and the heirs of Francois Barthlemy St. Marc Darby, which are the four children of this marriage to Constance DeBlanc. The widow and heirs were planning to sell the sugar plantation, which was situated in St. Mary Parish, to the two oldest sons for the purpose of cultivating sugar cane as a source of income.

At this family meeting, it was decided that the property belonging to the succession of Francois Barthlemy St. Marc Darby was to be disposed of at a one year credit from the day of the sale of the property. The purchasers of the property were to give their notes with good and sufficient security insolido to the satisfaction of those interested parties. The promissory notes were to bear interest at the rate of 8% a year from the date of the sale or the time due until the final payment. All mortgageable property will remain mortgaged by privilege of the vendor's lien until the full and final payment of the purchase price plus the interest accrued.

This succession sale has been previously advertised by written advertisement in the English and the French languages, which notices were posted on the door of the St. Martinville Courthouse, which is situated in the town of St. Martinville, in the Parish of St. Martin. Also, these notices were advertised by a similar advertisement, which was inserted in the Gazette, a weekly newspaper. The Gazette newspaper was published in the town of St. Martinville, Louisiana, in the 1840's. The advertisement stated that the property belonging to the succession of Francois Barthlemy St. Marc Darby was to be offered for sale at the coffee house of Mr. Louis Migues, which coffee house was situated in the town of New Iberia, Louisiana, and the sale would be held on January 30, 1847 at 1:00 P.M.

On January 30, 1847, at the coffee house of Mr. Louis Migues, in the town of New Iberia, Louisiana, Louis St. Marc Darby, of the City of New Orleans, Louisiana, and his brother, Francois Optat Darby, of the Parishes of St. Mary and St. Martin, purchased their late Father's, Francois St. Marc Darby, sugar plantation at an auction succession sale. The Darby brothers were the last and highest bidders. George Brians served at this sale as the auctioneer. The Darby sugar plantation was situated in the Parish of St. Mary. This plantation consisted of two separate tracts of land, one tract was located on the eastern side or the left bank of Bayou Teche, and the other tract was situated on the right bank or the west side of Bayou Teche. The tract of land that was situated on the east bank of Bayou Teche consisted of one thousand four hundred and eighty arpents of land. The tract of land, situated on the eastern bank of Bayou Teche, was over twice as large as the tract of land, that was situated on the west bank of Bayou Teche. The tract of land, which was situated on the west bank of Bayou Teche, measured twenty four arpents of land fronting Bayou Teche with a depth of forty arpents of land. The total arpentage of land of the tract on the west side of Bayou Teche was six hundred arpents of land. The Darby plantation was bounded in the 1840's on both the east and the west sides of Bayou Teche on the upperside or above by the lands of Julie Felicite Chretin, the second wife of Colonel Charles Napoleon Olivier, and on the lowerside or below by the lands of Thomas Houssan Thompson, or the Rosedale Plantation. Also, sold besides the land of the plantation were the buildings, which consisted of slave cabins, barns, sheds, out-houses, and store houses. On the sugar plantation along Bayou Teche, a ferry boat was used to transport the slaves, equipment, goods, workers, and other articles from one side of Bayou Teche to the other side of Bayou Teche.

The purchase price of the sugar plantation that belonged to the estate of Francois St. Marc Darby was $40,000.00, and the price of the ferry boat was $205.00. Louis St. Marc Darby and Francois Optat Darby, the purchasers, furnished their notes with good and sufficient security in soldio to the satisfaction of the interested parties.

The succession sale took place in the coffee house of Louis Migues, in the town of New Iberia, Louisiana. The witnesses for the sale of the sugar plantation of Francois St. Marc Darby were Louis Migues or Miguez, who was the owner of the coffee house where the sale took place, and Pierre Cabrolle.

Also, on January 1, 1847, Louis St. Marc Darby and his brother, Francois Optat Darby, formed a partnership for the purpose of operating the plantation that they had purchased. The Darby brothers planned to continue cultivating this sugar plantation as a sugar plantation. On this date, both Louis St. Marc Darby and Francois Optat Darby appeared before Barthlemy d'Aquesseau de la Houssaye, a notary public in the Parish of St. Mary, to form this partnership. The name of this partnership was Darby Brothers Planters. The Darby Brothers Planters partnership operated under the style of Darby Brothers Planters, and the name of Darby Brothers Planters for several years. Originally, the Darby Brothers Planters partnership was to commence, or to begin operations on January 30, 1847, and to continue in operation for a period of five years, or until January 30, 1852. The partnership, Darby Brothers Planters, could be dissolved by the mutual consent of either Louis St. Marc Darby, of New Orleans, Louisiana, or Francois Optat Darby, of St. Mary Parish.

Louis St. Marc Darby, a resident of New Orleans, Louisiana, was the party of the first part of the Darby Brothers partnership, and Francois Optat Darby, of St. Mary Parish, Louisiana, and the brother of Louis St. Marc Darby, was the party of the second part of the Darby Brothers partnership. Francois Optat Darby and Louis St. Marc Darby had purchased their late Father's, Francois St. Marc Darby, sugar plantation, jointly, with the intentions of carrying on the agricultural business of cultivating sugar cane, and processing the sugar cane into sugar and molasses. With half of the sugar plantation belonging to Louis St. Marc Darby, who was the party of the first part, and the other half belonging to Francois Optat Darby, who was the part of the second part.

The part of the first part, Louis St. Marc Darby, furnished to his partner, Francois Optat Darby, half of the sugar plantation, and twenty five slaves. The twenty five slaves were Chavetz Micheal, a man slave, age 24 years old, and valued at $1,000.00, Baptiste, a teenage boy slave, age 15 years old, and valued at $800.00, Jesse, a man slave, age 38 years old, and valued at $750.00, Jean Chavetz, a man slave, age 20 years old, and valued at $900.00, John, a man slave, age 25 years old, and valued at $650.00, Marvich, a man slave, age 40 years old, and valued at $400.00, Alictove, a teenage boy slave, age 15 years of age, and valued at $450.00, Dick, a boy slave, age 12 years of age, and valued at $350.00, Billy, a boy slave, age 8 years old, and valued at $300.00, Aimee', a girl slave, age 6 years of age, and valued at $200.00, Grace, a girl slave, age 3 years old, and valued at $425.00, Jane, a woman slave, age 27 years old, and valued at $400.00, Ann, a woman slave, age 30 years of age, and valued at $650.00, John, a boy slave, age 10 years old, and valued at $200.00, Martha, a girl slave, age 6 years old, and valued at $150.00, Maurich, a boy slave, age 3 years old, and valued at $100.00, Robert, a baby slave of 6 months of age, and worth $550.00, Henry, a boy slave, age 12 years old, and valued at

$300.00, George, a boy slave, age 9 years of age, and valued at $250.00, Joseph, a baby slave, age 9 months of age, and valued at $50.00, Noveris, a man slave, age 24 years of age, and valued at $550.00, Luimas, a boy slave, age 6 years old, and valued at $200.00, Levcade, a boy slave, age 3 years old, and valued at $150.00, and Casmir, a baby slave, age 6 months of age, and valued at $50.00.

The slaves were worth very little to their owners in value when the slaves were in babyhood, but as the slave child begin to grow its's value increased. Slaves were worth the most money during their teenage years, their 20's, their 30's and their 40's, or at the time when the most work could be accomplished by that slave.

The party of the second part of the Darby Brothers partnership was Francois Optat Darby, who put as his portion of the partnership these slaves, Nathan, a slave man, age 22 years old, and valued at $900.00, Tom, a slave man, age 25 years old, and valued at $900.00, Allen, a slave man, age 30 years of age, and valued at $800.00, Jack, a teenage slave, age 14 years old, and valued at $700.00, Jules, a boy slave, age 10 years old, and valued at $400.00, Harrison, a boy slave, age 9 years old, and valued at $300.00, Lusane, a child slave of six years of age, and valued at $200.00, Huson, a baby slave, age 14 months old, and valued at $50.00, Mariah, a slave, age 25 years of age, and valued at $550.00, Martha, a woman slave, age 25 years of age, and valued at $660.00, Alexis, a baby slave of 2 months of age, and valued at $25.00, Rose, a teenage girl slave, age 18 years of age, and valued at $700.00, and Avelina, a woman slave of 29 years of age, and valued at $800.00.

Also, Francois Optat Darby put up as his share of the Darby Brothers partnership, two mules, which were valued at $130.00 each, or the total valuation of the two mules was $260.00, and four horses. Apparently, Francois Optat Darby did not think the Darby Brothers partnership had enough land to cultivate sugar cane on, so he put up as his share into the he partnership a prescription right of sixty acres of land. These sixty acres of land were located in the rear of the lands of Charles Napoleon Olivier. This sixty acres of land was valued at $3.00 per acre.

Louis St. Marc Darby had invested $2,306.00 into the Darby Brothers partnership, which was a larger amount of money than the amount of money that was put into the partnership by his brother, Francois Optat Darby. Francois Optat Darby furnished his promissory notes for the sum of $1,513.00, and these promissory notes were conditioned to bear interest at 8% per year from the date of January 1, 1847, until paid. Also, these promissory notes were made payable to the order of Louis St. Marc Darby. It was well understood between Louis St. Marc Darby and Francois Optat Darby that after the satisfaction of the promissory notes and the interest, the whole property effects furnished by Louis St. Marc Darby and Francois Optat Darby for the benefit of the partnership shall be considered as belonging to the part of the first part, Louis St. Marc Darby, and the party of the second part, Francois Optat Darby, in the proportion of one half each.

Francois Optat Darby was to act as overseer of the sugar cane plantation, which was owned jointly by he and his brother, Louis St. Marc Darby. This sugar plantation was operated by the Darby Brothers partnership. Also, Francois Optat Darby received a salary of the sum of $1,200.00 over and above the portion coming to him as part owner of the plantation. Francois Optat Darby was called Optat by his family, friends, and associates. Also, Francois Optat Darby was to reside on the plantation, on the west side of Bayou Teche, in the plantation home, now known as Deepwood. Francois Optat Darby and his wife, Julie, and his children resided in this plantation home until January 1, 1857.

Francois Optat Darby was married in the month of July of the year of 1839 to Marie Julie Azemie Beauvois at the St. Martin of Tours Catholic Church in St. Martinville, Louisiana. Julie Beauvois was the daughter of Jean Baptiste Beauvois and his wife, Suzanne Ozenne Beauvois. After removing from Kaskaskias, Illinois Jean Baptiste Beauvois settled in the Bayou Teche area.
Jean Baptiste Beauvois died in the year of 1828, and his wife, Suzanne Ozonne Beauvois died on the date of July 17, 1826. Julie Beauvois Darby and her twin sister, Marie Marguerite Irma or Irene were born on the date of August 14, 1818. The twin sister of Julie Beauvois Darby, Marie Marguerite Irene or Irma Beauvois was married to Henri or Henry Camors.

The annual sugar crop was to be disposed of by the party of the first part, Louis St. Marc Darby. Louis St. Marc Darby acted as a consignor for sugar planters in that he would sell their harvest of sugar and molasses. Also, the sugar crop consigned to him, (Louis St. Marc Darby), as commission merchant, and the sugar crop was to be sold for the account of the Darby Brothers partnership. Louis St. Marc Darby was a partner in the firm of Darby and Tremount, a sugar factor business in the City of New Orleans, Louisiana.

The withdrawal of either Louis St. Marc Darby or Francois Optat Darby from the business of the Darby Brothers partnership could have resulted in the net proceeds from the sale being divided between the party of the first part, Louis St. Marc Darby, and the party of the second part, Francois Optat Darby. The debts contracted by either Louis St. Marc Darby or Francois Optat Darby, in the name of the Darby Brothers partnership shall bear on both Louis St. Marc Darby and his brother, Francois Optat Darby, and paid by the Darby Brothers in the amount of one half each.

By the year of 1854, Louis St. Marc Darby and Francois Optat Darby decided to sell their sugar plantation. The reason that the Darby brothers were desirous to sell the plantation was because the plantation was not profitable. In fact, the plantation was costing them more and more money to operate. So, the Darby brothers set out to look for a purchaser for their sugar plantation. They approached their Uncle Charles Napoleon Olivier, who was their uncle through marriage to their Aunt Celeste DeBlanc, and their Olivier first cousins, who were also engaged in the agricultural business of farming. Upon learning of the sale of the Darby Brothers sugar plantation, Charles Napoleon Olivier notified his youngest son by his first marriage, Joseph Devezin Olivier, who was residing in the island of Cuba, of the upcoming sale of the sugar plantation owned jointly by Louis St. Marc Darby and Francois Optat Darby.

On September 21, 1854, at the request of Francois Optat Darby, the party of the second part of the Darby Brothers partnership, a survey of the sugar plantation was made by the St. Mary Parish surveyor, Alexander L. Fields. The reason for this survey was because the Darby brothers were intending and making plans to sell their sugar plantation. From the land survey made by Alexander L. Fields, it was determined that the Darby Brothers sugar plantation consisted of one thousand, five hundred and thirty seven acres and seventy sevenths hundredths of an acre or one thousand, eight hundred and seventeen arpents and eight hundredths of an arpent of land on the east side of Bayou Teche. The tract of land of the Darby Brothers sugar plantation, which was located on the western side of Bayou Teche contained four thousand and twenty one arpents and eight hundredths of an arpent or two thousand and seven hundred and two acres and twenty seven hundredths of an acre in land on the western side of Bayou Teche.

At his sugar plantation on the island of Cuba, Joseph Olivier was notified by letter of the intention of the Darby Brothers to sell their plantation in the State of Louisiana. After residing on the island of Cuba since his marriage in the year of 1833 to Celeste Bellechase, Joseph Devezin Olivier was very desirous to return to the United States, and to his homeland, which was situated in the Southwest section on the edge of the bayou country of Louisiana and the Acadian country of Louisiana. Joseph Devezin Olivier was introduced to the region of the Gulf of Mexico through his tour of duty in the United States Navy, where he served as a midshipmen. Joseph Devezin Olivier was stationed on a ship in the Gulf of Mexico for most of his tour of duty. It was on the island of Cuba, which at that time was in the Dominica of Spain, that Joseph Devezin Olivier met his bride, Celeste Bellechase, who was the daughter of Joseph Bellechase of Cuba. They were married in the year of 1833.

After his marriage, Joseph Devezin Olivier was manager and administrator of three sugar estates or sugar plantations on the island of Cuba, which belonged to his wife, Celeste Bellechase. In Cuba, a sugar plantation is called a sugar estate. The three sugar estates managed by Joseph Devezin Olivier were Minivera, Monticello, and Rennion Desire. In 1837, Joseph D. Bellechase donated a sugar estate to his daughter, Celeste Bellechase, who was the wife of Joseph Devezin Olivier, and this sugar estate was named Monticello. Then in the year of 1839, Joseph D. Bellechase donated another sugar estate to his daughter, Celeste Bellechase Olivier, and this second sugar estate was named Minivera. The total value of Minivera was $50,000.00. Also, the sugar estate, Monticello, was where Joseph Devezin Olivier and Celeste Olivier made their home. The Monticello sugar estate was worth $7,000.00. The third sugar estate was named Rennion Desire. This plantation, Rennion Desire, was sold in the year of 1851, for $25,000.00. All of these sugar plantations were located on the island of Cuba, which at that time was situated in the Dominica of Spain. The total value of the three sugar estates owned in Cuba by Joseph Olivier and his wife, Celeste Bellechase Olivier amounted to the total amount of $82,000.00. While he (Joseph Devezin Olivier) was residing on the island of Cuba, whenever he had to have a notarial act, such as an act of conveyance or an act of mortgage passed, he had to go to the United States Consulate, which was situated in Havana, Cuba. While he was residing on the island of Cuba, Joseph Devezin Olivier was engaged in the practice of the sugar cane farming business and industry. Therefore, when the Darby Brothers plantation came up for sale, Joseph Devezin Olivier had plenty experience in the sugar business. There were many sugar planters from Louisiana, who had investments in sugar estates on the island of Cuba.

Joseph Devezin Olivier was the youngest son of Colonel Charles Napoleon Olivier by his first wife, Celeste Mathilda DeBlanc. Also, Constance DeBlanc, who was the widow of Francois St. Marc Darby, was the sister of Celeste DeBlanc Olivier, the first wife of Charles Napoleon Olivier. Therefore, the Darby Brothers, Louis St. Marc Darby and Francois Optat Darby, were the first cousins of Joseph Devezin Olivier. Joseph Devezin Olivier was called by the name, "Devezin," by his family, friends, and associates. The name, "Devezin," was the title in France of the Olivier family. In this time, of the 1850's, it was not unusual to approach cousins and other family members to purchase property.

On October 31, 1854, Joseph Devezin Olivier went from his residence, which was located on the sugar estate, called Monticello, which was situated near Mutanyan, on the island of Cuba, in the Dominica of Spain, to the United States Consulate in Havana, Cuba to execute an act of power of attorney. Joseph Devezin Olivier was granting power of attorney to his two half brothers, Jules Germain Olivier and Charles G. Olivier, of St. Mary Parish, Louisiana, to make a conveyance or a purchase in the name of Joseph Devezin Olivier of the Darby Brothers plantation. The reason that Joseph Devezin Olivier had decided to purchase the sugar plantation of the Darby Brothers was because he and his wife, Celeste

Bellechase Olivier, were desirous to return to the homeland of Joseph Devezin Olivier. Also, the Olivier's had decided to sell their sugar estates on the island of Cuba.

The Darby Brothers sugar plantation consisted of two thousand, nine hundred and nineteen arpents and thirty eight hundredths of an arpent of land. This sale is to be made with full warranty of title. Also, the plantation slaves are free from all liens, mortgages, or encumbrances. At this time, the purchase price of the plantation of the Darby brothers was agreed at $100,000.00. This was actually an intention to purchase the sugar plantation of Louis St. Marc Darby and Francois Optat Darby.

On Hallowen Day or October 31, 1854, Joseph D. Olivier went to the United States consulate in Havana, Cuba to execute an intent to purchase or the agreement to purchase the Darby Brothers plantation. In this act of intent to purchase, Joseph Devezin Olivier stated that he would assume the payment of the sum of $32,000.00 due in portions to Mistress Constance DeBlanc Darby, the mother of Louis St. Marc Darby and Francois Optat Darby. When the Darby Brothers purchased this sugar plantation at the succession sale of their late father's, Francois St. Marc Darby, estate, they borrowed $40,000.00 in notes owed to their mother, Constance DeBlanc Darby. From the time, the Darby Brothers purchased this plantation, in the year of 1847, Francois Optat Darby and Louis St. Marc Darby had only paid $8,000.00 on their loan. Also, Doctor Leonard Smith, the brother-in-law of the Darby Brothers, being the husband of their only sister, Mary Lodoskia Darby, and their younger brother, Joseph Alcide Darby, were also holders of this mortgage, which bore on the sugar plantation. The reason that Doctor Leonard Smith was considered a holder of the mortgage or the notes was because women, at this time in the 1850's in the State of Louisiana, could not make conveyances or mortgages without the assistance and authorization of their husbands. It was not until the 1970's, that women, in the State of Louisiana, were granted the right to make a deed of conveyance or mortgage or any other notarial act without the assistance and authorization of their husbands. Also, Joseph Devezin Olivier agreed to pay Louis St. Marc Darby and Francois Optat Darby, the vendors or the sellers in this conveyance, the sum of $25,000.00 on April 1, 1856 with interest at the rate of 8% on ($25,000.00) from the time Joseph Devezin Olivier is put in possession of the former Darby sugar plantation. For the balance of the purchase price, which amounts to $43,000.00, the attorneys for Joseph Devezin Olivier, Charles G. Olivier and Jules Germain Olivier will execute in the name of Joseph Devezin Olivier, eight promissory notes, which will be payable to Francois Optat Darby and Louis St. Marc Darby, on their order. Each one of the promissory notes was for the sum of $5,375.00 with interest at 8% per year from the first day of January or New Years Day of the year of 1855. The first payable note will be due on April 1, 1856, the second payable note will be due on April 1, 1857, the third payable note will be due on April

1, 1858, the fourth payable note will be due on April 1, 1859, the fifth payable note will be due on April 1, 1860, the sixth payable note will be due on April 1, 1861, the seventh payable note will be due on April 1, 1862, and the eighth payable note and final note will be due on April 1, 1863. Also, Joseph Devezin Olivier reserved the right to take up any of his promissory notes before the maturity of the note, and from paying the principal of such interest as may have accrued.

This sale was to include all the buildings, which were situated on the property, and the improvements made on the property by the Darby brothers, with the crop of sugar cane. The buildings on the plantation included barns, a residence house, now known as Deepwood, slave cabins, stables, a sugar house, warehouses, sheds, and other plantation buildings. There was a loading dock on Bayou Teche to ship out sugar and molasses, and to receive needed goods on the plantation.

Also, conveyed with this sale of the Darby Brothers plantation were seventy five slaves. These slaves were slaves for life, and were to continue working and residing on the plantation. These slaves were Henry, a slave man, age 25 years of age, Alphe, a slave man, age 34 years of age, Grace, a slave woman, age 55 years of age, Patimer, a salve, age 50 years old and crippled, Jessy, a slave, age 50 years of age, and crippled, Sissy, a teenage slave girl, age 18 years of age and her two children, Theresa, a girl of 4 years of age, and Emmanual, a slave toddler of one year old, Abralama, a slave woman, age 28 years old, and her three children, Uraine, a slave child, age 20 years old, Antoine, a slave child of 5 years old, and Catherine, a slave child of 3 years of age, Dulus, a slave, age 50 years old, and blind with only one eye, and a crippled wrist, Liz, a slave girl, of 12 years of age, John, a slave man, age 39 years old, who was a field hand, Charles, a slave man, age 37 years old, Tanchonmelto alias Time, a slave, age 23 years old, and a good house cook, Pauline, a slave, age 17 years old, and Pauline was a good field hand, Evariste, a boy slave of 14 years of age, Gulaume, a boy slave of 9 years old, Master Alexandrome, a slave, Ursule Martinagee, a slave, age 39 years old, and her two children, Zenon, a slave, age 20 years old, Seame, a slave, age 41 years old, and crippled, Parker, a slave, age 45 years old, Layunisu, a slave, age 54 years of age, Janus, a slave, age 35 years, and a mechanic, Janus has been having a lump on one of his knees, Rosa, a slave woman, age 23 years of age, and her three children, Manuel Shar, a slave, age 37 years of age, Aveline, a slave, age 37 years old, Ram, a slave, age 19 years old, an infant boy slave, Caroline, a slave woman, age 23 years of age, and her two children, William, a toddler slave 19 months old and Onezine H., a slave, Dick, a slave man, age 47 years old, Plomer, a slave of 16 years old, Philip, a slave man of 30 years old, Ann, a slave woman, age 31 years of age, and Prados, Ann's child of 18 months old, John, a slave teenage boy, age 15 years of age, Martha, a teenage girl slave, age 14 years old, Robert, a slave boy of 9 years old, Peter, a slave boy child of age 4 years, Peter and Robert are the children of a slave woman named Ann Mar, a slave of the age of 25 years old, and has runaway several times within the last eighteen months, Hunchun, a slave man, age 50 years old, Maria, a slave woman of 44 years old, Dick, a child slave of 4 years old, June, a teenage slave age 15 years of age, Tom, a slave man, age 30 years of age, Congo, a slave of 40 years of age, Harry, a slave man of 44 years of age, George, a slave man of 40 years of age, Louisiana, a slave woman age 42 years of age, Heaney, a slave girl, age 19 years old, and her two children, one child a toddler, who is 18 months old, George, a teenage boy slave age 17 years old, a child slave of the slave woman, Lucinda, who is 8 years of age, Huey, a slave man of 38 years old, Thomie, a slave of 27 years of age, and an unnamed slave woman and her four children, Leonuinde, a teenage girl salve, age 13 years of age, Cusmin, a child slave of age 9 years old, Filemin, a child slave of 3 years of age, and Salviz, a child slave of 2 years of age.

The livestock on the Darby plantation at this time of the conveyance consisted of sixty one mules, twenty work oxen, and a lot of hogs. Also, eight mules carts with iron axles, ten bagasse carts with iron axles, one pair of timber wheels, blacksmith tools, plows, hoes, axles, spades, and one draining wheel. All the farming utensils and the water cistern on the plantation were also sold.

The buildings on the plantation belonging to Dr. Leonard Smith, the brother-in-law of the Darby brothers, and occupied by Mrs. Constance DeBlanc Darby, the mother of the Darby brothers, which building was located on the lower line of the plantation are not to be included in the forthcoming sale of the Darby plantation. Also, eight mules, which were recently purchased by Louis St. Marc Darby and Francois Optat Darby for the partnership of the Darby Brothers was not included in the sale of the plantation.

It was understood between Louis St. Marc Darby and Francois Optat Darby, and Joseph Devezin Olivier that possession of the Darby plantation will be given to Joseph Devezin Olivier after Louis St. Marc Darby and Francois Optat Darby have taken off the crop of sugar and molasses off the plantation. The sugar and molasses was considered movable property.

Until, Joseph Devezin Olivier could take possession of his newly purchased sugar plantation, the movable property, which was situated on the plantation remained at risk of unintentionally being destroyed, by such forces as weather conditions, or a fire. In case of the unexpected or sudden death of any slave on the plantation, the value shall be determined by the parties (the Darby brothers and Joseph Devezin Olivier), themselves.

The attorney, Charles G. Olivier, agreed that Louis St. Marc Darby and Francois Optat Darby retain the use of the sugar house, warehouses, and loading docks to ship off this year's sugar crop of brown sugar and molasses. Also, the furniture and the baggage of Francois Optat Darby, his wife, Julie Beauvois Darby, and their children, Marie Suzanne, Marie Lucie, Francois Edgard, and Francois Optat, Junior, are to be shipped off from these loading docks, which are located on the plantation.

After the sugar and molasses are taken off the plantation, possession of the plantation is to be given or to pass on to Joseph Devezin Olivier, and this possession should be given presumably between now, (October 31, 1854) and the first of January of the year of 1855. Louis St. Marc Darby and Francois Optat Darby shall give reasonable notice of their vacating the plantation in the absence of Joseph Devezin Olivier to either Charles G. Olivier or to Eugene Barthlemy Olivier, both of St. Mary Parish, and the half brothers of Joseph Devezin Olivier.

Also, Joseph Devezin Olivier authorized his half brother, Charles G. Olivier, in the absence of Joseph Devezin Olivier to gain in the name of Joseph Devezin Olivier, to receive and acknowledge possession of the plantation conveyed to Joseph Devezin Olivier by Louis St. Marc Darby and Francois Optat Darby. No warranty was asked for the slave child born two or three days ago in the month of October of the year of 1854.

Also, on the date of October 31, 1854, Joseph Devezin Olivier agreed to the act of sale, which was in reality a sale and mortgage. Joseph Devezin Olivier agreed to pay the sum of $32,000.00 due in equal portions to Mrs. Constance DeBlanc Darby, Doctor Leonard Smith, and Joseph Alcide Darby. Also, it was agreed that Joseph Devezin Olivier would pay the sum of $25,000.00, which sum was conditioned to bear interest at 8% from the time that Joseph Devezin Olivier is put in possession of the plantation. For the balance of $43,000.00, Joseph Devezin Olivier had authorized his attorney to execute in the name of Joseph Devezin Olivier, eight promissory notes. Each promissory note or one of the eight notes was for the amount of $5,375.00, and conditioned to bear interest at 8% per year from January 1, 1855 until paid. The first note was payable on April 1, 1856, the second note was payable on April 1, 1857, and the eighth and final note was payable on April 1, 1863. At this time, Joseph Devezin Olivier reserved the right of taking up any of the notes before the maturity of the notes. For example, Joseph Devezin Olivier could pay the second note of the amount of $5,375.00 in the month of March of the year of 1858, and when the second note became due on April 1, 1858, Joseph Devezin Olivier would not have to pay the interest due. By reserving this right, Joseph Devezin Olivier saved himself from paying the interest due on this loan.

On November 16, 1854, Louis St. Marc Darby, of New Orleans, Louisiana, granted full authority to his brother, Francois Optat Darby, of St. Mary Parish, Louisiana, to make in the name of Louis

St. Marc Darby, an act of conveyance to Joseph Devezin Olivier of the State of Louisiana, but now residing in the Island of Cuba, to sell his title, rights, and interest in the sugar plantation that the Darby Brothers partnership had operated since 1847.

The sale from Louis St. Marc Darby and Francois Optat Darby to Joseph Devezin Olivier was made for the sum of $100,000.00, and this sale was actually a sale and mortgage. Joseph Devezin Olivier obligated himself to pay to Louis St. Marc Darby and Francois Optat Darby, the sum of $8,003.10, which was the sum due to Doctor Leonard Smith, who was married to the sister of the Darby brothers, Lodoskia Darby, and the amount was due on the date of January 3, 1855. This amount originated from an act of sale from Dr. Leonard Smith to the Darby brothers on September 27, 1847, and the act of sale was executed before Jean B. Darby, a notary public. Whereby, all of the mortgage is bearing on some of the slaves conveyed to Joseph Devezin Olivier from the Darby brothers. This mortgage was issued originally to secure the purchase money. Also, Joseph Devezin Olivier assumes to pay to Mrs. Constance DeBlanc Darby, two percent of the sum of $11,000.00 or $220.00 with the interest accruing from January 3, 1848. This note to Mrs. Constance DeBlanc Darby was originally issued by a virtue of an act of sale from Mrs. Constance DeBlanc Darby to her sons, Louis St. Marc Darby and Francois Optat Darby,on the date of September 27, 1847. Resulting from the act of sale on September 27 1847 was a mortgage securing the payment of the purchase money, and bearing upon some of the slaves conveyed in this act of sale to Joseph Devezin Olivier. A note in the amount of $10,756.69 and conditioned to bear interest at 8% per year is due to the minor child, Joseph Alcide Darby. Joseph Alcide Darby was the younger brother of Louis St. Marc Darby. Francois Optat Darby, and Lodoskia Darby Smith. The note was in favor of Joseph Alcide Darby, and this note bore upon some of the property conveyed by the act of sale to Joseph Devezin Olivier, which amount to be paid was $29,759.79

On the date of January 21, 1855, the act of sale or the act of conveyance of the Darby Brothers sugar plantation, which was situated on the Old Spanish Trail, was finalized, and the sale was passed. Also, at this time, possession of the plantation was given to Charles G. Olivier and Jules Barthlemy Olivier, the half brothers of the purchaser, Joseph Devezin Olivier. The reason that possession was not given to Joseph Devezin Olivier, was because Joseph Devezin Olivier was residing in the island of Cuba, on the sugar plantation, Mimosa, when possession of the Darby Brothers sugar plantation was given, so therefore his half brothers, Charles and Jules Olivier acted on his (Joseph Devezin Olivier's) behalf in accepting the land, slaves, and all other belonging situated on the plantation on this date. At this time, Francois Optat Darby and his wife, Julie, had vacated the plantation home of the Darby Brothers sugar plantation, which is now known as Deepwood.

Also, on this date of January 21, 1855, Madame Marie Azelie Julie Beauvois, the legal and lawful wife of Francois Optat Darby, took cognizance of the foregoing act of conveyance or the act of sale of the former Darby Brothers sugar plantation to the purchaser, Joseph Devezin Olivier. Mrs. Julie Darby was taken out of the physical and hearing presence of her husband, Francois Optat Darby, by the presiding notary public, the Honorable John A. Dumartrait, and the act of conveyance or the act of sale of the Darby Brothers plantation was read and explained very carefully to her. This reading and explaining of the act of conveyance to Mrs. Darby was required by the law of the State of Louisiana at this time in 1855. The reason for this was because Mrs. Darby, like all other married women in the State of Louisiana and those married women residing in other states with husbands owning property in the State of Louisiana, held a legal mortgage on the property of their legal and lawful husbands. The reason for this was for the restitution of the dotal, matrimonial, and paraphernal property of a wife, that her legal husband had legal control over. At this time, married women in the State of Louisiana, had no control over their own property without the consent and authorization of their husbands. If a husband sold any property that was either brought into the marriage by either the wife or the husband, or sold any property that was bought or purchased after the marriage, a wife had to be informed of the foregoing act of sale. A wife had the choice to either renounce or release her dotal, matrimonial, or paraphernal interests and rights. Also, a husband was unable to finalize or to legalize the foregoing act of sale without the consent of his wife. In this case of the Darby-Olivier conveyance, Mrs. Julie Darby did declare that she was aware of the act of sale, and she renounced and released her rights, interests, and privileges in the plantation granted to her by the laws of the State of Louisiana in favor of the purchaser, Joseph Devezin Olivier. Also, Mrs. Julie Darby renounced besides all the rights, interests, and privileges granted to her under the laws of the State of Louisiana, all the rights, interests, and privileges of any nature or kind to which she may be entitled to under the laws of the State of Louisiana in favor the purchaser, Joseph Devezin Olivier.

Francois Optat Darby acting as legal representative and attorney of his brother, Louis St. Marc Darby, took cognizance of the fact that Louis St. Marc Darby of the City of New Orleans, Louisiana, fully obligated and bounded himself, (Louis St. Marc Darby), to the purchaser, Joseph Devezin Olivier, by good authentic renunciation of the dotal, matrimonial, and paraphernal rights, interest, and privileges granted to his legal and lawful wife, Carmelite Coralie Fusilier. In the City of New Orleans, Louisiana, on the date of April 3, 1855, Coralie Fusilier Darby, the lawful and legal wife of Louis St. Marc Darby, declared that her husband, Louis St. Marc Darby and his brother, Francois Optat Darby, on the 27th day of the month of January of the year of 1855, by an act of conveyance or an act of sale, for the sum of $100,000.00, did convey and sell the Darby Brothers sugar planation to Joseph Devezin Olivier, who is residing on a sugar plantation, Mimosa, on the island of Cuba. Joseph Devezin Olivier was represented by his half brothers, Charles G. Olivier, and Jules Germain Olivier in this act of conveyance. Joseph Devezin Olivier did not come to the United States for the passing of the act of sale or the act of conveyance, because he was working in Cuba on his sugar plantations. Mrs. Coralie Fusilier Darby did declare and take cognizance of the fact of the foregoing act of sale, and she renounced and released all her dotal, matrimonial, and paraphernal rights, interest, and privileges to the Darby Brothers sugar plantation unto Joseph Devezin Olivier, the purchaser.

On March 25, 1835, the legislature of the State of Louisiana passed a law granting a legal and lawful mortgage on the property owned by their legal and lawful husbands. This laws was passed to protect the dotal, matrimonial, and paraphernal rights, interest, and privileges of a wife in the State of Louisiana. This legal mortgage was held on all property that was either brought into the marriage by either the husband or the wife, or purchased after marriage. Whenever a husband sold or mortgaged property or a plantation, the presiding notary public or attorney of the act of conveyance was required by the laws of the State of Louisiana to take the wife out of the physical and hearing presence of her husband, and inform and explain to the wife the details of the act of conveyance or the act of mortgage. This law was contained in the second section of this act of the Louisiana legislature. When the wife was informed of the act of sale, or the act of mortgage, she could make either a renunciation of her dotal, matrimonial, and paraphernal rights, interests, and privileges, or she could not release her rights, interests, and privileges. This part of an act of conveyance or an act of mortgage was known as renunciation. Today, in the State of Louisiana, this law is no longer required in an act of conveyance or mortgage.

Now on December 12, 1856, Francois Optat Darby went to the Fusilier plantation, which was situated on the Old Spanish Trail, in the present day town of Baldwin, Louisiana, to make a conveyance of the Fusilier sugar plantation, which was owned by Mrs. Louisa Dumartrait Fusilier. The money that Francois Optat Darby had to purchase the Fusilier plantation was the money he had received from his one half share of the sale of the Darby Brothers sugar plantation to Joseph Devezin Olivier. Francois Optat Darby was called Optat by his family, friends, and associates.

When Francois Optat Darby arrived at the Fusilier plantation, he was met by Alexander Fusilier, and Henry de la Houssaye, besides Mrs. Louisa Dumartrait Fusilier. George B. Shepherd, a notary public, was to preside at the execution of the act of conveyance. Francois Optat Darby was very well acquainted with the Fusilier plantation, because he and his wife, Julie, and the family had numerous visits with Louisa Dumartrait Fusilier.

Louisa Dumartrait Fusilier agreed to the sale and mortgage of her plantation for the purchase price of $30,552.00 to Francois Optat Darby.

The plantation home, where this act of conveyance was executed on the date of December 12, 1856 is still in existence today. This plantation house is located at 606 Main Street in the present day town of Baldwin, Louisiana, and it houses the Baldwin branch of the St. Mary Bank and Trust Company. After Francois Optat Darby and his family moved from their plantation home, which was situated on the Darby Brothers sugar plantation, the home is now known as Deepwood, they moved to this plantation home, now located in the present day town of Baldwin, Louisiana.

Also, on the date of December 12, 1856, with money from the sale of the Darby Brothers sugar plantation to Joseph Devezin Olivier, Francois Optat Darby purchased two tracts of land, which contained one thousand, and three hundred and eight six arpents of land and an undivided half of three hundred acres of swamp land. This swamp land was situated in the back of the Fusilier plantation, and this swamp land adjoined the Fusselier plantation. Also, sold to Francois Optat Darby was all the seed cane on the Fusilier plantation, besides the corn, hay, fodder, and sweet potatoes. Now, Francois Optat Darby, who was called Optat, was now once again a sugar planter.

Purchasing the Fusilier plantation was not enough, Francois Optat Darby also purchased from Henry Penn a tract of land, which measured one hundred and six acres of land and seventy six hundredths of an acre of land. This land was of timber, which was actually cypress swampland. This act of conveyance was also executed on the date of December 12, 1856, with the land selling at the purchase price of $25.00 an acre of land. The total sum of this sale and mortgage was $7,500.00. Besides being a sugar cane planter, Francois Optat Darby was now interested into going into the lumber business. The only reason that Francois Optat Darby considered purchasing the property from Henry Penn was for the cypress wood which was situated on the property. Henry Penn, Junior and his wife, Eugenia Deyris, (now Derise), were present at the execution of the act of conveyance of the cypress swampland, and the execution of the act of conveyance was held in St. Mary Parish, Louisiana.

Between the time span of January of the year of 1855 and the month of August of the year of 1857, Joseph Devezin Olivier, his wife, Celeste Bellechase Olivier, and their only daughter, Camille, returned to Louisiana to make their home. They moved into the plantation home, now known as Deepwood, of their recently purchased sugar plantation in St. Mary Parish.

On January 15, 1860, Doctor Leonard Smith, the brother-in-law of Francois Optat Darby and Louis St. Marc Darby, stated before

William Robertson, a notary public in the Parish of St. Martin, that Joseph Devezin Olivier had paid him (Doctor Leonard Smith) the full amount of the mortgage payment plus all the interest. The total amount of this mortgage in favor of Doctor Leonard Smith was the amount of $8,003.10, which was conditioned to bear interest at 8% from January 30, 1855 until paid in full. This mortgage resulted from the sale of the Darby Brothers sugar plantation to Joseph Devezin Olivier on the date of January 27, 1855. Joseph Devezin Olivier got the money to pay Doctor Leonard Smith from the proceeds of the cane crop of his newly purchased sugar plantation, and the sale of the sugar estate or plantation, Mimosa, which was situated on the island of Cuba. Therefore, the mortgage was fully cancelled in favor of Doctor Leonard Smith. Also, present at this acknowledgement were Messes Ashbusll Henshaw and Robert Marsh.

In the month of February of the year of 1868, Joseph Devezin Olivier made the first two sales of tracts of land that were situated on the original Gravier sugar plantation, which he had purchased from Louis St. Marc Darby and Francois Optat Darby. The Gravier sugar planation had retained its original acreage or arpentage through the ownership of Francois St. Marc Darby, and the ownership of his two sons, Louis St. Marc Darby and Francois Optat Darby. The probable reason that Joseph Devezin Olivier sought to sell these two tracts of land was due to the aftermath of the American Civil War. The Civil War had many effects that were even felt here in Louisiana by all the planters. The main hardship was that planters now had to pay the former slaves wages for the work on the plantations, and this resulted in monies being spent from the total proceeds of the year's harvest. Also, the Confederate money was no longer as valuable as it had been before the Civil War. Those planters that had some of their investments in banks in Confederate monies, probably suffered a financial hardship due to the devaluation of the confederate money against the American money. Another hardship suffered by the planters and landowners in the State of Louisiana was the land had been measured in arpents, and after the Civil War the measurement of land became more common in acreage. The planters and the landowners suffered the lost of their land not containing as many acres as arpents. This was because an arpent is 0.85 of an acre in measurement. Also, the arpents is the French term of measuring land, and the acre is the American term for land measurements.

On February 28, 1868, Joseph Devezin Olivier sold to Celeste Olivier Pellerin, the wife of Louis Pellerin, and a half sister of Joseph Devezin Olivier, a tract of land situated on the edge of his sugar plantation. This tract of land was on both the east and the west sides of Bayou Teche. This tract of land measured one arpent fronting Bayou Teche on both the east and the west sides of Bayou Teche, with a depth of forty arpents on both sides. Today, this tract of land comprises the land on the west side of Bayou Teche owned by Jessie Edgar Fleming, and Harry Rich, Senior, and on the east side of Bayou Teche the tract of land owned by Doctor and Mrs.

Thomas Borland, and the estate of J. Paulin Duhe. At the execution of this act of conveyance were Joseph Alcide Darby and Francois Dumartrait, who were serving as witnesses to the execution of this act of conveyance, and Barthlemy de la Houssaye, a notary public, who presided at the execution of the act of conveyance.

On March 2, 1868, Joseph Devezin Olivier sold to Joseph Alcide Darby, a first cousin of Joseph Olivier's, a tract of land on both the east and the west sides of Bayou Teche. At the execution of the act of conveyance were present Joseph Devezin Olivier, the vendor, Celeste Bellechase Olivier, the legal and lawful wife of Joseph Devezin Olivier, Joseph Alcide Darby, and Alexander Pellerin, and Francois M. Dumartrait, who were to serve as the subscribing witnesses to the execution of this act of conveyance. The notary public was Barthlemy de la Houssaye. Joseph Alcide Darby was a resident of St. Mary Parish, Louisiana. Also, Joseph Devezin Olivier was known as, "Devezin," by his family, friends, and associates.

This tract of land, that was sold to Joseph Alcide Darby, consisted of five hundred arpents of land on both the east and the west sides of Bayou Teche. It was bounded on the upperside or above by the property of Joseph Devezin Olivier, and Celeste Bellechase Olivier, and on the east side by the land belonging to the succession of Charles Olivier, and on the west side by the land of Edouard Simon and Zeon Olivier, both free men of color. Also, conveyed or sold to Joseph Alcide Darby were two pairs of beef cows and three milk cows.

Celeste Bellechase Olivier, the legal and lawful wife of Joseph Devezin Olivier, renounce her dotal, matrimonial, and paraphernal rights, privileges, and titles to these two tracts of land. Celeste Bellechase Olivier stated that she was fully aware and took cognizance of the sale of both the these tracts of land. The reason that Mrs. Olivier had to do this reununication was required by the law of the State of Louisiana at this time. The execution of these two conveyances was held at the residence home of Mr. and Mrs. Joseph Devezin Olivier, which home is now known as Deepwood.

On January 7, 1869, Joseph Devezin Olivier mortgaged his sugar plantation. Joseph Devezin Olivier mortgaged his sugar plantation to Mistress Marion Courtois for the sum of $4,000.00. Mistress Marion Courtois loaned Joseph Devezin Olivier $4,000.00 in United States money at 8% interest per year from January 7, 1869. Joseph Devezin Olivier executed his promissory notes in favor of Mistress Marion Courtois, and these notes were payable four years after January 7, 1869. The 8% interest was due and payable annually at or before the end of the year of 1869, 1870, 1871, and 1872.

The sugar plantation that was mortgaged by Joseph Devezin Olivier was situated on both the east and the west sides of Bayou Teche, and had an area of two thousand, three hundred and thirty nine arpents of land. This sugar plantation was bounded above or on the upperside by land presently belonging to Joseph Alcide Darby, and below or on the lowerside by land belonging to Thomas H. Thompson. Joseph Devezin Olivier resided on the sugar plantation in the plantation home, and the plantation home of the Olivier plantation is now known as Deepwood. Also, on the sugar plantation was a sugar house, cabins, and barns. All the buildings and structures that were situated on the planation at this time were embraced in this mortgage in favor of Mistress Marion Courtois.

Celeste Bellechase Olivier, the lawful wife of Joseph Devezin Olivier, was taken out of the hearing and physical presence of her husband, Joseph Devezin Olivier, by the notary public, Patrick Burke. The reason for this was for Patrick Burke to read and carefully explain to Celeste Bellechase Olivier the terms and conditions of the mortgage made by her husband, Joseph Devezin Olivier, in favor of Mistress Marion Courtois. After a due reading and explanation, Celeste Bellechase Olivier approved and ratified the foregoing act of mortgage. In the signing of signatures on the act of mortgage, Mrs. Marion Courtois made her mark. Mrs. Marion Courtois did not know how to write her name in longhand.

On September 30, 1869, Joseph Devezin Olivier acknowledged by a statement of amounts due by Joseph Devezin Olivier to his legal and lawful wife, Celeste Bellechase Olivier. This was a conventional mortgage on the immovable property of Joseph Devezin Olivier. In the year of 1833, Joseph Devezin Olivier had received $50,000.00 from Celeste Bellechase Olivier for his half interest in the sugar estate, Mimosa, which was owned by Celeste Bellechase Olivier on the island of Cuba. This $50,000.00 was appropriated to the own use of Joseph Devezin Olivier. In the year of 1837, Joseph Devezin Olivier received $7,000.00 from his wife, Celeste Bellechase Olivier. This seven thousand dollars was derived from the sale of the sugar estate, Monticello, which was situated on the island of Cuba, and was the separate property of Celeste Bellechase Olivier. The seven thousand dollars that was derived from the sale of this sugar estate was appropriated to the own use of Joseph Devezin Olivier by his wife, Celeste Bellechase Olivier. Then in 1851, Celeste Bellechase Olivier sold the sugar estate, Rennion Desire, which was her separate property for the amount of $25,000.00. This twenty-five thousand dollars derived from the sale of the sugar estate, Rennion Desire, was appropriated to the own use of Joseph Devezin Olivier. The total amount of monies owed by Joseph Devezin Olivier to his wife, Celeste Bellechase Olivier, came to the total amount of $82,000.00.

On October 4, 1869, in the presence of William Robertson, a notary public, Joseph Devezin Olivier gave a true and correct statement of the total amounts of monies owed to his legal and lawful wife, Celeste Bellechase Olivier, who was originally from the island of Cuba. Celeste Bellechase Olivier had been reared on the island of Cuba. Also, Celeste Bellechase Olivier was able to converse in both the French language and the Spanish language.

On April 20, 1870, by act of Dation en paienment, Joseph Devezin Olivier declared that he was indebted to his wife, Celeste Bellechase Olivier, in the full sum of $82,000.00 for monies loaned to him by her since their marriage. This $82,000.00 had been appropriated to Joseph Devezin Olivier for his own use. To satisfy the claims of Celeste Bellechase Olivier against Joseph Devezin Olivier, the sugar plantation was conveyed to Celeste Bellechase Olivier.

This sugar plantation was situated in the newly created Parish of Iberia, or the former Parish of St. Mary. Also, this sugar plantation was known as the residence of Mr. and Mrs. Joseph Devezin Olivier. The sugar plantation had a frontage along Bayou Teche of thirty arpents on the west side of Bayou Teche, and a frontage of twenty four and eighty eight hundredths of an arpent on the east side of Bayou Teche. It contained a total of two thousand, nine hundred and nineteen arpents and eighty eight hundredths of an arpent of land. This was the total arpentage after the sale of the two tracts of land, one tract was sold to Joseph Alcide Darby, and the second tract was sold to Celeste Olivier Pellerin.

From the amount of land that is reserved and deducted out of the sale of the Dation en paienment is the land sold to Celeste Olivier Pellerin. The lower and upper boundaries of the Joseph Alcide Darby tract of land ran back on both the east and the west sides of Bayou Teche in parallel lines to the full depth of the Olivier sugar plantation. This tract of land consisted of five hundred arpents of land. Therefore, five hundred and eight arpents of land were deducted from the Dation en paienment, which gave the Olivier sugar plantation a total arpentage of two thousand, three hundred and thirty nine arpents and eighty eight hundredths of an arpent of land to be sold to Celeste Bellechase Olivier.

Also, included with the two thousand, three hundred thirty nine, and eight hundredths of an arpent of land, were four mules, four horses, a herd of twenty head of gentle creole cattle, the farm utensils of all kinds, which were used in the cultivation of the sugar cane crop, and other crops grown on the plantation. Also, one third of the growing crop of sugar cane that was on the plantation was also sold.

Also, included in this Dation en paienment were all the hereditary rights of Joseph Devezin Olivier in and to the succession of his late Father, Colonel Charles Napoleon Olivier and Celeste DeBlanc Olivier.

This Dation en paienment sold all the real estate, movable property, and improvements to Celeste Bellechase Olivier with full warranty of title and subrogation of all the rights and actions in warranty, which Joseph Devezin Olivier had against all previous owners of the land.

This Dation en paienment was made in consideration of the price of $47,000.00. This price of $47,000.00 was the amount at which all the property had been estimated and valued. For the consideration of the Dation en paienment, Celeste Bellechase Olivier released her husband, Joseph Devezin Olivier, for a portion of her claim equal to the $47,000.00 or the purchase price of the sugar plantation. Now, the balance due to Celeste Bellechase Olivier was $35,000.00 in principal for which she reserved her mortgage against her husband, Joseph Devezin Olivier. Celeste Bellechase Olivier accepts the present sale of her husband's sugar plantation to her, and acknowledges that the property has been delivered to her on the 20th day of April of the year of 1870.

Also, Celeste Bellechase Olivier did dispense with the introduction of the certificate required by Article 3328 of the Louisiana Civil Code. This Article 3328 of the Louisiana Civil Code releases the notary public or the recorder of an act of conveyance from all responsibility in the premises. Also, the United States Internal Revenue stamps in the amount of $47.00 were affixed and cancelled.

On May 13, 1870, Celeste Bellechase Olivier acted under the authority granted to her by Judge Emmanuel J. Etie, who was serving as Judge of the Parish of Iberia, to mortgage her sugar plantation. Celeste Bellechase Olivier was indebted to Daniel B. O'Connell, of the City of New Orleans, Louisiana, for the sum of $2,500.00. Celeste Bellechase Olivier subscribed four promissory notes to her own order and endorsed by her. One promissory note was for the amount of $1,000.00, and it was due on May 4, 1871. The other three promissory notes were in the amount of $500.00 each, and these three promissory notes were due on December 1, 1870. All these promissory notes were payable at the office of Daniel B. O'Connell in the City of New Orleans, Louisiana. Daniel B. O'Connell was the sugar factor or the man who sold the sugar and sugar products that was harvested and made on the Olivier sugar plantation.

Celeste Bellechase Olivier applied to Judge Emmanuel J. Etie, of Iberia Parish, for permission to mortgage her sugar plantation, which was situated in the newly created parish of Iberia, formerly a part of St. Mary Parish. The reason that Celeste Bellechase Olivier sought this mortgage was to enhance the value of her property. Celeste Olivier was mortgaging the sugar plantation with all the machinery in the sugar house, the engine in the sugar house, the sugar house or the sugar mill, the agricultural implements, and all the livestock and cattle on the plantation. The reason that Celeste Olivier had to apply to Judge Emmanuel J. Etie was because of the absence of her husband, Joseph Devezin Olivier, who was attending to business on the Island of Cuba. She was authorized by Joseph Devezin Olivier to make a mortgage in his absence. Judge Emmanuel J. Etie granted Celeste Bellechase Olivier permission after having found that the mortgage was to accrue to the benefit of Celeste Bellechase Olivier.

Celeste Bellechase Olivier bounded herself to consign her farm to Daniel B. O'Connell, of the City of New Orleans, Louisiana. All of Celeste Bellechase Olivier's portion of the sugar cane crop proceeds were to be applied to the extinguishment of the mortgage. Daniel B. O'Connell is entitled to his usual commission of 2 1/2% of the proceeds of all the sugar cane crop made on the Olivier plantation and consigned to Daniel B. O'Connell.

It was well understood between Daniel B. O'Connell and Celeste Bellechase Olivier that the present mortgage granted is the second mortgage bearing on the Olivier plantation. The first mortgage is in favor of Mistress Marion Courtois. The second mortgage will not impair on the first mortgage.

On February 19, 1872, Celeste Bellechase Olivier made a donation of land to her nieces and nephews, who were the children of her husband's half brother, Alexander Olivier and his wife, Amanda Lastrapes Olivier. This donation was made as a wish and intention of Celeste Bellechase Olivier, because of her love and affection for her nieces and her nephews. The nieces of Celeste Bellechase Olivier were Alice, Annette, Aimee, and Celeste, and the nephews were Albert, Agricole, George, and Jules.

This was the piece of property that Celeste Olivier had donated to these minor children of Alexander and Amanda Olivier, a certain tract of land, which was situated on the east side of Bayou Teche, in the Parish of Iberia. This tract of land measured one and one half arpents fronting the east side of Bayou Teche with a depth of forty arpents of land. This tract of land was bounded on the north side by the lands of Celeste Bellechase Olivier, on the south and the west side by Bayou Teche, and on the east side by the lands belonging to the estate of Thomas H. Thompson, or the Rosedale Plantation.

The property that was donated was subject to the usufruct by Amanda Lastrapes Olivier, during her lifetime. This was in view of the great love and affection which Celeste Bellechase Olivier has for and toward Mistress Amanda Lastrapes Olivier,who was the legal wife of Alexander Olivier. Also, such being the intention of Celeste Bellechase Olivier, that she does reserve the usufruct of the property during the life of Amanda Lastrapes Olivier. This usufruct is based on account of the consideration of the love and affection, which Celeste Bellechase Olivier has for Amanda Lastrapes Olivier.

The value of this tract of land was the price of $500.00. Also, this act of donation was executed at the home of Celeste Bellechase Olivier, which is now known as Deepwood. The witnesses to the execution of this act of donation were Alexander Pellerin, and Alfred de la Houssaye. P. E. Burke was the recorder of this act of donation, whereas Leonce de la Houssaye served in his capacity as the presiding notary public.

On February 3, 1875, Celeste Bellechase Olivier, leased her plantation to Edmund Lapeyrouse of St. Martin Parish, Louisiana. The terms of the lease were for the period of five years, beginning on January 1, 1875, and ending on January 1, 1880.

The plantation of Celeste Bellechase Olivier contained sixteen hundred arpents of land, which were situated on both the east and the west sides of Bayou Teche. The Olivier plantation was bounded in the year of 1875, on the north side by lands of Edward Kissack, which lands was known as the Olive Branch Plantation, on the south and the east sides by lands formerly belonging to the succession of Thomas Houssan Thompson, or the Rosedale Plantation, on the west side by lands belonging to Louisa Dumartrait. Also, leased to Edmund Lapeyrouse was all the agricultural implements on the plantation.

Celeste Olivier also leased to Edmund Lapeyrouse the steam cotton gin, one hundred and seventeen sacks, in which to carry the cotton in, one pair of bellows, which is an instrument used to stimulate a strong a blast of air, such as to stimulate a fire, two awits, a lot of blacksmith tools, six mules, two horses, five work oxen, four shovels, three axes, five hoes, three spades, three cross cut saws, three saws, six ploughs or plows with harnesses for four, three harrows, six carts, one wagon, two flooks, two yokes, one corn mill, one corn sheller, and one ferry boat or flat boat with chain attached. The ferry boat or the flat boat was used to go from the east side of Bayou Teche to the west side of Bayou Teche. Also, Edmund Lapeyrouse had seed cane or plant cane to plant about four or five hundred arpents of land in sugar cane on the Olivier plantation. Later on, it was erased the agreement to release to Edmund Lapeyrouse seed cane to plant four or five hundred arpents in sugar cane. A bellow was an instrument that was used to stimulate a strong blast of air, such as to stimulate a fire.

Edmund Lapeyrouse had the right to take out of the woods on the Olivier plantation, chopped wood for his own use, such as heating in the wintertime, and for cooking and heating water. Also, Edmund Lapeyrouse had the use of the plantation woods to obtain chopped wood that was necessary for the use in the plantation sugar house. The chopped wood that was used in the plantation sugar house was used to fuel the fire that was boiling the sugar cane juice in the making of brown sugar and molasses.

Celeste Bellechase Olivier and her husband, Joseph Devezin Olivier, reserved from the lease agreement with Edmund Lapeyrouse, the use of the dwelling house, now known as Deepwood, and six buildings, which buildings were situated in the yard of the dwelling house. Also, there was a fence enclosing the dwelling house from the areas of land of the plantation.

Edmund Lapeyrouse had the use of all the other buildings, which included a house that he was permitted to use as his dwelling house. Also, there were the slave cabins, which Edmund Lapeyrouse could utilize for the use as the dwelling house for his laborers. According to the lease agreement, Edmund Lapeyrouse had the use of the sugar house, the cotton gin, and the docks for the shipping of his sugar that was processed on the Olivier plantation.

Also, Edmund Lapeyrouse had to put in good repair the sugar house, the engine in the sugar house, and the sugar kettles, the cotton gin, and all the other plantation buildings that may be in need of repairs. Edmund Lapeyrouse had a claim for the amount the repairs, which was stated in the price of the lease agreement. This claim was to facilitate Edmund Lapeyrouse in making those needed repairs.

In the lease agreement, Celeste Bellechase Olivier authorizes Edmund Lapeyrouse that this lease was made and accepted on these following terms and conditions. Edmund Lapeyrouse had to pay Celeste Bellechase Olivier and her husband, Joseph Devezin Olivier, at the end of each year of the lease agreement the following amounts, at the end of the year of 1875, a total amount of $400.00, at the end of 1876, the total amount of $600.00, at the end of 1877, the total amount of $800.00, at the end of 1878, the total amount of $1,000.00, and at the end of the year of 1879, the total amount of $1,000.00. It was also agreed that Edmund Lapeyrouse had to deliver to Celeste Bellechase Olivier and her husband, Joseph Devezin Olivier, fifty barrels of clean corn, twenty five cords of wood, fodder for two horses, and ten barrels of potatoes.

It was agreed by the terms of the lease between Celeste Bellechase Olivier, and her husband, Joseph Devezin Olivier, the grantors, and Edmund Lapeyrouse, the grantee, that if during the time of the lease agreement, the net proceeds of the crop should be over and above the annual payment of the lease, and the expenses necessary to carry on the planting of the sugar cane crop, and the cotton crop, then Edmund Lapeyrouse will apply the surplus of the net proceeds to the payment of the debt of the plantation, which is a mortgage. Also, Celeste Bellechase Olivier, and her husband, Joseph Devezin Olivier, are to have control of the plantation, and they are responsible for all the costs and expenses for running the plantation.

Edmund Lapeyrouse was aware that there is a mortgage bearing on the Olivier plantation. This mortgage was in favor of Mistress Marion Courtois of the town of New Iberia, Louisiana. Edmund Lapeyrouse accepted the lease agreement of the Olivier plantation, knowing that there was a mortgage bearing on the plantation.

In the lease agreement, Edmund Lapeyrouse bounds and makes himself responsible for the administration of the Olivier plantation, and to protect the Olivier plantation's buildings, tools, and crops. Also, Edmund Lapeyrouse is to manage the Olivier plantation, and to return the Olivier plantation at the end of the lease agreement in good condition.

The lease agreement between Mrs. Celeste Bellechase Olivier and her husband, Joseph Devezin Olivier, with Edmund Lapeyrouse was executed at the Olivier plantation home on the Olivier plantation, now known as Deepwood. The lease agreement was executed by Leonce de la Houssaye, who was a notary public in the town of New Iberia, Louisiana. Also, Leonce de la Houssaye went to the Olivier plantation home on the Olivier plantation to execute this legal act of a lease agreement. The two witnesses to the execution of this lease agreement were Alexander Pellerin and Alexander Olivier.

On April 11, 1875, Celeste Bellechase Olivier, who was the legal wife of Joseph Devezin Olivier, made a conveyance to Alexander Olivier. Alexander Olivier was the half brother of the husband of Celeste Bellechase Olivier. Leonce de la Houssaye served as the notary public at the execution of this act of conveyance.

The tract of land that was conveyed measured one half of an arpent fronting Bayou Teche, with a forty arpent depth. Also, this tract of land was situated on the east bank of Bayou Teche, in the Parish of Iberia. The boundaries of this tract of land were on the north side by the lands of Octave Reggio, on the south side by lands of the children of Alexander and Amanda Olivier, on the east side by the lands of Celeste Bellechase Olivier, and on the west side by Bayou Teche. The purchase price of this tract of land was $300.00, which was paid in cash.

On the date of April 11, 1875, Celeste Bellechase Olivier conveyed at a cash sale, a tract of land to Octave Reggio. This tract of land that was sold measured three arpents fronting Bayou Teche, by a depth of forty arpents or thirty-eight American acres of land. This tract of land was bounded on the north side by the lands this day sold to N. Reggio, on the south side by the lands this day sold to Amanda Lastrapes, the legal and lawful wife of Alexander Olivier, on the west side by Bayou Teche, and on the east side by the lands of Celeste Bellechase Olivier. The purchase price of this tract of land was $1,200.00, which amount was paid in cash.

This act of conveyance was executed by Leonce de la Houssaye, a notary public in the town of New Iberia, Louisiana. Also, the two subscribing witnesses to the execution of this act of conveyance were Gabriel Olivier and Alexander Olivier. Since the year of 1870, Celeste Bellechase Olivier had been separated from her husband, Joseph Devezin Olivier, in property, and she was able to sell various tracts of land from her plantation.

On April 11, 1875, Celeste Bellechase Olivier conveyed a tract of land to her sister-in-law, Amanda Lastrapes, who was the wife of Alexander Olivier. Both, Celeste Bellechase Oliver and Amanda Lastrapes Olivier were separated from their respective husbands in property. A wife and a husband could be separated in property by an act of Dation en paienment, in which one spouse could convey all the immovable and movable property to the other spouse.

The tract of land that was sold to Amanda Lastrapes Olivier measured one half an arpent of land fronting Bayou Teche, with a depth of fifty arpents or forty two American acres and five hundredths of an American acre of land. Also, this tract of land was bounded on the north side by the land of Octave Reggio, on the south side by the land of the children of Alexander and Amanda Oliver, on the west side by Bayou Teche, and on the east side by the lands of Celeste Bellechase Olivier.

At the execution of the act of conveyance, Amanda Lastrapes Olivier paid Celeste Bellechase Olivier the purchase price of $300.00 for the tract of land. This act of conveyance was executed by Leonce de la Houssaye, with L. Alexander Pellerin serving in his capacity as a witness to the execution of this act of conveyance.

Also, on the date of April 11, 1875, Celeste Bellechase Olivier conveyed a tract of land to Armand N. Reggio, of the Parish of Lafourche. At the execution of this act of conveyance, Armand N. Reggio was represented by Octave Reggio, who acted as the attorney in fact. The reason that Octave Reggio was able to represent Armand N. Reggio was because Octave Reggio had the power of attorney vested in him, by a legal act of power of attorney, which act was executed on the date of March 25, 1875.

The tract of land that was sold to Armand N. Reggio measured three arpents fronting Bayou Teche, with a depth of forty arpents or thirty eight American acres of land. This tract of land was bounded on the north side by the lands of Celeste Bellechase Olivier, on the south side by the lands this day sold to Octave Reggio, on the west side by the Bayou Teche, and on the east side by the lands of Celeste Bellechase Olivier. This tract of land was sold for the purchase price of $800.00, which was paid in cash money at the execution of this act of conveyance.

On the date of April 16, 1876, Celeste Bellechase Olivier sold to Octave Reggio another tract of land from her plantation. This second tract of land measured three arpents fronting Bayou Teche, with a depth of forty arpents. Also, this tract of land was bounded on the north side by the lands of Celeste Bellechase Olivier, on the south side by the land sold to Octave Reggio by Celeste Bellechase Olivier, on the west side by Bayou Teche, and on the east side by the lands sold by Celeste Bellechase Olivier to Octave Reggio.

Celeste Bellechase Olivier had sold a tract of land, which measured three arpents fronting Bayou Teche, with a depth of forty arpents to her sister-in-law, Amanda Lastrapes Olivier, who was the wife of Alexander Olivier. Alexander Olivier was the half brother of Joseph Devezin Olivier, who was the husband of Celeste Bellechase Olivier. Also, Alexander Olivier was the son of Charles Napoleon Olivier by his second wife, Julie Felicite Chretin, who was from St. Landry Parish, whereas, Joseph Devezin Olivier was the son of Charles Olivier by his first wife, Celeste Mathilda DeBlanc, who was from St. Martin Parish. This tract of land was bounded on the north side by the lands sold on the date of May 26, 1876 to Amanda Lastrapes Olivier, the wife of Alexander Olivier, and west by Bayou Teche, and east by the lands of Celeste Bellechase Olivier. This tract of land was originally sold to Octave Reggio on the date of April 11, 1875.

On December 27, 1877, Celeste Bellechase Olivier was assisted and authorized by her husband, and also by a certificate of the Judge of the 3rd Judicial District Court to contract a mortgage with Mrs. Felicie Indest. Celeste Bellechase Olivier was examined by the Judge out of the presence of her husband, Joseph Devezin Olivier, regarding the object and purpose of this mortgage, and also the manner and the reason of contracting this debt. Celeste Bellechase Olivier was indebted to Felicite Indest for the sum of $1,000.00. For the reimbursement of this debt, Celeste Bellechase Olivier had subscribed her certain promissory note, and this note is dated December 27, 1877. Also, this promissory note is made to the order of Felicite Indest. This promissory note is stipulated to bear interest at the rate of 8% from the date of July 2, 1877. Also, this promissory note is payable on the date of January 2, 1878. This promissory note was delivered to Felicite Indest through her attorney, P. Laurent Renoudet.

For the reimbursement of this debt, Celeste Bellechase Olivier specially mortgaged and hypothecated unto Felecite Indest, the Olivier sugar plantation, which was situated on the Old Spanish Trail, in the newly created Parish of Iberia. This plantation was also known as the residence of Celeste Olivier. The Olivier sugar plantation had a frontage of thirty arpents on the west side of Bayou Teche, with a depth of forty arpents, and on the east side of Bayou Teche the frontage was twenty four arpents and eighty eight hundredths of an arpent of land. The total area contained in the Olivier sugar plantation was two thousand, nine hundred and nineteen arpents and eighty eight hundredths of an arpent of land. This plantation was bounded on the upperside by the northwest line on both sides of Bayou Teche and by land on the west side of Bayou Teche formerly belonging to Joseph Devezin Oliver, but now belonging to Henry Train, and by land on the east side of Bayou Teche formerly belonging to Samuel Andrus, and later belonging to the estate of Charles Napoleon Olivier, in the rear on the east side by land formerly belonging to the estate of Samuel Andrus, and on the lower side or the southeast side on both sides of Bayou Teche by land formerly belonging to the widow and heirs of Thomas Houssan Thompson, or the Rosedale Plantation, and in the rear on the western side of Bayou Teche by lands formerly belonging to the estate of Edward or Edouard Simon, and now of Zenon Decuir. The dividing line bounding the Olivier sugar plantation has the directions and relative positions as shown by a sketch of survey made by Boyd Warner, as copied by Alexander L. Fields, on the date of September 21, 1854.

The money that Celeste Bellechase Olivier borrowed from Felcite Indest was solely for the benefit of Celeste Bellechase Olivier, and for the advantages of her separate paraphernal property. This money was not to be used for the separate use and benefit of Joseph Devezin Olivier, the husband of Celeste Bellechase Olivier, nor for the benefit of the separate estate of Joseph Devezin Olivier. Also, this money was not to be used for the community property of acquets and gains of Joseph Devezin Olivier and Celeste Bellechase Olivier. The Judge found that all the requirements of Article 128 of the Civil Code of the State of Louisiana existed in this case. Therefore, Celeste Bellechase Olivier was granted the right to contract a mortgage with Felcite Indest for the separate property of Celeste Bellechase Olivier.

On April 21, 1884, Joseph Devezin Olivier granted a right of way to Morgan's Louisiana & Texas Railroad and Steamship Company to put a railroad track through his sugar plantation, which was situated on the west side of Bayou Teche in Iberia Parish. Now, the railroad track that once was the Morgan's Louisiana & Texas Railroad & Steamship Company is now the railroad track of the Southern Pacific Railroad Company.

By the year of 1888, Celeste Bellechase Olivier began to sell tracts of land of her sugar plantation to different purchasers. Now, the former Gravier and Darby sugar plantation was to be owned by separate owners of different tracts of land on both the west and the east sides of Bayou Teche. It became harder and harder to make ends meet after the Civil War, and Celeste Bellechase Olivier did not want to mortgage her sugar cane farm any more than she could pay off. So by the year of 1888, she began to seriously think about selling various tracts of land to several different persons who were interested in purchasing land in the country as an investment.

On February 17, 1888, Celeste Bellechase Olivier sold a tract of land, which was situated on the east side of Bayou Teche to William Bayne. This tract of land measured five arpents and sixty three and a half feet fronting Bayou Teche on the east side, with a forty arpent depth. The total arpentage contained in this tract of land was two hundred and thirteen arpents of land. This tract of land was bounded above or on the upperside by a road, which was forty feet wide, and below or on the lowerside by the lands of the widow of Augustus Reggio, in front by Bayou Teche, and in the rear by lands of Celeste Bellechase Olivier. Celeste B. Olivier also grants and sells a right of way over the forty foot wide road that bounded this tract on the upperside. Also, Celeste Bellechase Olivier grants and sells a right of way over another road, which is forty feet wide. This second road is situated across the Bayou Teche, and opposite the first road. This second road leads from Bayou Teche to the public road, now known as the Old Jeanerette Road, leading to New Iberia, Louisiana.

This conveyance to William Bayne by Celeste Bellechase Olivier was made in consideration of the amount of $1,704.00. This total amount of $1,704.00 was paid in cash at the execution of the act of conveyance by William Bayne to Celeste Bellechase Olivier. Also, Celeste Olivier acknowledges the receipt of the purchase money and granted full acquittance.

It was understood by William Bayne, the purchaser, and Celeste Bellechase Olivier, the seller, that the right of way over the two roads, that are sold with the land are established as a servitude in favor of William Bayne, or the owner of the land and benefits the tract of land. This servitude or right of way will remain forever for the benefit of the present purchaser, William Bayne.

Also, on February 17, 1888, at the residence on the Olivier plantation, now known as Deepwood, Celeste Bellechase Olivier conveyed another tract of land to Augustus Packard. This tract of land was also located on the east side of Bayou Teche. This tract of land measured five arpents and sixty three and a half arpents fronting Bayou Teche on the east side, with the depth of forty arpents. The total area of this tract of land measured two hundred and thirteen arpents. This tract of land was bounded above or on the upperside by the land of Edward Kissack or Olive Branch Plantation, below or on the lowerside by a road, which was forty feet wide, in front by Bayou Teche, in the rear by lands belonging to Celeste Bellechase Olivier. Also, Celeste Bellechase Olivier sells and grants a right of way along the road, and the right of way along the road bounds this tract on the lowerside.

Celeste Bellechase Olivier also sells and grants a right of way over another road, which is forty feet wide, and situated across Bayou Teche. This road is just opposite the first road, and the second road is now known as the Old Jeanerette Road.

This sale was made for the purchase price of $1,704.00. Augustus Packard paid the purchase price of $1,704.00 to Celeste Bellechase Olivier at the execution of this act of conveyance. Also, Celeste Bellechase Olivier granted full acquittal and acknowledged the receipt of having received the money due her.

Then on the date of February 26, 1888, Celeste Bellechase Olivier conveyed a tract of land to Frank Welch, an attorney, who resided in the town of Jeanerette, Louisiana. Frank Welch served the town of Jeanerette, Louisiana as its first mayor. The execution of this act of conveyance was held at the residence of Celeste Bellechase Olivier, on the Olivier plantation, now known as Deepwood.

The tract of land sold to Frank M. Welch was situated on the west side of Bayou Teche, in Iberia Parish. This tract of land measured nine arpents fronting Bayou Teche, with a forty arpent depth. This sale also included the railroad track of the Southern Pacific Railroad Company, which was formerly Morgan's Louisiana & Texas Railroad & Steamship Company. This tract of land was bounded above or on the upperside by the lands of Perry, below or on the lowerside by the lands of Celeste Bellechase Olivier, in front by Bayou Teche, and in the rear by the lands of Zenon Decuir. This tract of land now comprises the Cloverdale Plantation, which is owned by McDaniel Jeffery.

The railroad track is sold subject to the right of way granted to Augustus Packard and William Bayne, by act of conveyance of their tracts of land, which comprised formerly the Olivier sugar plantation. The right of way on the road, which is forty feet wide, and leads from Bayou Teche to the public road, leading to New Iberia, now known as the Old Jeanerette Road. The lower line of the railroad track begins on Bayou Teche at a point fifteen arpents distance from the upper line of Alfred Gonsoulin's land.

Also, Celeste B. Olivier sold to Frank M. Welch on this same day, another tract of land, which was situated on the eastern side of Bayou Teche. This tract of land contained a total of one thousand and twenty four arpents of land. This tract of land was bounded on the north and the eastern sides by land belonging to the estate of Charles Napoleon Olivier, and on the south side by the lands of Alfred Gonsoulin, and on the west side by the lands of Augustus Packard, William Bayne, the widow of Augustus Reggio, Octave Reggio, and the heirs of Alexander Olivier.

The purchase price of this tract of land was the sum of $3,500.00, which was paid in installments. Frank M. Welch paid in current money the sum of $3,000.00 to Celeste B. Olivier at the execution of this act of conveyance. Celeste B. Olivier acknowledged the receipt of the $3,000.00 in part payment, and granted acquittal. For the payment of the balance of the purchase price of $500.00, Frank M. Welch has furnished two promissory notes. These two promissory notes were endorsed in the blank, and made to the order of Frank M. Welch. The first promissory note is payable on the date of March 1, 1889, and the second promissory note is payable on the date of September 1, 1889. Each of the two promissory notes was for the amount of $250.00. These two promissory notes were conditioned to bear interest at 8% per year from maturity until paid in full.

Also, on the date of February 27, 1888, Celeste Bellechase Olivier conveyed a tract of land to Edwin O. Faulkner, who was a resident of Walsh County, Dakota. This tract of land conveyed to Edwin O. Faulkner was situated on the western side of Bayou Teche, in the Parish of Iberia. This tract measured four and one half arpents fronting Bayou Teche on the west side, with a depth of forty arpents. Also, included was a deduction of the Southern Pacific Railroad Company, formerly Morgan's Louisiana & Texas Railroad & Steamship Company, of one hundred and seventy eight arpents of land, which was granted as a right of way by Joseph Devezin Olivier. This tract of land was bounded above or on the upperside by the lands of Celeste Bellechase Olivier, below or on the lowerside by the lands of Eliza Brown, in front by Bayou Teche, and in the rear by the lands of Zenon Decuir. This tract of land now forms a part of Ravenswood Plantation, which is owned by Frank R. Courts.

This conveyance between Edwin O. Faulkner and Celeste Bellechase Olivier was made and accepted for the purchase price of $1,424.00. Edwin O. Faulkner paid in current money the sum of $600.00 to Celeste Bellechase Olivier. Then for the balance of $800.00 of the purchase price, Edwin O. Faulkner had made and subscribed a promissory note. This promissory note is to the order of Edwin O. Faulkner, endorsed in the blank, and dated the date of February 27, 1888. This promissory note is payable in three months from the date of February 27, 1888, or on the date of May 27, 1888. Also, this promissory note was conditioned to bear interest at the rate of 8%, which interest will accrue from maturity until paid in full.

Edwin O. Faulkner and Celeste Bellechase Olivier agree to dispense with the production of the certificate, which was required by Article 3364 of the Civil Code of the State of Louisiana. This certificate will exonerate the notary public from all responsibility and liability in these premises. Also, Celeste Bellechase Olivier declared that all the taxes have been paid on the tract of land conveyed to Edwin O. Faulkner.

Also, on the date of February 27, 1888, at the residence home of the Olivier's, which was located on the Olivier plantation, Celeste Bellechase Olivier met with Eliza Brown, who was the wife of George Brown, to make a conveyance of a tract of land. This land that was conveyed to Eliza Brown was situated on the western side of Bayou Teche, in the Parish of Iberia.

The tract of land conveyed to Eliza Brown by Celeste Bellechase Olivier measured four and a half arpents fronting Bayou Teche, on the west side, with a depth of forty arpents. Also, after the deduction of the railroad track of the Southern Pacific Railroad Company, which was conveyed in a right of way by Joseph Devezin Olivier, and this deduction amounted to one hundred and seventy-eight arpents of land. This tract of land was bounded above or on the upperside by the lands of Edwin O. Faulkner, below or on the lowerside in part by the lands of Alfred Gonsoulin, in the front by the public road, which leads from New Iberia to Jeanerette, and is the Old Spanish Trail, and in front by Bayou Teche, and in the rear by the lands of Zenon Decuir. Also, Celeste Bellechase Olivier declared that all the taxes assessed against the property have been paid. This tract of land is now presently a part of Ravenswood Plantation,which is owned by Frank R. Courts.

This sale was made for the purchase price of $1,424.00. Eliza Brown paid the sum of $824.00 in current money to Celeste Bellechase Olivier, at the execution of this act of conveyance. For the balance of the purchase price of $600.00, Eliza Brown has furnished her promissory note. This promissory note is endorsed in the blank, dated the date of February 27, 1888, and is for the sum of $600.00. Also, this promissory note bears interest at 8% per annum from the date of February 27, 1888 or until paid in full.

Both, Celeste Bellechase Olivier and Eliza Brown agreed to dispense with the production of the certificate required by Article 3364 of the Civil Code of the State of Louisiana. This certificate exonerates the notary public from all responsibility and liability in the premises conveyed to Eliza Brown.

On the date of December 6, 1889, Celeste Bellechase Olivier made a mortgage in favor of Pierre Le Bron. This mortgage was for the sum of $700.00. Celeste Bellechase Olivier gave her promissory note, which note was endorsed in the blank, and made to the order of Celeste Bellechase Olivier or her own order. This promissory note was due after one year or on the date of December 6, 1890.

Celeste Bellechase Olivier mortgaged her homeplace, now known as Deepwood, and this was all the property she now owned. The tract of land that was mortgaged was situated on the west side of Bayou Teche, and it was about six miles below the town of New Iberia, Louisiana. This tract of land had a frontage of six arpents along the west side of Bayou Teche, with a depth of forty arpents or thirty four American acres of land. This tract of land was bounded in the front by Bayou Teche, in the rear by the lands of Zenon Decuir, above by the lands of Alexander Stewart, or the Cloverdale Plantation, and below by the lands of Edwin O. Faulkner. The total area of this tract of land was two hundred and forty arpents of land or one hundred and seventy three arpents of land.

The tract of land conveyed to Eliza Brown by Celeste Bellechase Olivier measured four and a half arpents fronting Bayou Teche, on the west side, with a depth of forty arpents. Also, after the deduction of the railroad track of the Southern Pacific Railroad Company, which was conveyed in a right of way by Joseph Devezin Olivier, and this deduction amounted to one hundred and seventy-eight arpents of land. This tract of land was bounded above or on the upperside by the lands of Edwin O. Faulkner, below or on the lowerside in part by the lands of Alfred Gonsoulin, in the front by the public road, which leads from New Iberia to Jeanerette, and is the Old Spanish Trail, and in front by Bayou Teche, and in the rear by the lands of Zenon Decuir. Also, Celeste Bellechase Olivier declared that all the taxes assessed against the property have been paid. This tract of land is now presently a part of Ravenswood Plantation,which is owned by Frank R. Courts.

This sale was made for the purchase price of $1,424.00. Eliza Brown paid the sum of $824.00 in current money to Celeste Bellechase Olivier, at the execution of this act of conveyance. For the balance of the purchase price of $600.00, Eliza Brown has furnished her promissory note. This promissory note is endorsed in the blank, dated the date of February 27, 1888, and is for the sum of $600.00. Also, this promissory note bears interest at 8% per annum from the date of February 27, 1888 or until paid in full.

Both, Celeste Bellechase Olivier and Eliza Brown agreed to dispense with the production of the certificate required by Article 3364 of the Civil Code of the State of Louisiana. This certificate exonerates the notary public from all responsibility and liability in the premises conveyed to Eliza Brown.

On the date of December 6, 1889, Celeste Bellechase Olivier made a mortgage in favor of Pierre Le Bron. This mortgage was for the sum of $700.00. Celeste Bellechase Olivier gave her promissory note, which note was endorsed in the blank, and made to the order of Celeste Bellechase Olivier or her own order. This promissory note was due after one year or on the date of December 6, 1890.

Celeste Bellechase Olivier mortgaged her homeplace, now known as Deepwood, and this was all the property she now owned. The tract of land that was mortgaged was situated on the west side of Bayou Teche, and it was about six miles below the town of New Iberia, Louisiana. This tract of land had a frontage of six arpents along the west side of Bayou Teche, with a depth of forty arpents or thirty four American acres of land. This tract of land was bounded in the front by Bayou Teche, in the rear by the lands of Zenon Decuir, above by the lands of Alexander Stewart, or the Cloverdale Plantation, and below by the lands of Edwin O. Faulkner. The total area of this tract of land was two hundred and forty arpents of land or one hundred and seventy three arpents of land.

The act of mortgage was executed by Leonce de la Houssaye, a notary public in the town of New Iberia, Louisiana. The two witnesses to the execution of this act of mortgage were Armand Cousin, Junior and Charles Eugene Reggio, who was the son-in-law of Celeste Bellechase Olivier. Charles Eugene Reggio was married to Camille Olivier, who was the only daughter of Joseph Devezin Olivier and Celeste Bellechase Olivier.

On the date of April 1, 1890, Celeste Bellechase Olivier executed an act of mortgage, which mortgage was in favor of Mrs. William Dillion, of New Iberia, Louisiana. The sum of this mortgage was $1,300.00. Celeste Bellechase Olivier furnished her promissory notes, which notes were drawn and subscribed to her own order. Also, the promissory note was endorsed in the blank by Celeste Bellechase Olivier. These promissory notes were conditioned to bear 8% interest per annum from the date of maturity. Celeste Bellechase Olivier had mortgaged her homeplace, now known as Deepwood.

On the date of July 1, 1896, Celeste Bellechase Olivier, who was the widow of Joseph Devezin Olivier, was indebted to Mrs. Elisa Fraber, who was the widow of Aristede Renaud. The amount of this mortgage was $2,500.00. Celeste Bellechase Olivier had given her promissory note, which note was endorsed in the blank. Also, the promissory note was due on July 1, 1898. This promissory note was handed to Albert Decuir, who was the agent and representative of Elisa Fraber Renaud. This promissory note should be paid at maturity, but the yearly interest shall be paid in advance.

Celeste Bellechase Olivier mortgaged her homeplace, which is now known as Deepwood, and which is situated on the Old Spanish Trail. This tract of land was located on the west side of Bayou Teche, about six miles below the town of New Iberia, Louisiana. It measured six arpents fronting the west side of Bayou Teche by a depth of forty arpents or thirty four American acres of land. This tract of land was bounded in front by Bayou Teche, in the rear by the lands of Zenon Decuir, above by the lands of Alexander Stewart, or the Cloverdale Plantation, below or on the lowerside by the lands of Edwin O. Faulkner, and just beyond the Grand Marais or the Great Swamp. This tract of land comprised all the lands owned by Celeste Bellechase Olivier.

Also, Celeste Bellechase Olivier agreed to keep the land and the buildings insured for the sum of $1,000.000. This insurance policy had to be with a good and solvent insurance company, which was doing business in the State of Louisiana. Also, this insurance policy on the property and the buildings should be carried until the final payment of the mortgage.

THE WELCH AND THE STEWART FARM

On the date of May 7, 1888, Frank M. Welch conveyed or sold a part of the original tract of land, which he (Frank M. Welch) had purchased from Celeste Bellechase Olivier, who was the wife of Joseph Devezin Olivier, on the date of February 27, 1888, to Alexander Stewart. Frank M. Welch only kept ownership of the tract of land, which was situated on the western side of Bayou Teche, in the Parish of Iberia, formerly St. Mary Parish, for only two and one half months, before reselling this tract of land. Frank M. Welch had purchased this tract of land for investment reasons. Also, Frank M. Welch was an attorney, and a resident of the town of Jeanerette, Louisiana.

This tract of land that was conveyed on May 7, 1888 to Alexander Stewart was situated on the western side of Bayou Teche, and it contained two hundred and forty arpents of land in area. Also, this tract of land had a frontage of six arpents fronting Bayou Teche on the western bank of the bayou, and a depth of forty arpents. The boundaries of this tract of land were on the upperside or above were the lands of Mr. Perry, on the lowerside or below were the lands of Frank M. Welch, in the rear were the lands of Zenon Decuir, and the boundary in the front was Bayou Teche. All the buildings and improvements were sold with the tract of land, except one cabin, which was reserved by Frank M. Welch, the vendor or seller, in this act of conveyance.

The purchase price of this tract of land was $2,640.00, which was paid in cash money at the execution of the act of conveyance. Frank M. Welch acknowledged the receipt and granted Alexander Stewart acquittal for the indebtness of the purchase price of $2,640.00. Also, Frank M. Welch had paid all the taxes that were assessed against the property.

The act of conveyance was executed in the law office of P. Laurent Renoudet, a notary public and a lawyer in the town of New Iberia, Louisiana. Also, William F. Weeks and James P. Brown served in their capacity as competent witnesses to the execution of this act of conveyance between the parties of Frank M. Welch and Alexander Stewart. Also, Frank M. Welch and Alexander Stewart exonerated the notary public, P. Laurent Renoudet, from all liability in the premises of the land being conveyed.

On June 2, 1888, Frank M. Welch conveyed the remainder of the lands on the west side of Bayou Teche, which were originally purchased from Celeste Bellechase Olivier to Alexander Stewart. Alexander Stewart was planning on building a house to reside in, and also on going into the business of farming sugarcane.

This tract of land that was sold to Alexander Stewart was also situated on the west side of Bayou Teche, in the Parish of Iberia. This tract of land contained seventy eight arpents in area. Also, this tract of land was bounded above or on the upperside by the lands of Frank M. Welch, below or on the lowerside by the lands of Mrs. Joseph Devezin Olivier (Celeste Bellechase Olivier), in front by Bayou Teche, and in the rear by the lands of Zenon Decuir.

At the execution of the act of conveyance, Alexander Stewart paid the purchase price of $858.00 in cash money to Frank M. Welch. Frank M. Welch acknowledged the receipt of the payment of the purchase price, and granted Alexander Stewart acquittal for the indebtness of the purchase price of $858.00

Now, Alexander Stewart had purchased for himself a plantation or a farm, which he (Alexander Stewart) would began in the business of farming. Also, Alexander Stewart built the family home on the tract of land. Alexander Stewart named his newly purchased plantation, Cloverdale Plantation. The Cloverdale Plantation was once a part of the original Olivier, Darby, and Gravier sugar plantations.

Alexander McDonald Stewart was the owner of the Cloverdale Plantation, which he purchased from Celeste Bellechase Olivier. Also, Alexander Stewart was married to the former Martha Peoples Stewart. Alexander Stewart died at his residence on Cloverdale Plantation, on the date of May 6, 1937.

Alexander Stewart left a last will and testament, which was made in the nuncupative form by private act on the date of April 20, 1933. Alexander McDonald Stewart wrote his last will and testament at his plantation home on the Cloverdale Plantation.

The law firm that probated this last will and testament of Alexander McDonald Stewart was Weeks and Weeks, of the City of New Iberia, Louisiana. Alexander Stewart bequeathed by his wish to his daughter, Margaret Ann Stewart, the home place of the Cloverdale Plantation. Alexander Stewart resided at this home place on the Cloverdale Plantation. The home place of the Cloverdale Plantation contained one hundred and thirty seven acres of land. His son, Robert J. Stewart, was bequeathed one dollar, because he had already received his share of the estate of his father, Alexander McDonald Stewart.

Also, Isabella Stewart Summarll was bequeathed eleven shares of paid stock in the Calcasieu Building and Loan Association. She had received property belonging to the Cloverdale Plantation during the lifetime of Alexander McDonald Stewart. Harry Alexander Stewart received twenty shares of paid stock in the Iberia Building and Loan Association, which was situated in the City of New Iberia, Louisiana. Martha Stewart Jeffery received eight shares of paid stock in the Iberia Building and Loan Association, as well as four shares of paid stock in the Calcasieu Building and Loan Association of Lake Charles, Louisiana. Also,

Martha Stewart Jeffery received $500.00 in Liberty Bonds from her Father, Alexander McDonald Stewart.

Alexander Stewart left belonging to his estate one lot of household furniture, which consisted of the furniture in three bedrooms, one living room, the stove and cooking utensils in the kitchen, the household linen, the chinaware, and all the furniture and furnishing on the Cloverdale Plantation. The funeral services for Alexander McDonald Stewart were conducted by the Dauterive Funeral Home of New Iberia, Louisiana.

THE OLIVIER HOME, THE REGGIO HOME, THE BUSSY INVESTMENT, THE LEONARD HOME, THE PATOUT HOME, THE HEBERT INVESTMENT, THE PATOUT HOME, THE AMMERMAN HOME, AND THE LABICHE HOME

On August 3, 1893, Camille Olivier, the wife of Charles Eugene Reggio, inherited the home and property that belonged to her mother, Celeste Bellechase Olivier. Camille and her husband, Charles Reggio moved into and lived in the house, now known as Deepwood. Since their return from the island of Cuba, the home, now known as Deepwood, was the residence home of Celeste Bellechase Olivier and her husband, Joseph Devezin Olivier, who were the parents of Camille Olivier Reggio.

In either the month of February or March of the year of 1900, Camille Olivier Reggio sold to John C. Bussy, of the town of Jeanerette, Louisiana, the home, now known as Deepwood, and the property at a cash sale.

On January 14, 1905, John C. Bussy sold the home, Deepwood, and the property to J. A. and Henry Leonard at a sale and mortgage. The tract of land measured six arpents fronting Bayou Teche on the west side, with a depth between parallel lines of forty arpents. This property was bounded on the north side by the lands of Alexander McDonald Stewart, on the south side by the lands of John C. Bussy, on the east side by Bayou Teche, and on the western side by Decuir. Also, included in the sale were three hundred barrels of rice, twelve acres of seed sugar cane, a half stack of hay, six mules, one cart, four plows or ploughs, and other agricultural implements.

This sale was made and accepted for the consideration of the purchase price of $8,500.00. J. A. and Henry Leonard paid $500.00 in cash money at the execution of the act of conveyance. The balance of the purchase price was the sum of $8,000.00 was to be paid in installments. J. A. and Henry Leonard furnished their eight promissory notes, which promissory notes was made to their own order and endorsed in the blank. The eight promissory notes were dated the date of June 14, 1906. Each one of the eight promissory notes were for the sum of $1,000.00. The eight promissory notes were conditioned to bear interest at the rate of 8%, and the interest was payable annually. These eight promissory notes were payable in one, two, three, four, five, six, seven, and eight years after the date of the sale. Because J. A. and Henry Leonard were unable to pay the eight promissory notes, they were forced to sell the home, now known as Deepwood, and the property to John C. Bussy.

On the date of July 2, 1895, Camille Olivier Reggio granted a right of way to Morgan's Louisiana & Texas Railroad, & Steamship Company. Camille Olivier was the legal and lawful wife of Charles Eugene Reggio. Also, she was the only child and heir of her deceased parents, Celeste Bellechase Olivier and Joseph Devezin Olivier. The two witnesses to the execution of this deed of right of way were James G. Parkerson and D. A. Webster.

The strip of land that Camille Olivier Reggio was conveying to Morgan's Louisiana & Texas Railroad & Steamship Company was forty feet wide in diameter on the south side adjoining the right of way of Morgan's Louisiana & Texas Railroad & Steamship Company. Also, this strip of land was nine hundred and twenty eight feet in length, which was to increase the right of way to seventy five feet in depth on the south side from the central line of the present railroad track of Morgan's Louisiana & Texas Railroad & Steamship Company. This railroad track ran over and across the lands of Camille Olivier Reggio. This strip of land was located on the west side of Bayou Teche. Also, the boundaries of this strip of land were on the upperside by the lands of Alexander McDonald Stewart, which was known as the Cloverdale Plantation, and on the lowerside by the lands of John C. Bussy. This strip of land commences at station number 6299+925 and extends to station number 6278+20.

The purchase price of this strip of land was the sum of $120.00. Also, this purchase price was to cover compensation for loss of any crops. It was stipulated further that any additional culverts or drains were to be made at a point to be designated by Charles Eugene Reggio.

At the signing of this deed of right of way, Camille Olivier Reggio was unable to sign her name due to paralysis. Camille Olivier Reggio made her mark on this document, which constituted for her legal signature. Also, at the execution of this deed of right of way, Charles Eugene Reggio assisted his wife, Camille Olivier Reggio in granting this servitude.

On the date of May 19, 1919, Sidney Patout, and his wife, Qunita Mestayer Patout, purchased the home, now known as Deepwood and the property from John C. Bussy. This tract of land was located in the second ward of Iberia Parish, Louisiana, and it measured one hundred and seventy five arpents of land. This sale was made and accepted for the purchase price of $13,125.00. Sidney Patout paid John C. Bussy the sum of the purchase price or $13,125.00, in cash money at the execution of this act of conveyance.

THE FAULKNER FARM, THE BUSSY FARM, THE REGGIO FARM, AND THEN RAVENSWOOD PLANTATION

On the date of June 1, 1889, Edwin O. Faulkner, and his wife, Elizabeth Faulkner, conveyed their farm to John C. Bussy, of Jeanerette, Louisiana. Edwin O. Faulkner and Elizabeth Faulkner were residents of the City of Chicago, Illinois, which is situated in the County of Cook, in the State of Illinois. Edwin O. Faulkner was originally from Walsh County, which is situated in the State of North Dakota. The reason that Edwin Faulkner purchased this farm in the State of Louisiana was for investment reasons. Also, at this same time, Peter Sandager, and Carl Hendrickson, both from the State of North Dakota, had purchased the Rosedale Plantation, which was in the same vicinity as the farm that Edwin Faulkner had purchased, and then James Petris Brown and his wife, Avery Mary Brown, had purchased the Pine Grove Plantation, now known as Beau Pre', was located in the same vicinity as the farm that Edwin Faulkner had purchased and the Rosedale Plantation that was purchased by Peter Sandager and Carl Hendrickson. This was the same farm that Edwin O. Faulkner had purchased from Celeste Bellechase Olivier on the date of the 27th of February of the year of 1888.

The farm that was sold to John C. Bussy was located on the west side of Bayou Teche, and it had a depth of forty arpents of land. There was to be considered a deduction of land, which formed the Southern Pacific Railroad track, and this railroad track comprised one hundred and seventy acres of land.

This act of conveyance was made and accepted for the price and sum of $1,500.00. At the execution of the act of conveyance, John C. Bussy paid Edwin O. Faulkner and his wife, Elizabeth Faulkner, the purchase price of $1,500.00 in cash current money. Edwin O. Faulkner granted full acquittal and acknowledged the receipt of the purchase price.

This tract of land at one time had been a part of the land purchased by Joseph Devezin Olivier from Francois Optat Darby and Louis St. Marc Darby. Also, it was part of the sugar plantation that was purchased by Francois St. Marc Darby from Jacques Gravier. In the Faulkner / Olivier deed this was put into the deed concerning the boundaries of the tract of land that was conveyed by Celeste Bellechase Olivier to Edwin O. Faulkner. The tract of land was bounded above or on the upperside by the lands of Celeste

Bellechase Olivier, and below by the land this day sold to Eliza A. Brown, in front by Bayou Teche, and in the rear by the lands of Zenon Decuir.

On the date of March 3, 1905, Camille Olivier Reggio, the wife of Charles Eugene Reggio, acquired the Faulkner/Bussy farm from the seller, John C. Bussy, of Jeanerette, Louisiana, by a sale and mortgage. Actually, the tract of land that Camille Olivier Reggio was purchasing was a part of her parents, Celeste Bellechase Olivier and Joseph Devezin Olivier's sugar plantation. Camille Olivier Reggio was born on the island of Cuba at Mimosa, the sugar estate or sugar plantation where her parents were residing prior to purchasing the Darby Brothers sugar plantation., and moving to the State of Louisiana.

This tract of land that was being conveyed to Camille Reggio was situated on the west side of Bayou Teche, in the Parish of Iberia. It measured four and one half arpents fronting Bayou Teche, and the depth of the tract of land was forty arpents. There was a deduction of land, which comprised a right of way of the Morgan's Louisiana & Texas Railroad & Steamship Company, now the railroad track belongs to the Southern Pacific Railroad. This tract of land was bounded above or on the upperside by the property of J. A. and Henry Leonard, now known as Deepwood, and below or on the lowerside by the property of Mrs. Alfred Gonsoulin, recently purchased by Sarah Lettia Kemp Fitzhugh, who was the widow of George Nicholas Fitzhugh, and on the east by Bayou Teche, and in the rear or on the west side by the lands of Zenon Decuir.

This conveyance or sale was made and accepted for the price and sum of $3,5000.00 and it was a sale and mortgage. John C. Bussy acknowledged and accepted the payment of the purchase price in this manner and form: $1,000.00 is to be paid in cash at the execution of the act of conveyance, and the balance of the purchase price of $2,500.00 will be paid in installments. John C. Bussy granted full acquittance and discharge for the $1,000.000 paid in cash money at the execution of the act of conveyance. Camille Olivier Reggio has furnished her five promissory notes, which promissory notes were made to the order of herself, and by her endorsed in the blank. These five promissory notes were dated the 3rd day of March in the year of 1902. Each of the five promissory notes were for the sum of $500.00. Each one of the promissory notes were due in one year from the date of sale, two years from the date of sale, three years from the date of sale, four years from the date of sale, and five years from the date of sale. The interest on these five promissory notes was 8%, and the interest had to be paid annually. Camille Olivier Reggio, the purchaser, resigns the right to pay any of the five promissory notes before the note matured. Also, John C. Bussy declared that all the taxes were paid on the farm. The execution of the act of sale was passed at the home of Camille Olivier Reggio, which is now known as Deepwood.

After she had purchased this tract of and from John C. Bussy, Camille Olivier Reggio and her husband, Charles Eugene Reggio, built a home on the property. This home is now occupied by the Frank Courts family at 6707 Old Spanish Trail. Charles Reggio farmed sugar cane on this property.

On the date of February 18, 1908, Charles Eugene Reggio granted a right of way to the Bayou Teche Railway and Light Company through this property.

On the date of March 19, 1910, Charles Eugene Reggio acquired the tract of land from the succession of Camille Olivier Reggio, his deceased wife. This farm was located near Olivier Station, and it contained one hundred and forty arpents of land. Besides inheriting the real estate property, Charles Reggio acquired four American horses, two John Deere plows, one mower and harrow, and two milk cows. The farm was appraised at $2,950.00, and the four American mules were each valued at $100.00, one / 3 wheel mule cart was appraised at $50.60, one Avery cultivator was appraised at $15.00, one buggy was appraised at $40.00, one old American mare was appraised at $50.00, two John Deere plows were appraised at $10.00, one mower and harrow was appraised at $6.00, and the two milk cows were appraised at $30.00. The total appraisement was $3,551.60. The property was exempted from the payment of inheritance taxes.

Camille Olivier Reggio departed her life at her home on the plantation,and this home is now occupied by the Frank Courts family at 6707 Old Spanish Trail. Before her death, Camille Olivier Reggio was paralyzed and unable to write, but she was of sound mind, memory, and understanding. She requested that L. Pellerin, a notary public, receive her last will and testament, which she dictated to L. Pellerin in the presence of witness, Bob Edgar. By her last will and testament, Camille Reggio left her husband, Charles Eugene Reggio all the property that she possessed at the time of her death. In the event that Charles Eugene Reggio should die, then all of the property would be left to the sister-in-law of Camille Olivier Reggio, Mathlida Reggio. Mathlida Reggio was the sister of Charles Eugene Reggio, and they were the children of Augustus N. Reggio. Also, Mathlida Reggio was residing with her brother, Charles Reggio and his wife, Camille Olivier Reggio, her sister-in-law. Mathilda Reggio was helping out with the household chores, and taking care of Camille Olivier Reggio during her infirm time. Also, Camille Olivier Reggio wished that the amount of $30.00 be sustained from her possessions for masses for the repose of the soul of Camille Olivier Reggio, and that of her departed parents, Joseph Devezin Olivier and Celeste Bellechase Olivier. Camille Olivier Reggio died in the month of December of the year of 1908 in her bedroom, which was situated in her country plantation home on the Reggio farm.

After the death of his wife, Charles Eugene Reggio continued to live in the house on the farm, and he continued to cultivate his land in sugar cane. Charles Reggio was involved in the sugar industry as his father-in-law, Joseph Devezin Olivier, had been.

On August 2, 1911, Charles Reggio granted a right of way through his property to the Iberia, St. Mary, and Eastern Railroad Company. Also, on December 4, 1911, Charles Reggio granted a right of way through his property to the Southwestern Traction and Power Company.

On the date of January 21, 1914, John C. Bussy, of the town of Jeanerette, Louisiana, acquired the Reggio farm from Charles Eugene Reggio by an act of retrocession. Charles Eugene Reggio stated that for the consideration of the five mortgage notes, in the amount of $2,500.00, or each note was for the amount of $500.00, he (Charles Eugene Reggio) was unable to pay the notes. As a result of being unable to pay the notes, Charles Eugene Reggio retroceded his farm to John C. Bussy. These mortgage notes originated from the sale of the property by John C. Bussy to Camille Olivier Reggio. Also, these five mortgage notes were identified with this act of sale by paraphering the notes, "Ne Varieteur," by the notary public or attorney. This retroaction of the Reggio farm was passed in the town of Jeanerette, Louisiana, and E. J. LeJeune and E. C. Edwards served as witnesses to this act of retroaction. Also, all the taxes which were assessed against the Reggio farm have been paid.

On the date of October 19, 1918, John C. Bussy, of the town of Jeanerette, Louisiana, sold the Reggio farm to Sarah Lettia Kemp Fitzhugh, who was the widow of George Nicholas Fitzhugh. Sarah Fitzhugh was a resident of the town of New Iberia, Louisiana, and she lived at 221 East Main Street. Sarah Fitzhugh purchased the Reggio farm for investment reasons. Also, included in the sale of the Reggio farm were four mules, three mule carts, a mowing machine, a hay stack, and all the plows and farming implements. The movable property, the mules, the mule carts, the mowing machine, the hay stack, and the plows and farming implements were all in the possession of Charles Eugene Reggio. After the act of retroaction, Charles Eugene Reggio continued to live on the farm, and farm the land in sugar cane. It was well understood between Sarah Fitzhugh, Charles Eugene Reggio, and John C. Bussy that possession of the former Reggio farm would be given to Sarah Fitzhugh after Charles Reggio had made and taken off his crop of sugar cane. The property could not be surrendered by Charles Reggio before February 1, 1919.

The sale of the Reggio farm to Sarah Fitzhugh was made for the amount of $13,068.75. Sarah Fitzhugh paid $5,000.00 at the execution of the act of conveyance, and John C. Bussy acknowledged the receipt and approved of the payment of the $8,068.75, which was the balance of the purchase price. Sarah Fitzhugh had furnished her two promissory notes to the order of John C. Bussy, and these notes were dated October 19, 1918. Each note was for the sum of $4,043.37, and payable on or before one or two years after October 19, 1918. These two promissory notes were conditioned to bear interest at 8% per annum from the date of October 19, 1918. The two promissory notes furnished by Sarah Fitzhugh were paraphed, "Ne Variteur," by the notary public, Ventruss J. Smith, and identified with this act of conveyance.

In order to secure the punctual payment of the two promissory notes at maturity, as well as, of all the interest accrued on the two promissory notes, and to secure the payment of all reimbursement of all fees by any lawyers, who may be employed to collect the principal and interest in part payment, the property (the Reggio farm) will remain specially mortgaged until the full and the final payment of all promissory notes and interest. Also, Sarah Fitzhugh binded herself not to sell or to encumber the property, or even have a vendor's lien until the mortgage was paid in full and cancelled.

Sarah Fitzhugh named her newly purchased farm, Ravenswood Plantation, for the manor home of the Fitzhugh family in England. Her husband, George Nicholas Fitzhugh, was a direct descendant of the Fitzhugh family in England, and the Fitzhugh manor home in England was named Ravenworth. Also, Sarah Fitzhugh decided to call her newly purchased plantation, Ravenswood, for the name of the town of Ravenswood, West Virginia, were the Fitzhugh family in America had settled. The husband of Sarah Fitzhugh, George Nicholas Fitzhugh was from the town of Ravenswood, West Virginia. The town of Ravenswood, West Virginia was named by the Fitzhugh ancestors of George Nicholas Fitzhugh for the ancestral home of the Fitzhugh family in England, Ravensworth.

On the date of April 25, 1925, by an act of donation, Sarah Lettia Kemp Fitzhugh donated and gave her only daughter, Kate Fitzhugh Russell, the plantation, Ravenswood.

On the date of November 10, 1926, Kate Fitzhugh Russell, who was the widow of Joseph Pierre Russell sold to her only daughter, Heneritta Sarah Russell, who was the wife of Doctor Frank Field Courts, the plantation, Ravenswood. This plantation was considered the paraphernal property and separate property of Henreritta Russell Courts, and not the community property of acquets and gains of her marriage with Doctor Frank Field Courts.

THE BROWN FARM, THE GONSOULIN FARM AND THEN RAVENSWOOD PLANTATION

On the date of August 3, 1889, Eliza A. Brown, and her husband, George Brown, met with P. Laurent Renoudet, a notary public in the town of New Iberia, Louisiana, for the purpose of executing an act of conveyance between the Browns and Alfred Gonsoulin. Alfred Gonsoulin was a resident of the town of Jeanerette, Louisiana. Also, Alfred Gonsoulin was interested in purchasing the Brown tract of land for farming and investment reasons.

This tract of land that was being conveyed was situated on the west side of Bayou Teche, in the Parish of Iberia, and it was located about six miles below the town of New Iberia, Louisiana. It measured four and one half arpents fronting Bayou Teche, on the west side, and it had a depth of forty arpents. Also, the right of way of the Southern Pacific Railroad passed through this property, and this right of way consisted of one hundred and seventy arpents of land. This tract of land was bounded above or on the upperside by the lands of Edwin O. Faulkner, below or on the lowerside in part by the lands of Peter E. Sandager and Carl Hendrickson or the Rosedale Plantation, and in part by the public road leading from New Iberia to Jeanerette, Louisiana, and this public road is known as the Old Spanish Trail, and in the rear by the lands of Zenon Decuir. This was the same property that was purchased by Eliza A. Brown from Celeste Bellechase Olivier, on the date of February 27, 1888. Also, this property was purchased with the paraphernal funds or the own separate funds of Eliza Brown. Actually, this was Eliza Brown's personal and separate investment, not an investment of the community of acquets and gains, which existed between Eliza A. Brown, and George Brown.

Also, conveyed to Alfred Gonsoulin besides the real estate property, was one American horse, one American mare, two plows, gearing for the plows, and all the agricultural implements, such as hoes, shovels, harrows, picks, rakes, and all other instruments used in the processing of cultivating crops. Also, all the crop of sugar cane that was growing on the farm.

This conveyance was made and accepted for the purchase price of $1,500.00. Alfred Gonsoulin paid the purchase price of $1,500.00 in cash or current money, on the date of August 3, 1889, at the execution of this act of conveyance. Eliza Brown granted full acquittance and acknowledged the receipt. Also, Eliza Brown stated that all the taxes assessed against the farm were paid.

On October 9, 1894, Suzette Decuir Gonsoulin, the widow of Alfred Gonsoulin, acquired the farm by an act of cash sale from Anatole Gonsoulin and others. It was well understood that the sale and transfer covers all the movable and immovable property on the farm. This farm was the same farm that was purchased by Alfred Gonsoulin from Eliza A. Brown, and this farm contained one hundred and seventy eight American acres of land.

The ownership of the Alfred Gonsoulin farm is a naked ownership, and was inherited by the children of Alfred Gonsoulin. The Alfred Gonsoulin farm comprises three sevenths (3/7) of Alfred Gonsoulin's share of the community of acquets and gains, which existed between Alfred Gonsoulin, who was deceased and his wife, Suzette Decuir Gonsoulin.

This conveyance was made and accepted for the consideration of $8,215,80. This purchase price made $2,738.60 for each child of Alfred Gonsoulin's to receive as part of their inheritance from their late father, Alfred Gonsoulin. Suzette Decuir Gonsoulin paid the purchase price of $2,378.60 in cash or current money to each vendor, who was a child of herself and Alfred Gonsoulin from their marriage. The payment of the purchase price was made at the execution of this act of conveyance. All the vendors acknowledge the receipt and grant full acquittal.

On December 24, 1910, Suzette Decuir Gonsoulin, the surviving widow of Alfred Gonsoulin, conveyed the Gonsoulin farm to her daughter, Eva Gonsoulin, who was the wife of A. Damas Moresi, and the brother of Eva Gonsoulin, Alfred Gonsoulin, a single man. This was the same farm that Suzette Decuir Gonsoulin had purchased from Anatole Gonsoulin. The execution of this act of conveyance was presided by Jules P. Robin, who was a notary public, and served in this capacity at this time.

On November 14, 1918, Eva Gonsoulin Moresi, the wife of A. Damas Moresi of Jeanerette, Louisiana, and her brother, Alfred Gonsoulin, sold the farm to Sarah Lettia Kemp Fitzhugh, the widow of George Nicholas Fitzhugh. The Fitzhugh family were originally from Ravenswood, West Virginia, and they had moved down south in the aftermath of the Civil War. This act of conveyance was passed in the town of Jeanerette, Louisiana, in the presence of L. T. Dulany, a notary public, who served in his capacity.

At the time of this conveyance, Sarah Fitzhugh, the purchaser, was aware that there was a mortgage on the farm in the amount of $1,000.00. This mortgage originated from the act of conveyance between Eva Gonsoulin and Alfred Gonsoulin from Suzette Decuir Gonsoulin.

This sale was made and accepted for the consideration of the price and sum of $8,500.00. Sarah Fitzhugh paid the purchase price of $8,500.00 in cash or current money, at the execution of this act of conveyance, which was held in the town of Jeanerette, Louisiana. The vendors, Eva Gonsoulin Moresi and Alfred Gonsoulin gave full acquittal and acknowledged the receipt. Sarah Fitzhugh assumed the sum of $1,000.00, which was a mortgage held on the farm, together with all the interest on the principal amount. Therefore, the note of $1,000.00 was assumed by the purchaser, Sarah Fitzhugh.

The taxes assessed against the farm for the year of 1918 are to be paid by the vendors, Eva Gonsoulin Moresi and Alfred Gonsoulin. The tax rolls have been completed and the amount of taxes owed on this farm have been assessed. The tax receipt is to be turned over to Sarah Fitzhugh after the taxes are paid by the vendors, Eva Gonsoulin Moresi and Alfred Gonsoulin.

After purchasing the Gonsoulin farm, Sarah Fitzhugh, who was the widow of George Nicholas Fitzhugh, renamed the farm Ravenswood Plantation, for the Fitzhugh family homeland in the State of West Virginia.

THE EDGAR FARM

On March 8, 1878, Celeste Olivier Pellerin, the widow of Louis Pellerin, conveyed a tract of land to Joseph Cyr, of Jeanerette, Louisiana. This tract of land was situated in the Parish of Iberia, on the west side of Bayou Teche. It had two and one fourth arpent frontage on the west side of Bayou Teche, and the depth of the tract of land was forty arpents deep. This tract of land was bounded on the north and west sides by the lands of Henry Train, on the south side by the lands of Mrs. Louisa Dumartrait, and east by Bayou Teche. Also, included were all the buildings and improvements.

The purchase price of this tract of land was $1,200.00. Celeste Pellerin, the vendor or the seller, of the tract of land, bounded herself to reimburse Joseph Cyr in three years from the date of March 6, 1878, for the amount of the purchase price of the tract of land of $1,200.00. It was understood between Joseph Cyr and Celeste Olivier Pellerin that at the expiration of the time agreed upon for the redemption of the tract of land, the sum of $1,200, if not paid to Joseph Cyr, the purchaser, by Celeste Pellerin, the seller, then the title, rights, interests, and privileges of ownership in the tract of land will revert to Joseph Cyr. The expiation was due on the date of March 6, 1881. If the property should revert to Joseph Cyr, then he will have the right to use the property with complete title, interest, and privileges of ownership.

Then on March 18, 1878, at the residence of Celeste Olivier Pellerin, Celeste Pellerin, the seller of the tract of land, met with Joseph Cyr, the purchaser, to amend and annull the act passed on the date of March 8, 1878, which act was passed in front of the presiding notary public, Barthlemy de la Houssaye. Celeste Olivier

Pellerin was desirous of the terms and conditions of the sale, so therefore, she granted, bargained, sold, conveyed, and set over the tract of land to Joseph Cyr. The purchase price of this sale was the sum of $1,200.00, which was paid in cash at the execution of the act of conveyance, on the date of March 18, 1878, by Joseph Cyr.

On May 11, 1879, Joseph Cyr sold the tract of land, which he had purchased from Celeste Olivier Pellerin, for the sum of $1,200.00 to the Queryrouze & Bois Wholesale of the City of New Orleans, Louisiana. This tract of land was probably used for the growing of vegetable crops, because Joseph Cyr was the owner of a grocery store in the town of Jeanerette, Louisiana. It is possible that the fresh vegetables sold in the grocery store owned by Joseph Cyr were raised, grown, and harvested on this tract of land. It is also possible that Joseph Cyr had an account with the Queryrouze & Bois Wholesale, and the reason that he sold this tract of land to that company was due to a debt owed by Joseph Cyr to the Queryrouze & Bois Wholesale. Oscar Bois represented the firm of Queryrouze & Bois Wholesale and the interests of Leon Queryrouze, at the execution of the act of conveyance. Joseph Cyr reserves the right during the months to redeem the property back by paying the taxes. It was understood that the right to redeem the property would expire on the date of May 10, 1880. At the time of the execution of the act of sale, all the taxes assessed against the land were paid.

The wholesale company of Queryrouze & Bois was a wholesale grocery, that specialized in all kinds of wines and liquors, and all kinds of western produce. The wholesale company was located at the corner of Old Levee and Bienville Streets in the City of New Orleans, Louisiana.

Joseph Cyr died by June of the year of 1880. Therefore, his widow, Emilie J. Hoffner, purchased the tract of land back from the wholesale company of Queryrouze & Bois on the date of June 1, 1880, by an act of conveyance. J. A. Breaux acted as agent for the Queryrouze & Bois Wholesale Company, of the City of New Orleans, Louisiana.

On April 18, 1883, Emilie Julie Hoffner Cyr, the widow of Joseph Cyr, conveyed the tract of land to Francois and Remi A. Porche. Emilie Cyr had agreed to deliver the property the same day (April 18, 1878) as the execution of the act of conveyance.

This tract of land that was conveyed by Emilie Julie Hoffner Cyr, the widow of Joseph Cyr, was situated on the west side of Bayou Teche, in the Parish of Iberia. This tract of land had a frontage of two and one fourth arpents on the west side of Bayou Teche. The boundaries of this tract were on the north side by the lands of Henry Train, on the south side by the lands of Alcee

Garginee, on the east side by Bayou Teche, and on the west side by the lands of Henry Train.

On January 11, 1876, Francois Porche and Remi A. Porche, both residents of Iberia Parish, Louisiana, conveyed through an act of exchange this tract of land to Alexander Edgar, a resident of St. Mary Parish.

On April 24, 1889, Alexander Edgar conveyed this tract of land to Edward T. Weeks, who was an agent. This was a cash sale, and the purchase price was $500.00, which was paid by Edward T. Weeks, at the execution of the act of conveyance. This tract of land was acquired by Edward T. Weeks for the present vendor or seller, Alexander W. Edgar.

Also, on April 14, 1889, Edward T. Weeks, the agent for Alexander W. Edgar, conveyed the tract of land back to Alexander W. Edgar, at a sale and mortgage. It was well understood between Alexander Edgar, the purchaser of this tract of land, and Edward T. Weeks, the vendor or seller of this tract of land, that Edward T. Weeks in his own capacity as agent for Alexander W. Edgar only warrants the title of the property conveyed against his (Edward T. Weeks) own acts, while in the possession of Edward T. Weeks, and no further. Also, Alexander W. Edgar, the purchaser, waives the restitution by Edward T. Weeks, the agent or holder or holders of the promissory note in the amount of $500.00, and all amounts paid under or in accordance with the contract evidenced by this deed in the event of eviction or disturbances in the property or the rights of the property. Alexander W. Edgar further declared that he waives and renounces all rights, homesteads, and all exemptions under which he has or may have in the tract of land, under Article 244 of the Constitution of the State of Louisiana. This waiver and renouncing of homestead is made for the full amount of homestead rights and exemptions, but is special only and made in favor of no debt, save the debt recognized and contracted. All the taxes are paid on the tract of land being conveyed.

This tract of land was situated on the west side of Bayou Teche, and it had a depth of forty five arpents. This is the same tract of land that was conveyed to Edward T. Weeks, this 19th day of April of the year of 1889, and then the tract of land was reconveyed back to Alexander W. Edgar, on the 19th day of April of the year of 1889.

On September 4, 1948, the succession of Alexander W. Edgar, and his wife, Louise Smith Edgar, was opened in Iberia Parish, Louisiana, and this tract of and was inherited by John M. Edgar, Louise Edgar, Robert Edgar, May Edgar, who was the wife of Harry Rich, Senior, and Jessie Edgar, who was the wife of Richard Fleming.

This tract of land was situated on the south side of

Bayou Teche, in the second ward of Iberia Parish, Louisiana. Also, this tract of land had a frontage of two and one fourth arpents fronting Bayou Teche, or four hundred and sixty four feet along Bayou Teche. This tract of land contained a total of four thousand, nine hundred and forty five feet of land, without the land deductions for the three right of ways that cross the Edgar tract of land. The three right of ways that cross the Edgar tract of land are the right of way for the Louisiana State Highway #182, formerly United States Highway #90, the right of way for the Missouri Pacific Railroad track, and the right of way for the Southern Pacific Railroad track, formerly Morgan's Louisiana & Texas Railroad & Steamship Company. Together these three right of ways contained a total of two hundred and forty five feet, which leaves a net total of four thousand, six hundred and ninety five feet. The Edgar tract of land is bounded on the north side by Bayou Teche, on the south side by the Grand Marais or the Great Swamp drainage ditch, on the east side by the lands of Robert Stewart or Cloverdale Plantation, and on the west side by the lands of Mrs. Fannie R. Kling.

A view of a plantation barn, on the west side of Bayou Teche. This barn is situated on the former Gravier sugar plantation, the plantation of Francois St. Marc Darby, the Darby Brothers plantation, the plantation of Joseph Devezin Olivier.

A view on the west side of Bayou Teche of the former Gravier sugar plantation, the plantation of Francois St. March Darby, the Darby Brothers plantation, and the plantation of Joseph Devezin Olivier.

Glossary

Appolina or Appoline	a French name for women, which is the femmine for the masculine name Napoleon
arpent	a French measure of land, which is equivalent to .85 of an acre of land
double concession	a grant of land granted to two or more people
fils	a French term meaning junior or younger
free man of color	a free man of color, or a man of color who is not a slave, and is able to own property, slaves, and other property
free people of color	a term used to distinguish people of color who were free to live as they chose and not slaves, these free people of color were able to own property, slaves, were allowed to pursue an education, some owned their own businesses
free woman of color	a woman, who is not a slave, and is free and able to own property, slaves, and other property
griffe	a term used to refer to an American born black slave
Le Jeune	a French surname meaning Young
moiety	a French term meaning half
mouton	a French term meaning sheep
paraphed	a notary public or a lawyer puts a seal or paraphes a legal document or instrument, such as

document or instrument, such as a mortgage, act of sale, will, etc to identify it with another legal document

pere — a French term meaning senior or elder or older

pieux — a French term for the pointed or triangle shaped uppper portion
256 of a fence post

sugar factor — a firm or company engaged in the business of selling sugar and its products to other food companies, restauraunts, sugar refining companies, and other companies buying sugar and its products. Also, a sugar factor acted as a bank, in that the farmers could borrow money against their next year's crop proceeds. The sugar factor held the money or the proceeds of the sugar cane crop in accounts in the name of the farmer, and this account acted like a modern day bank account.

INDEX

A

B

C

D

E

F

G

H

I

J

K

L

M

N

O

P

Q

R

S

T